D0142719

Applied Probability and Stochastic Processes

RICHARD M. FELDMAN
Industrial Engineering Department
Texas A & M University

CIRIACO VALDEZ-FLORES
Sielken, Incorporated

PWS PUBLISHING COMPANY

I(T)P ▪ An International Thomson Publishing Company

Boston ▪ Albany ▪ Bonn ▪ Cincinnati ▪ Detroit ▪ London
Madrid ▪ Melbourne ▪ Mexico City ▪ New York ▪ Paris
San Francisco ▪ Singapore ▪ Tokyo ▪ Toronto ▪ Washington

Library of Congress Cataloging-in-Publication Data
Feldman, Richard M. (Richard Martin).
Applied probability and stochastic processes / Richard M. Feldman, Ciriaco Valdez-Flores.
p. cm.
Includes index
ISBN 0-534-93921-X
1. Stochastic processes. I. Valdez-Flores, Ciriaco. II. Title.
QA274.F45 1995 95-38395
519.2-dc20 CIP

Sponsoring Editor *Jonathan Plant*
Production/Composition/Art *Publication Services*
Developmental Editor *Mary Thomas Stone*
Manufacturing Coordinator *Wendy Kilborn*
Text Printer/Binder *R. R. Donnelley & Sons/ Crawfordsville*
Production Coordinator *Elise S. Kaiser*
Marketing Manager *Nathan Wilbur*
Cover Designer *Elise S. Kaiser*
Cover Printer *John Pow Company*

Printed and bound in the United States of America.
95 96 97 98 99 — 10 9 8 7 6 5 4 3 2 1

This book is printed on recycled, acid-free paper.

For more information, contact:
PWS Publishing Company
20 Park Plaza
Boston, MA 02116

International Thomson Publishing Europe
Berkshire House 168-173
High Holborn
London WC1V 7AA
England

Thomas Nelson Australia
102 Dodds Street
South Melbourne, 3205
Victoria, Australia

Nelson Canada
1120 Birchmont Road
Scarborough, Ontario
Canada M1K 5G4

International Thomson Editores
Campos Eliseos 385, Piso 7
Col. Polanco
11560 Mexico D.F., Mexico

International Thomson Publishing GmbH
Königswinterer Strasse 418
53227 Bonn, Germany

International Thomson Publishing Asia
221 Henderson Road
#05-10 Henderson Building
Singapore 0315

International Thomson Publishing Japan
Hirakawacho Kyowa Building, 31
2-2-1 Hirakawacho
Chiyoda-ku, Tokyo 102
Japan

Contents

Preface

This book is a result of teaching stochastic processes at various levels to junior and senior undergraduates and beginning graduate students during a several-year period. In teaching such a course, we have realized a need to furnish students with material that gives a mathematical presentation while at the same time providing proper foundations to allow students to build an intuitive feel for probabilistic reasoning. We have tried to maintain a balance in presenting advanced but understandable material that sparks the interest and challenges students, without the discouragement that often comes as a consequence of not understanding the material. Our intent in this text is to develop stochastic processes in an elementary but mathematically precise style and to provide sufficient examples and homework exercises that will permit students to understand the range of application areas for stochastic processes.

The only prerequisites for an undergraduate course using this textbook are a previous course covering calculus-based probability and statistics and familiarity with basic matrix operations. Access to, and familiarity with, software that enables easy matrix manipulation would be helpful but not necessary. Use of software such as Mathematica® or Maple® or familiarity with APL would be especially helpful when covering the chapter on Markov chains.

This book could also be used for an introductory course to stochastic processes at the graduate level, in which case an additional prerequisite of linear programming should be required if the chapter on Markov decision theory is to be covered. It would also be helpful to expect graduate students to be competent programmers in some scientific programming language. Two chapters cover advanced topics that would be skipped in an undergraduate course: Chapter 9, Markov Decision Processes, and Chapter 10, Advanced Queues. Knowledge of linear programming is necessary for Chapter 9, and a programming language would be very helpful in understanding the concepts in Chapter 10.

The first chapter is a review of probability. It is intended simply as a review; the material is too terse if students have not previously been exposed to probability. However, our experience is that most students do not learn probability until after two or three exposures to it, so this chapter should serve as an excellent summary and review for most students. The second chapter begins the introduction to random processes and covers Markov chains. The approach in this chapter and in Chapter 4 is similar to the approach taken by Çınlar (*Introduction to Stochastic Processes,* Prentice-Hall, 1975). The homework problems cover a wide variety of modeling situations in an attempt to begin the development of "modelers."

Chapter 3 is a very brief introduction to simulation. It is somewhat unusual to place the chapter so early in the text, but our feeling is that it is beneficial for students to use simulation to help them develop an intuitive understanding of the processes being studied. Chapter 3 covers only enough material to enable students to simulate Markov chains and Markov processes. The basics of event-driven simulation are covered in Chapter 6. There will be no loss in continuity if the material in Chapter 3 is ignored until just before Chapter 6. Another advantage of briefly introducing simulation early is that it enables the student to begin a programming problem (e.g., Exercises 3.8, 3.9) early in the semester so they are not rushed in completing it. Then, when simulation is covered in greater depth, the students will have some idea and experience in simulation that should make understanding Chapter 6 easier.

Chapter 4 is an introduction to continuous-time Markov processes. The major purpose of the chapter is to provide the tools necessary for the development of queueing models; therefore, the emphasis in the chapter is on steady-state analyses. The final section of Chapter 4 is an extremely brief treatment of the time-dependent probabilities for Markov processes. This section should be skipped for most undergraduate classes. Queueing theory is introduced in Chapter 5 and continued in the final chapter of the book. The intent of Chapter 5 is to develop the student's appreciation for queueing processes and, as in Chapter 2, an attempt has been made to develop a wide variety of modeling situations through the homework problems. Consistent with some of the modern trends in queueing theory, we have included sections in Chapter 5 introducing queueing networks and some approximations.

Chapter 6 has two sections: the first deals with the specifics of event-driven simulations, the second introduces some of the statistical issues for output analysis. If the mechanical details of simulation (such as future events lists) are not of interest to the instructor, the first section of Chapter 6 can be skipped with no loss of continuity. Chapter 3 together with the second section of Chapter 6 should yield an excellent introduction to simulation. No programming language is assumed since our purpose is not to produce experts in simulation, but simply to introduce the concepts and develop an interest in simulation within the student. If simulation is covered adequately by other courses, both Chapters 3 and 6 can be skipped.

Chapters 7 and 8 introduce a change in tactics in that they deal with specific problem domains: the first is inventory and the second is replacement. Applied probability can be taught as a collection of techniques useful for a wide variety of applications, or it can be taught as various application areas for which randomness plays an important role. The first six chapters focus on particular techniques with some applications being emphasized through examples and the homework problems. Chapters 7 and 8,. however, focus on two problem domains that have been historically important in applied probability and stochastic processes. It was difficult to decide on the proper location for these two chapters. There are some Markov chain references in the last section of the inventory chapter; therefore, it is best to start with Chapters 1 and 2 for most courses. After covering the second chapter, it would be appropriate and easily understood if the next chapter taught were Chapter 7, 8, 3, or 4. It simply depends on the inclination of the instructor.

Chapters 9 and 10 are included for advanced students. Chapter 9 covers Markov decision processes, and Chapter 10 is a presentation of phase-type distributions and the matrix geometric approach to queueing systems adopted from the work of M. F. Neuts (*Matrix-Geometric Solutions in Stochastic Models,* Johns Hopkins University Press, 1981).

Acknowledgments

We are indebted to many of our colleagues for their invaluable assistance and professional support. In particular, we thank G. L. Curry; B. L. Deuermeyer; G. L. Hogg; W. Kuo; R. L. Sielken, Jr.; M. L. Spearman; and M. A. Wortman for their help and encouragement. In addition, Mingchih Chen has been an invaluable assistant in proofreading and helping with the exercises at the end of the chapters.

We are grateful to the following reviewers for their suggestions and comments:

Mary R. Anderson, *Arizona State University;* Rajan Batta, *State University of New York at Buffalo;* M. Jeya Chandra, *The Pennsylvania State University;* S. Kumar, *Rochester Institute of Technology;* Gang Li, *University of North Carolina;* Jye-Chyi Lu, *North Carolina State University;* Ditlev Monrad, *University of Illinois at Urbana-Champaign;* Marcel F. Neuts, *The University of Arizona;* Paul L. Schillings, *Montana State University;* and G. Don Taylor, *University of Arkansas at Fayetteville.*

We also wish to express our appreciation to our wives, Alice Feldman and Nancy Vivas-Valdez, for their patience and loving support. Finally, we acknowledge our thanks through the words of the psalmist, "Give thanks to the Lord, for He is good; His love endures forever." (Psalms 107:1, NIV)

Richard M. Feldman
Ciriaco Valdez-Flores

1

Basic Probability

The random phenomena inherent in most processes point to the need for stochastic modeling and statistical analyses. Before discussing modeling or statistical analysis, it is necessary to review the fundamental concepts of basic probability. This chapter is not intended to teach probability theory, but it is used for review and to establish a common ground for the notation and definitions used throughout the book. Therefore, we briefly review some fundamentals of probability while assuming a previous exposure to probability.

1.1 BASIC DEFINITIONS

To understand probability, it is best to envision an experiment for which the result is unknown. All possible outcomes must be defined, and the collection of the outcomes is called the sample space. Probabilities are assigned to subsets of the sample space, called events. We shall give the rigorous definition for probability. However, the reader should not be discouraged if an intuitive understanding is not immediately acquired. This takes time and the best way (only way) to understand probability is by working problems.

Definition 1.1 *An element of a* SAMPLE SPACE *is an* OUTCOME. *A set of outcomes, or equivalently a subset of the sample space, is called an* EVENT.

Definition 1.2 *A* PROBABILITY SPACE *is a three-tuple* $(\Omega, \mathcal{F}, \Pr)$ *where* Ω *is a sample space,* $\mathcal{F}$ *is a collection of events from the sample space, and* $\Pr$ *is a probability law that assigns a number to each event contained in* $\mathcal{F}$. *Furthermore,* $\Pr$ *must satisfy, for* $A, B \in \mathcal{F}$, *the following conditions:*

- $\Pr(\Omega) = 1$,
- $\Pr(A) \geq 0$,

- $\Pr(A^c) = 1 - \Pr(A)$, *where* A^c *is the complement of* A,
- $\Pr(A \cup B) = \Pr(A) + \Pr(B)$ *if* $A \cap B = \emptyset$, *where* $\emptyset$ *denotes the empty set.*

Note that the collection of events, $\mathcal{F}$, in the definition of a probability space must satisfy some technical mathematical conditions that are not worth our time to describe in this text. If the sample space contains a finite number of elements, then $\mathcal{F}$ usually consists of all the possible subsets of the sample space. The four conditions on the probability measure Pr should appeal to a person's intuitive concept of probability. The first condition indicates that something from the sample space must happen, the second condition indicates that negative probabilities are illegal, the third condition indicates that the probability of an event is equal to one minus the probability of its complement, and the fourth condition indicates that the probability of the union of two disjoint (or mutually exclusive) events is the sum of their individual probabilities. Actually, the third condition is redundant but it is listed in the definition because of its usefulness.

A probability space is the full description of an experiment; however, it is not always necessary to work with the entire space. One possible reason for working within a restricted space is because certain facts about the experiment are already known. For example, suppose a dispatcher at a refinery has just sent a barge containing jet fuel to a terminal 800 miles downriver. Personnel at the terminal would like a prediction on when the fuel will arrive. The experiment consists of all possible weather, river, and barge conditions that would affect the travel time downriver. However, when the dispatcher looks outside rain is seen. Thus, the original probability space can be restricted to include only rainy conditions. Probabilities thus restricted are called conditional probabilities according to the following definition.

DEFINITION 1.3 *Let* $(\Omega, \mathcal{F}, \Pr)$ *be a probability space where* A *and* B *are events in* $\mathcal{F}$ *with* $\Pr(B) \neq 0$. *The* CONDITIONAL PROBABILITY *of* A GIVEN B, *denoted* $\Pr(A \mid B)$, *is*

$$\Pr(A \mid B) = \frac{\Pr(A \cap B)}{\Pr(B)}.$$

Venn diagrams are sometimes used to illustrate relationships among sets. In the diagram of Figure 1.1, assume that the probability of a set is proportional to its area. Then the value of $\Pr(A \mid B)$ is the proportion of the area of set B that is occupied by the set $A \cap B$.

EXAMPLE 1.1 A telephone manufacturing company makes radio phones and plain phones and ships them in boxes of two (same type in a box). Periodically, a quality control technician randomly selects a shipping box, records the type of phone in the box (radio or plain), and then tests the phones and records the number that were defective. The sample space is

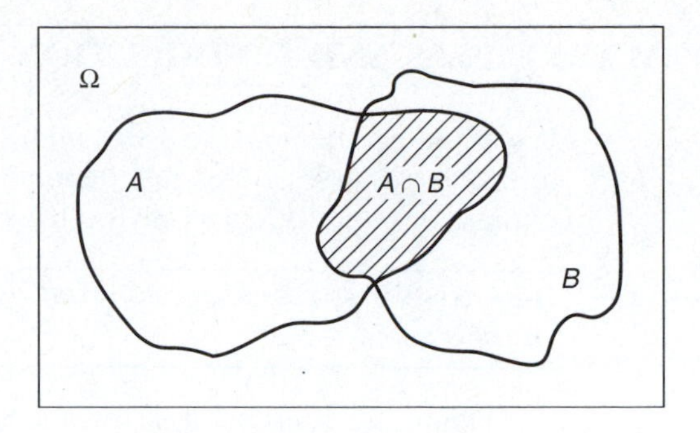

FIGURE 1.1 Venn diagram illustrating events A, B, and $A \cap B$.

$$\Omega = \{(r,0),(r,1),(r,2),(p,0),(p,1),(p,2)\},$$

where each outcome is an ordered pair; the first component indicates whether the phones in the box are the radio type or plain type and the second component gives the number of defective phones. The set $\mathcal{F}$ is the set of all subsets, namely,

$$\mathcal{F} = \{\emptyset,\{(r,0)\},\{(r,1)\},\{(r,0),(r,1)\},\cdots,\Omega\}.$$

There are many legitimate probability laws that could be associated with this space. One possibility is

$$\begin{aligned}
\Pr\{(r,0)\} &= 0.45, & \Pr\{(p,0)\} &= 0.37,\\
\Pr\{(r,1)\} &= 0.07, & \Pr\{(p,1)\} &= 0.08,\\
\Pr\{(r,2)\} &= 0.01, & \Pr\{(p,2)\} &= 0.02.
\end{aligned}$$

By using the last property in Definition 1.2, the probability measure can be extended to all events; for example, the probability that a box is selected that contains radio phones and at most one phone is defective is given by

$$\Pr\{(r,0),(r,1)\} = 0.52.$$

Now let us assume that a box has been selected and opened. We observe that the two phones within the box are radio phones, but no test has yet been made on whether or not the phones are defective. To determine the probability that at most one phone is defective in the box containing radio phones, define the event A to be the set $\{(r,0),(r,1),(p,0),(p,1)\}$ and the event B to be $\{(r,0),(r,1),(r,2)\}$. In other words, A is the event of having at most one defective phone, and B is the event of having a box of radio phones. The probability statement can now be written as

$$\Pr\{A \mid B\} = \frac{\Pr(A \cap B)}{\Pr(B)} = \frac{\Pr\{(r,0),(r,1)\}}{\Pr\{(r,0),(r,1),(r,2)\}} = \frac{0.52}{0.53} = 0.991. \quad \blacksquare$$

▶ *Suggestion: Do Exercises 1.1 and 1.2.*

1.2 RANDOM VARIABLES AND DISTRIBUTION FUNCTIONS

It is often cumbersome to work with the outcomes directly in mathematical terms. Random variables are defined to facilitate the use of mathematical expressions and to focus only on the outcomes of interest.

Definition 1.4 *A* Random Variable *is a function that assigns a real number to each outcome in the sample space.*

Figure 1.2 presents a schematic illustration of a random variable. The name "random variable" is actually a misnomer, because it is not random and is not a variable. As illustrated in the figure, the random variable simply maps each point (outcome) in the sample space to a number on the real line.[1]

Revisiting Example 1.1, let us assume that management is primarily interested in whether or not at least one defective phone is in a shipping box. In such a case a random variable D might be defined such that it is equal to zero if all the phones within a box are good and equal to 1 otherwise; that is,

$$D(r,0) = 0, \quad D(p,0) = 0,$$
$$D(r,1) = 1, \quad D(p,1) = 1,$$
$$D(r,2) = 1, \quad D(p,2) = 1.$$

The set $\{D = 0\}$ refers to the set of all outcomes for which $D = 0$ and a legitimate probability statement would be

$$\Pr\{D = 0\} = \Pr\{(r,0),(p,0)\} = 0.82.$$

To aid in the recognition of random variables, the notational convention of using only capital Roman letters (or possibly Greek letters) for random

FIGURE 1.2 A mapping from the sample space to the real numbers.

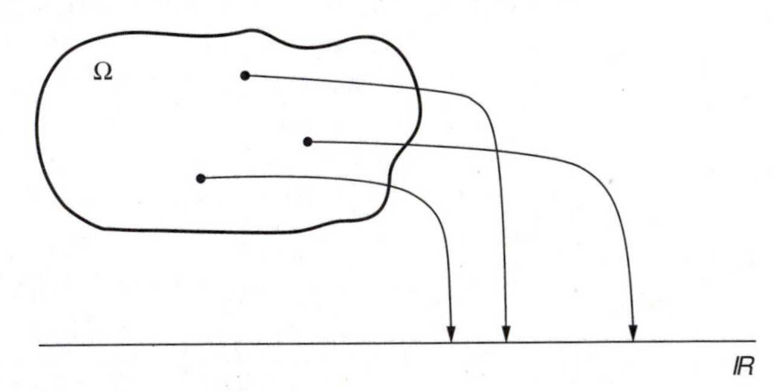

[1]Technically, the space into which the random variable maps the sample space may be more general than the real number line, but for our purposes, the real numbers will be sufficient.

variables is followed. Thus, if you see a lowercase Roman letter, you know immediately that it cannot be a random variable.

Random variables are either discrete or continuous depending on their possible values. If the possible values can be counted, the random variable is called discrete; otherwise, it is called continuous. The random variable D defined in the previous example is discrete. To give an example of a continuous random variable, define T to be a random variable that represents the length of time that it takes to test the phones within a shipping box. The range of possible values for T is the set of all positive real numbers, and thus T is a continuous random variable.

A cumulative distribution function (CDF) is often used to describe the probability law underlying the random variable. The cumulative distribution function (usually denoted by a capital Roman letter or a Greek letter) gives the probability accumulated up to and including the point at which it is evaluated.

DEFINITION 1.5 *The function F is the* CUMULATIVE DISTRIBUTION FUNCTION (CDF) *for the random variable X if*

$$F(a) = \Pr\{X \le a\}$$

for all real numbers a.

The CDF for the random variable D defined above is

$$F(a) = \begin{cases} 0 & \text{if } a < 0 \\ 0.82 & \text{if } 0 \le a < 1 \\ 1.0 & \text{if } a \ge 1 \end{cases}. \tag{1.1}$$

Figure 1.3 gives the graphical representation for F. The random variable T defined to represent the testing time for phones within a randomly chosen box is continuous and there are many possibilities for its probability law since we have not yet defined its probability space. As an example, the function G (see the right-hand graph in Figure 1.10) is the cumulative distribution function describing the randomness that might be associated with T:

$$G(a) = \begin{cases} 0 & \text{if } a < 0 \\ 1 - e^{-2a} & \text{if } a \ge 0 \end{cases}. \tag{1.2}$$

FIGURE 1.3 Cumulative distribution function for Equation (1.1) for the discrete random variable D.

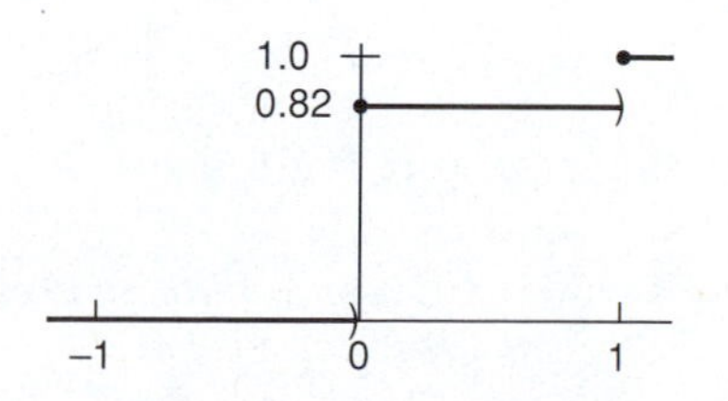

Property 1.6 A cumulative distribution function F has the following properties:

- $\lim_{a \to -\infty} F(a) = 0$,
- $\lim_{a \to +\infty} F(a) = 1$,
- $\lim_{a \to b^+} F(a) = F(b)$,
- $F(a) \le F(b)$ if $a < b$.

The first and second properties indicate that the graph of the cumulative distribution function always begins on the left at zero and limits to one on the right. The third property indicates that the cumulative distribution function is right-continuous. The fourth property indicates that the function is nondecreasing.

It is possible to describe the random nature of a discrete random variable by indicating the size of jumps in its cumulative distribution function. Such a function is called a probability mass function (denoted by a lowercase letter) and gives the probability of a particular value occurring.

DEFINITION 1.7 *The function f is the* PROBABILITY MASS FUNCTION *(pmf) of the discrete random variable X if*

$$f(k) = \Pr\{X = k\}$$

for every k in the range of X.

If the pmf is known, then the cumulative distribution function is easily found by

$$F(a) = \sum_{k \le a} f(k). \tag{1.3}$$

The situation for a continuous random variable is not quite as easy because the probability that any single given point occurs must be zero. Thus, we talk about the probability of an interval occurring. With this in mind, it is clear that a mass function is inappropriate for continuous random variables; instead, a probability density function (denoted by a lowercase letter) is used.

DEFINITION 1.8 *The function g is called the* PROBABILITY DENSITY FUNCTION *(pdf) of the continuous random variable Y if*

$$\int_a^b g(u)\, du = \Pr\{a \le Y \le b\}$$

for all a, b in the range of Y.

From Definition 1.8 we see that the pdf is the derivative of the cumulative distribution function and

$$G(a) = \int_{-\infty}^{a} g(u)\,du. \tag{1.4}$$

The cumulative distribution functions for the example random variables D and T are defined in Eqs. (1.1) and (1.2). We complete that example by giving the pmf for D and the pdf for T as follows:

$$f(k) = \begin{cases} 0.82 & \text{if } k = 0 \\ 0.18 & \text{if } k = 1 \\ 0 & \text{otherwise} \end{cases} \tag{1.5}$$

and

$$g(a) = \begin{cases} 2e^{-2a} & \text{if } a \geq 0 \\ 0 & \text{otherwise} \end{cases}. \tag{1.6}$$

EXAMPLE 1.2 Discrete random variables need not have finite ranges. A classical example of a discrete random variable with an infinite range is from 1920 due to Rutherford, Chadwick, and Ellis.[2] An experiment was performed to determine the number of α particles emitted by a radioactive substance in 7.5 seconds. The radioactive substance was chosen to have a long half-life so that the emission rate would be constant. After many experiments, it was found that the number of emissions in 7.5 seconds was a random variable, N, whose pmf could be described by

$$\Pr\{N = k\} = \frac{(3.87)^k e^{-3.87}}{k!} \quad \text{for } k = 0, 1, \cdots.$$

The discrete random variable N has a countably infinite range and the infinite sum of its pmf equals one. In fact, this distribution is fairly important and will be discussed later under the heading of the Poisson distribution. Figure 1.4 shows its pmf graphically.

The notion of independence is very important when dealing with more than one random variable. Although we postpone the discussion on multi-

FIGURE 1.4 The Poisson probability mass function of Example 1.2.

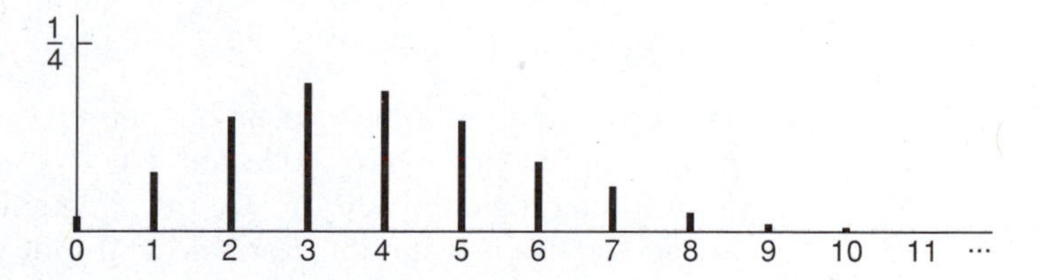

[2] See N. A. Rahman, *A Course in Theoretical Statistics* (New York: Hafner Publishing Co., 1968) pp. 209–210.

variate distribution functions until Section 1.5, we introduce the concept of independence at this point.

DEFINITION 1.9 *The random variables* $X_1, \cdots, X_n$ *are* INDEPENDENT *if*

$$\Pr\{X_1 \leq x_1, \cdots, X_n \leq x_n\} = \Pr\{X_1 \leq x_1\} \times \cdots \times \Pr\{X_n \leq x_n\}$$

for all possible values of $x_1, \cdots, x_n$.

Conceptually, random variables are independent if knowledge of one (or more) random variable does not "help" in making probability statements about the other random variables. Thus, an alternative definition of independence could be made using conditional probabilities (see Definition 1.3) where the random variables X_1 and X_2 are called independent if $\Pr\{X_1 \leq x_1 \mid X_2 \leq x_2\} = \Pr\{X_1 \leq x_1\}$ for all values of x_1 and x_2.

For example, suppose that T is a random variable denoting the length of time it takes for a barge to travel from a refinery to a terminal 800 miles downriver, and R is a random variable equal to 1 if the river condition is smooth when the barge leaves and 0 if the river condition is not smooth. After collecting data to estimate the probability laws governing T and R, we would not expect the two random variables to be independent because knowledge of the river conditions would help in determining the length of travel time.

One advantage of independence is that it is easier to obtain the distribution for sums of random variables if they are independent than if they are not. When the random variables are continuous, the pdf of the sum involves an integral called a *convolution*.

Property 1.10 Let X_1 and X_2 be independent continuous random variables with pdf's given by $f_1(\cdot)$ and $f_2(\cdot)$. Let $Y = X_1 + X_2$, and let $h(\cdot)$ be the pdf for Y. The pdf for Y can be written, for all y, as

$$h(y) = \int_{-\infty}^{\infty} f_1(y - x) f_2(x)\, dx.$$

Furthermore, if X_1 and X_2 are both nonnegative random variables, then

$$h(y) = \int_{0}^{y} f_1(y - x) f_2(x)\, dx.$$

EXAMPLE 1.3 Our electronic equipment is highly sensitive to voltage fluctuations in the power supply so we have collected data to estimate when these fluctuations occur. After much study, it has been determined that the time between voltage spikes is a random variable with pdf given by Eq. (1.6), where the unit of time is hours. Furthermore, it has been determined that the random variables describing the time between two successive voltage spikes are independent. We have just turned the equipment on and would like to know the probability that within the next 30 minutes at least two spikes will occur.

Let X_1 denote the time interval from when the equipment is turned on until the first voltage spike occurs, and let X_2 denote the time interval from when the first spike occurs until the second spike occurs. The question of interest is to find $\Pr\{Y \leq 0.5\}$, where $Y = X_1 + X_2$. Let the pdf for Y be denoted by $h(\cdot)$. Property 1.10 yields

$$h(y) = \int_0^y 4e^{-2(y-x)}e^{-2x}\,dx$$

$$= 4e^{-2y}\int_0^y dx = 4ye^{-2y},$$

for $y \geq 0$. The pdf of Y is now used to answer our question, namely,

$$\Pr\{Y \leq 0.5\} = \int_0^{0.5} h(y)\,dy = \int_0^{0.5} 4ye^{-2y}\,dy = 0.264.$$

It is also interesting to note that the convolution can be used to give the cumulative distribution function if the first pdf in the preceding property is replaced by the CDF; in other words, for *nonnegative* random variables we have

$$H(y) = \int_0^y F_1(y-x)f_2(x)\,dx. \tag{1.7}$$

Applying Eq. (1.7) to our voltage fluctuation question yields

$$\Pr\{Y \leq 0.5\} \equiv H(0.5) = \int_0^{0.5} [1 - e^{-2(0.5-x)}]2e^{-2x}\,dx = 0.264.$$ ■

We rewrite the convolution of Eq. (1.7) slightly to help in obtaining an intuitive understanding of why the convolution is used for sums. Again, assume that X_1 and X_2 are independent, nonnegative random variables with pdf's f_1 and f_2, then

$$P\{X_1 + X_2 \leq y\} = \int_0^y F_2(y-x)f_1(x)\,dx.$$

The interpretation of $f_1(x)\,dx$ is that it represents the probability that the random variable X_1 falls in the interval $(x, x + dx)$ or, equivalently, that X_1 is approxmately x. Now consider the time line in Figure 1.5. For the sum to be less than y, two events must occur: first, X_1 must be some value (call it x) less than y; second, X_2 must be less than the remaining time which is

FIGURE 1.5 Time line illustrating the convolution.

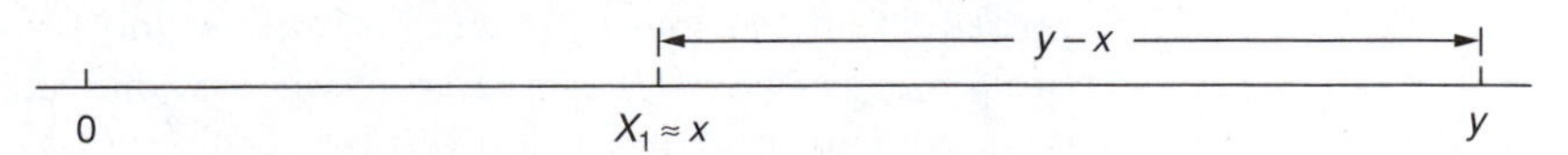

$y - x$. The probability of the first event is approximately $f_1(x)\,dx$, and the probability of the second event is $F_2(y - x)$. Because the two events are independent, they are multiplied together; and because the value of x can be any number between 0 and y, the integral is from 0 to y.

► *Suggestion: Do Exercises 1.3–1.6.*

1.3 MEAN AND VARIANCE

Many random variables have complicated distribution functions and it is therefore difficult to obtain an intuitive understanding of the behavior of the random variable by simply knowing the distribution function. Two measures, the mean and variance, are defined to aid in describing the randomness of a random variable. The *mean* equals the arithmetic average of infinitely many observations of the random variable and the *variance* is an indication of the variability of the random variable. To illustrate this concept, we use the square root of the variance, which is called the *standard deviation*. In the nineteenth century, the Russian mathematician P. L. Chebyshev showed that for any given distribution, *at least* 75% of the time the observed value of a random variable will be within two standard deviations of its mean and *at least* 93.75% of the time the observed value will be within four standard deviations of the mean. These are general statements; specific distributions will give much tighter bounds. (For example, a commonly used distribution is the normal "bell shaped" distribution. With the normal distribution, there is a 95.44% probability of being within two standard deviations of the mean.) Both the mean and variance are defined in terms of the expected value operator, which we now define.

DEFINITION 1.11 *Let h be a function defined on the real numbers and let X be a random variable. The* EXPECTED VALUE *of $h(X)$ is given, for X discrete, by*

$$E[h(X)] = \sum_k h(k)f(k)$$

where f is its pmf, and for X continuous, by

$$E[h(X)] = \int_{-\infty}^{\infty} h(s)f(s)\,ds$$

where f is its pdf.

EXAMPLE 1.4 A supplier sells eggs by the carton containing 144 eggs. There is a small probability that some eggs will be broken and he refunds money based on broken eggs. We let B be a random variable indicating the number of broken eggs per carton with a pmf given by

k	$f(k)$
0	0.779
1	0.195
2	0.024
3	0.002

A carton sells for \$4, but a refund of 5 cents is made for each broken egg. To determine the expected income per carton, we define the function h as follows

k	$h(k)$
0	4.00
1	3.95
2	3.90
3	3.85

Thus, $h(k)$ is the net revenue obtained when a carton is sold containing k broken eggs. Since it is not known ahead of time how many eggs are broken, we are interested in determining the *expected* net revenue for a carton of eggs. Definition 1.11 yields

$$\begin{aligned} E[h(B)] &= 4.00 \times 0.779 + 3.95 \times 0.195 \\ &\quad + 3.90 \times 0.024 + 3.85 \times 0.002 = 3.98755. \end{aligned}$$ ■

The expected value operator is a linear operator, and it is not difficult to show the following property.

Property 1.12 Let X and Y be two random variables with c being a constant, then

- $E[c] = c$,
- $E[cX] = cE[X]$,
- $E[X + Y] = E[X] + E[Y]$.

In the egg example, since the cost per broken egg is a constant ($c = 0.05$), the expected revenue per carton could be computed as

$$\begin{aligned} E[4.0 - 0.05B] &= 4.0 - 0.05E[B] \\ &= 4.0 - 0.05(0 \times 0.779 + 1 \times 0.195 + 2 \times 0.024 + 3 \times 0.002) \\ &= 3.98755. \end{aligned}$$

The expected value operator provides us with the procedure to determine the mean and variance.

DEFINITION 1.13 *The* MEAN, μ, *and* VARIANCE, σ^2, *of a random variable X are defined as*

$$\mu = E[X], \qquad \sigma^2 = E[(X - \mu)^2],$$

respectively. The STANDARD DEVIATION *is the square root of the variance.*

Property 1.14 The following are often helpful as computational aids:

- $\text{var}(X) = \sigma^2 = E[X^2] - \mu^2$,
- $\text{var}(cX) = c^2\,\text{var}(X)$,
- If $X \geq 0$, then $E[X] = \int_0^\infty [1 - F(s)]\,ds$ where $F(x) = \Pr\{X \leq x\}$,
- If $X \geq 0$, then $E[X^2] = 2\int_0^\infty s[1 - F(s)]\,ds$ where $F(x) = \Pr\{X \leq x\}$.

EXAMPLE 1.5 The mean and variance calculations for a discrete random variable can be easily illustrated by defining the random variable N to be the number of defective phones within a randomly chosen box from Example 1.1. In other words, N has the pmf given by

$$\Pr\{N = k\} = \begin{cases} 0.82 & \text{if } k = 0 \\ 0.15 & \text{if } k = 1 \\ 0.03 & \text{if } k = 2 \end{cases}.$$

The mean and variance are therefore given by

$$\begin{aligned} E[N] &= 0 \times 0.82 + 1 \times 0.15 + 2 \times 0.03 \\ &= 0.21, \\ \text{var}(N) &= (0 - 0.21)^2 \times 0.82 + (1 - 0.21)^2 \times 0.15 + (2 - 0.21)^2 \times 0.03 \\ &= 0.2259. \end{aligned}$$

Or, an easier calculation for the variance (Property 1.14) is

$$\begin{aligned} E[N^2] &= 0^2 \times 0.82 + 1^2 \times 0.15 + 2^2 \times 0.03 \\ &= 0.27, \\ \text{var}(N) &= 0.27 - 0.21^2 \\ &= 0.2259. \end{aligned}$$

EXAMPLE 1.6 The mean and variance calculations for a continuous random variable can be illustrated with the random variable T whose pdf was given by Eq. (1.6). The mean and variance are therefore given by

$$E[T] = \int_0^\infty 2se^{-2s}\,ds = 0.5,$$

$$\text{var}(T) = \int_0^\infty 2(s - 0.5)^2 e^{-2s}\,ds = 0.25.$$

Or, an easier calculation for the variance (Property 1.14) is

$$E[T^2] = \int_0^\infty 2s^2 e^{-2s}\,ds = 0.5,$$

$$\text{var}(T) = 0.5 - 0.5^2 = 0.25.$$

▶ *Suggestion: Do Exercises 1.7–1.13.*

1.4 IMPORTANT DISTRIBUTIONS

Many distribution functions are used so frequently that they have become known by special names. In this section, some of the major distribution functions are given. The student will find it helpful in years to come if these distributions are committed to memory. Several textbooks[3] are available that give more complete descriptions of distributions, and we recommend gaining a familiarity with a variety of distribution functions before any serious modeling is attempted.

Uniform-discrete. The random variable N has a discrete uniform distribution if there are two integers a and b such that the pmf of N can be written as

$$f(k) = \begin{cases} 1/(b-a+1) & \text{for } k = a, a+1, \cdots, b \\ 0 & \text{otherwise} \end{cases}. \tag{1.8}$$

Then,

$$E[N] = \frac{a+b}{2}, \qquad \text{var}(N) = \frac{(b-a+1)^2 - 1}{12}.$$

EXAMPLE 1.7 Consider rolling a fair die. Figure 1.6 shows the uniform pmf for the "number of dots" random variable. Notice in the figure that, as the name "uniform" implies, all the probabilities are the same. ■

Bernoulli. The random variable N has a Bernoulli distribution if there is a number $0 < p < 1$ such that the pmf of N can be written as

$$f(k) = \begin{cases} 1-p & \text{for } k = 0 \\ p & \text{for } k = 1 \\ 0 & \text{otherwise} \end{cases}. \tag{1.9}$$

Then,

$$E[N] = p, \qquad \text{var}(N) = p(1-p).$$

FIGURE 1.6 A discrete uniform probability mass function.

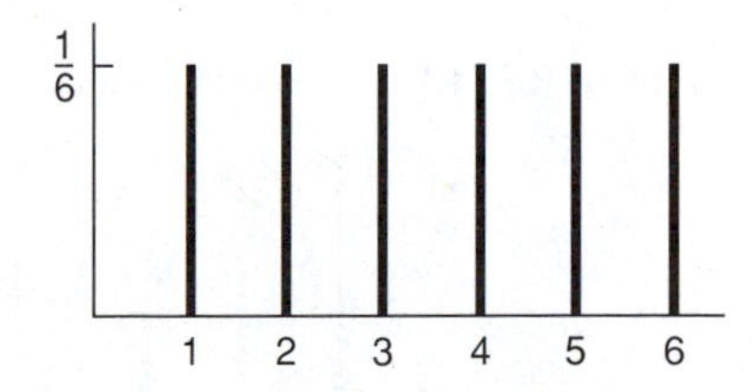

[3] Chapter 6 of A. M. Law and W. D. Kelton, *Simulation Modeling and Analysis,* 2nd ed. (New York: McGraw-Hill Book Company, 1991) contains a good summary of several distributions.

Binomial. (By James Bernoulli, 1654–1705; published posthumously in 1713.) The random variable N has a binomial distribution if there is a number $0 < p < 1$ and a positive integer n such that the pmf of N can be written as

$$f(k) = \begin{cases} \frac{n!}{k!(n-k)!} p^k (1-p)^{n-k} & \text{for } k = 0, 1, \cdots, n \\ 0 & \text{otherwise} \end{cases}. \tag{1.10}$$

Then,

$$E[N] = np, \qquad \text{var}(N) = np(1-p).$$

The value of p is often thought of as the probability of a success. The binomial pmf evaluated at k thus gives the probability of k successes occurring out of n trials. The binomial random variable with parameters p and n is the sum of n (independent) Bernoulli random variables each with parameter p.

EXAMPLE 1.8 We are monitoring calls at a switchboard in a large manufacturing firm and have determined that one-third of the calls are long distance and two-thirds of the calls are local. We have decided to pick four calls at random and would like to know how many calls in the group of four will be long distance. In other words, let N be a random variable indicating the number of long-distance calls in the group of four. Thus, N is binomial with $n = 4$ and $p = \frac{1}{3}$. It also happens that in this company, half of the individuals placing calls are women and half are men. We would also like to know how many of the group of four were calls placed by men. Let M denote the number of men placing calls; thus, M is binomial with $n = 4$ and $p = \frac{1}{2}$. The pmf's for these two random variables are shown in Figure 1.7. Notice that for $p = 0.5$, the pmf is symmetric, and as p varies from 0.5, the graph becomes skewed. ■

Geometric. The random variable N has a geometric distribution if there is a number $0 < p < 1$ such that the pmf of N can be written as

$$f(k) = \begin{cases} p(1-p)^{k-1} & \text{for } k = 1, 2, \cdots \\ 0 & \text{otherwise} \end{cases}. \tag{1.11}$$

FIGURE 1.7 Two binomial probability mass functions.

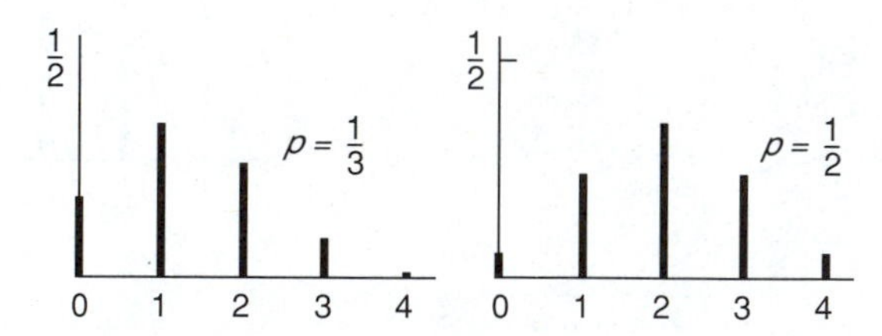

Then,

$$E[N] = \frac{1}{p}, \qquad \text{var}(N) = \frac{1-p}{p^2}.$$

The idea behind the geometric random variable is that it represents the number of trials until the first success occurs. In other words, p is thought of as the probability of success for a single trial, and we continually perform the trials until a success occurs. The random variable N is then set equal to the number of trials that we had to perform. Note that although the geometric random variable is discrete, its range is infinite.

EXAMPLE 1.9 A car saleswoman has made a statistical analysis of her previous sales history and determined that each day there is a 50% chance that she will sell a luxury car. After further careful analysis, it is also clear that a luxury car sale on one day is independent of the sale (or lack of it) on another day. On New Year's Day (a holiday in which the dealership is closed) the saleswoman is contemplating when she will sell her first luxury car of the year. If N is the random variable indicating the day of the first luxury car sale ($N = 1$ implies the sale was on January 2), then N is distributed according to the geometric distribution with $p = 0.5$, and its pmf is shown in Figure 1.8. Notice that theoretically the random variable has an infinite range, but for all practical purposes the probability of the random variable being larger than seven is negligible. ■

Poisson. (By Simeon Denis Poisson, 1781–1840; published in 1837.) The random variable N has a Poisson distribution if there is a number $\lambda > 0$ such that the pmf of N can be written as

$$f(k) = \begin{cases} (\lambda^k e^{-\lambda})/k! & \text{for } k = 0, 1, \cdots \\ 0 & \text{otherwise} \end{cases}. \tag{1.12}$$

Then,

$$E[N] = \lambda, \qquad \text{var}(N) = \lambda.$$

The Poisson distribution is the most important discrete distribution in stochastic modeling. It arises in many different circumstances. One use is as an

FIGURE 1.8 A geometric probability mass function.

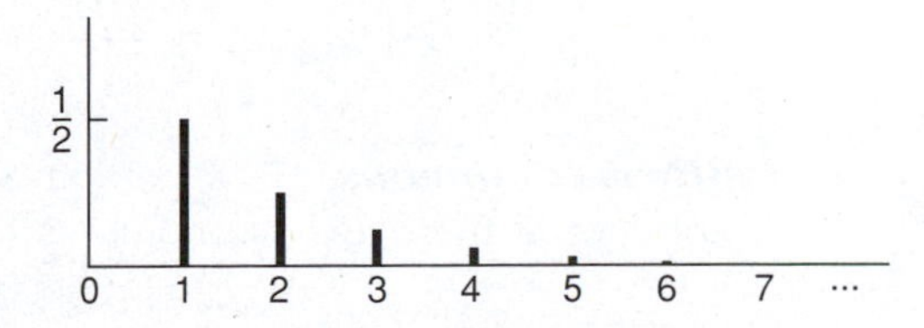

approximation to the binomial distribution. For n large and p small, the binomial is approximated by the Poisson by setting $\lambda = np$. For example, suppose we have a box of 144 eggs and there is a 1% probability that any one egg will break. Assuming that the breakage of eggs is independent of other eggs breaking, the probability that exactly 3 eggs will be broken out of the 144 can be determined using the binomial distribution with $n = 144$, $p = 0.01$, and $k = 3$; thus,

$$\frac{144!}{141!\,3!}(0.01)^3(0.99)^{141} = 0.1181,$$

or by the Poisson approximation with $\lambda = 1.44$, which yields

$$\frac{(1.44)^3 e^{-1.44}}{3!} = 0.1179.$$

In 1898, L. V. Bortkiewicz[4] reported that the number of deaths due to horse-kicks in the Prussian army was a Poisson random variable. Although this seems like a silly example, it is very instructive. The reason that the Poisson distribution holds in this case is due to the binomial approximation feature of the Poisson. Consider the situation: There would be a small chance of death by horse-kick for any one person (i.e., p small) but a large number of individuals in the army (i.e., n large). There are many analogous situations in modeling that deal with large populations and a small chance of occurrence for any one individual within the population. In particular, arrival processes (such as arrivals to a bus station in a large city) can often be viewed in this fashion and thus described by a Poisson distribution. Another common use of the Poisson distribution is in population studies. The population size of a randomly growing organism can often be described by a Poisson random variable. W. S. Gosset, using the pseudonym of Student, showed in 1907 that the number of yeast cells in 400 squares of haemocytometer followed a Poisson distribution. Radioactive emissions are also Poisson as indicated in Example 1.2. (Figure 1.4 also shows the Poisson pmf.)

Many arrival processes are well approximated using the Poisson probabilities. For example, the number of arriving telephone calls to a switchboard during a specified period of time or the number of arrivals to a teller at a bank during a fixed period of time are often modeled as a Poisson random variable. Specifically, we say that an arrival process is a Poisson process with mean rate λ if arrivals occur one at a time and the number of arrivals during an interval of length t is given by the random variable N_t where

$$\Pr\{N_t = k\} = \frac{(\lambda t)^k e^{-\lambda t}}{k!} \quad \text{for } k = 0, 1, \cdots. \tag{1.13}$$

Uniform-continuous. The random variable X has a continuous uniform distribution if there are two numbers a and b with $a < b$ such that the pdf

[4] Rahman, p. 206.

of X can be written as

$$f(s) = \begin{cases} 1/(b-a) & \text{for } a \leq s \leq b \\ 0 & \text{otherwise} \end{cases}. \tag{1.14}$$

Then its cumulative probability distribution is given by

$$F(s) = \begin{cases} 0 & \text{for } s < a \\ \dfrac{s-a}{b-a} & \text{for } a \leq s < b \\ 1 & \text{for } s \geq b \end{cases}$$

and

$$E[X] = \frac{a+b}{2}, \qquad \text{var}(X) = \frac{(b-a)^2}{12}.$$

The graphs for the pdf and CDF of the continuous uniform random variables are the simplest of the continuous distributions. As shown in Figure 1.9, the pdf is a rectangle and the CDF is a "ramp" function.

Exponential. The random variable X has an exponential distribution if there is a number $\lambda > 0$ such that the pdf of X can be written as

$$f(s) = \begin{cases} \lambda e^{-\lambda s} & \text{for } s \geq 0 \\ 0 & \text{otherwise} \end{cases}. \tag{1.15}$$

Then its cumulative probability distribution is given by

$$F(s) = \begin{cases} 0 & \text{for } s < 0 \\ 1 - e^{-\lambda s} & \text{for } s \geq 0 \end{cases}$$

and

$$E[X] = \frac{1}{\lambda}, \qquad \text{var}(X) = \frac{1}{\lambda^2}.$$

The exponential distribution is an extremely common distribution in probabilistic modeling. One very important feature is that the exponential distribution is the only continuous distribution that contains no memory. Specifically, an exponential random variable X has the property that

$$\Pr\{X > t + s \mid X > t\} = \Pr\{X > s\}.$$

FIGURE 1.9 The probability density function and cumulative distribution function for a continuous uniform distribution between 1 and 3.

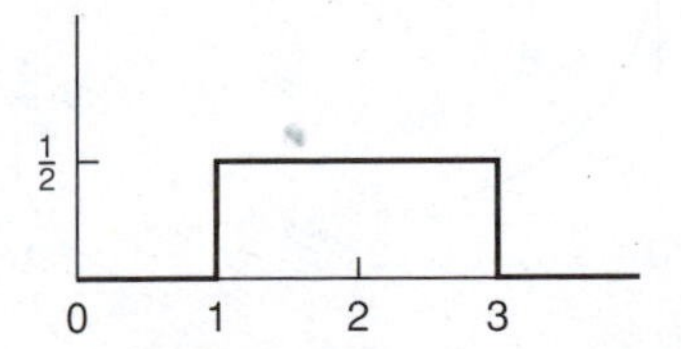

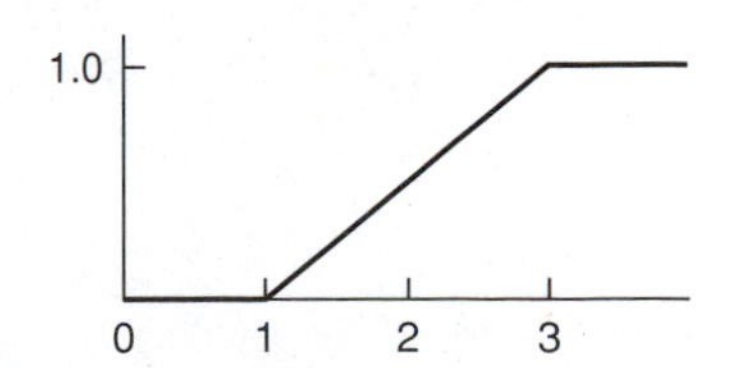

That is, if, for example, a machine's failure time is due to purely random events (such as voltage surges through a power line), then the exponential random variable would properly describe the failure time. However, if failure is due to the wearing out of machine parts, then the exponential distribution would not be suitable (see Exercise 1.23).

As a result of this lack of memory, a very important characteristic is that if the number of events within an interval of time occurs according to a Poisson random variable, then the time between events is exponential (and vice versa). Specifically, if an arrival process is a Poisson process [Eq. (1.13)] with mean rate λ, the times between arrivals are governed by an exponential distribution with mean $1/\lambda$. Furthermore, if an arrival process is such that the times between arrivals are exponentially distributed with mean $1/\lambda$, the number of arrivals in an interval of length t is a Poisson random variable with mean λt.

EXAMPLE 1.10 A software company has received complaints regarding their responsiveness for customer service. They have decided to analyze the arrival pattern of phone calls to customer service and have determined that the arrivals form a Poisson process with a mean of 120 calls per hour. Because a characteristic of a Poisson process is exponentially distributed interarrival times, we know that the distribution of the time between calls is exponentially distributed with a mean of 0.5 minutes. Thus, the graphs of the pdf and CDF describing the randomness of interarrival times are shown in Figure 1.10. ■

Erlang. (Named after the Danish mathematician A. K. Erlang for his extensive use of it and his pioneering work in queueing theory in the early 1900s.) The random variable X has an Erlang distribution if there is a positive integer m and a positive number λ such that the pdf of X can be written as

$$f(s) = \begin{cases} [m\lambda(m\lambda s)^{m-1}e^{-m\lambda s}]/(m-1)! & \text{for } s \geq 0 \\ 0 & \text{otherwise} \end{cases}. \qquad \textbf{(1.16)}$$

Then,

$$E[X] = \frac{1}{\lambda}, \qquad \text{var}(X) = \frac{1}{m\lambda^2}.$$

FIGURE 1.10 The probability density function and cumulative distribution function for an exponential distribution with mean 0.5.

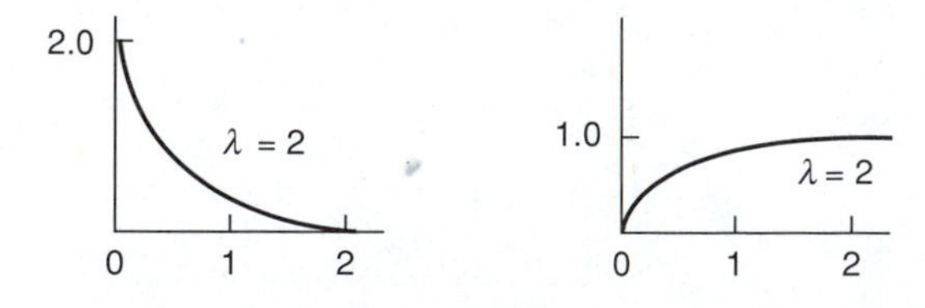

The Erlang distribution is a special case of a more general family of distributions called *gamma distributions,* which are not presented here. The usefulness of the Erlang is due to the fact that an Erlang random variable with parameters m and λ is the sum of m (independent) exponential random variables each with parameter $m\lambda$. In modeling process times, the exponential distribution is often inappropriate because the standard deviation is as large as the mean. Engineers usually try to design systems that yield mean process times significantly smaller than the standard deviation. Notice that for the Erlang distribution, the standard deviation decreases as the square root of the parameter m increases so that processing times with a small standard deviation can often be approximated by an Erlang random variable.

Figure 1.11 illustrates the effect of the parameter m by graphing the pdf for a type-2 Erlang and a type-10 Erlang. (The parameter m establishes the "type" for the Erlang distribution.) Notice that a type-1 Erlang is an exponential random variable so its pdf would have the form shown in Figure 1.10.

Weibull. In 1939, W. Weibull[5] developed a distribution for describing the breaking strength of various materials. Since that time, many statisticians have shown that the Weibull distribution can often be used to describe failure times for many different types of systems. The Weibull distribution has two parameters: a scale parameter, λ, and a shape parameter, α. Its cumulative distribution function is given by

$$F(s) = \begin{cases} 0 & \text{for } s < 0 \\ 1 - e^{-(\lambda s)^{\alpha}} & \text{for } s \geq 0 \end{cases}.$$

Both scale and shape parameters can be any positive number. The shape parameter determines the general shape of the pdf (see Figure 1.12), and the scale parameter either expands or contracts the pdf. The moments of the Weibull are a little difficult to express because they involve the gamma function. The gamma function is defined, for $x > 0$, as

$$\Gamma(x) = \int_0^{\infty} s^{x-1} e^{-s}\, ds,$$

FIGURE 1.11 Two Erlang probability density functions with mean 1.

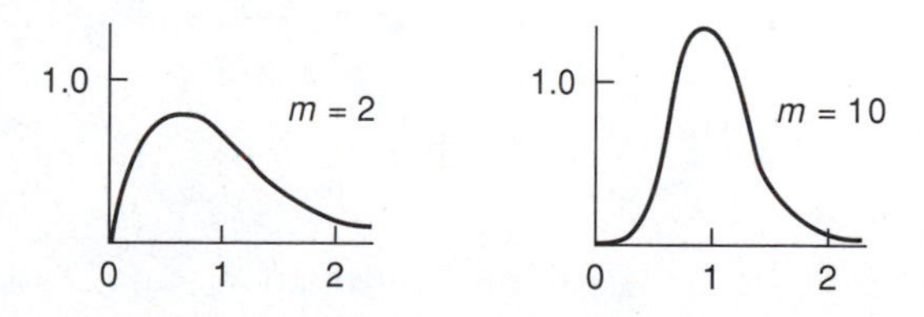

[5]R. E. Barlow and F. Proschan, *Statistical Theory of Reliability and Life Testing* (New York: Holt, Rinehart and Winston, 1975), p. 73.

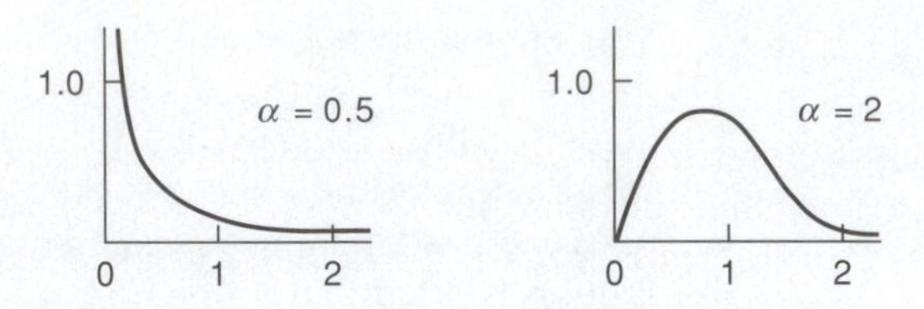

FIGURE 1.12 Two Weibull probability density functions with mean 1.

which implies the relationship $\Gamma(x+1) = x\Gamma(x)$, for $x \geq 1$. Thus, if x is a positive integer, $\Gamma(x) = (x-1)!$. If x is not an integer, and you are doing a problem by hand, your best option is to look up the gamma function in a mathematical tables book (see Appendix C). If it is necessary to incorporate the gamma function into a computer program, there are excellent approximations (see Appendix B.2.1).

The mean and variance for the Weibull distribution are

$$E[X] = \frac{1}{\lambda}\Gamma\left(1 + \frac{1}{\alpha}\right), \qquad \text{var}(X) = \frac{1}{\lambda^2}\left[\Gamma\left(1 + \frac{2}{\alpha}\right) - \Gamma\left(1 + \frac{1}{\alpha}\right)^2\right].$$

When the shape parameter is greater than 1, the shape of the Weibull pdf is unimodal, similar to the Erlang with its type parameter greater than 1. When the shape parameter equals 1, the Weibull pdf is an exponential pdf. When the shape parameter is less than 1, the pdf is similar to the exponential except that the graph is asymptotic to the y-axis instead of hitting the y-axis.

Figure 1.12 provides an illustration of the effect that the shape parameter has on the Weibull distribution. Because the mean values were held constant for the two pdf's shown in the figure, the value for λ varied. The left-hand pdf in the figure has $\lambda = 2$, which, together with $\alpha = 0.5$, yields a mean of 1 and a standard deviation of 2.236; the right-hand pdf has $\lambda = 0.886$, which, together with $\alpha = 2$, yields a mean of 1 and a standard deviation 0.523.

Normal. (Discovered by A. de Moivre, 1667–1754, but usually attributed to Carl Gauss, 1777–1855.) The random variable X has a normal distribution if there are two numbers μ and σ with $\sigma > 0$ such that the pdf of X can be written as

$$f(s) = \frac{1}{\sigma\sqrt{2\pi}} e^{-(s-\mu)^2/2\sigma^2} \quad \text{for } -\infty < s < \infty. \tag{1.17}$$

Then,

$$E[X] = \mu, \qquad \text{var}(X) = \sigma^2.$$

The normal distribution is the most common distribution and is recognized by most people by its bell-shaped curve. Its pdf and CDF are shown in Figure 1.13 for a normally distributed random variable with mean 0 and standard deviation 1.

Although the normal distribution is not widely used in stochastic modeling, it is, without question, the most important distribution in statistics. The normal distribution can be used to approximate both the binomial and Poisson distributions. A common rule of thumb is to approximate the binomial whenever n (the number of trials) is larger than 30. If $np < 5$, then use the Poisson for the approximation with $\lambda = np$. If $np \geq 5$, then use the normal for the approximation with $\mu = np$ and $\sigma^2 = np(1-p)$. Furthermore, the normal can be used to approximate the Poisson whenever $\lambda > 30$. When using a continuous distribution (like the normal) to approximate a discrete distribution (like the Poisson or binomial), the interval between the discrete values is usually split halfway. For example, if we desire to approximate the probability that a Poisson random variable will take on the values 29, 30, or 31 with a continuous distribution, then we would determine the probability that the continuous random variable is between 28.5 and 31.5.

EXAMPLE 1.11 The software company mentioned in the previous example has determined that the arrival process is Poisson with a mean arrival rate of 120 calls per hour. The company would like to know the probability that in any one hour 140 or more calls arrive. To determine that probability, let N be a Poisson random variable with $\lambda = 120$, let X be a random variable with $\mu = \sigma^2 = 120$, and let Z be a standard normal random variable (i.e., Z is normal with mean 0 and variance 1). Our question is answered as follows:

$$\begin{aligned}\Pr\{N \geq 140\} &\approx \Pr\{X > 139.5\} \\ &= \Pr\{Z > (139.5 - 120)/10.95\} \\ &= \Pr\{Z > 1.78\} = 1 - 0.9625 = 0.0375.\end{aligned}$$

The importance of the normal distribution is due to its property that sample means from almost any practical distribution will limit to the normal; this property is called the *Central Limit Theorem*. We state this property now even though it needs the concept of statistical independence, which is not yet defined. However, because the idea should be somewhat intuitive, we state the property at this point because it is so central to the use of the normal distribution.

FIGURE 1.13 The probability density function and cumulative distribution function for a standard normal distribution.

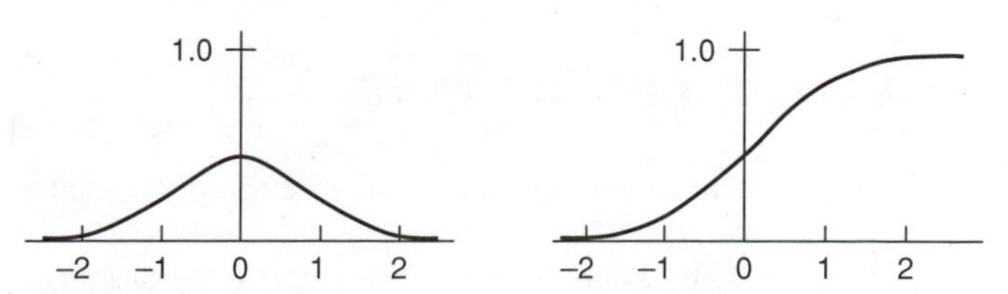

Property 1.15 ***Central Limit Theorem*** Let $\{X_1, X_2, \cdots, X_n\}$ be a sequence of n independent random variables each having the same distribution with mean μ and (finite) variance σ^2, and define

$$\overline{X} = \frac{X_1 + X_2 + \cdots + X_n}{n}.$$

Then, the distribution of the random variable Z defined by

$$Z = \frac{\overline{X} - \mu}{\sigma/\sqrt{n}}$$

approaches a normal distribution with mean 0 and standard deviation 1 as n gets large.

Skewness. Before moving to the discussion of more than one random variable, we mention an additional descriptor of distributions. The first moment gives the central tendency for random variables, and the second moment is used to measure variability. The third moment, which was not discussed previously, is useful as a measure of skewness (i.e., nonsymmetry). Specifically, the coefficient of skewness, γ, for a random variable T with mean μ and standard deviation σ is defined by

$$\gamma = \frac{E[(T - \mu)^3]}{\sigma^3}, \tag{1.18}$$

and the relation to the other moments is

$$E[(T - \mu)^3] = E[T^3] - 3\mu E[T^2] + 2\mu^3.$$

A symmetric distribution has $\gamma = 0$; if the mean is to the left of the mode, $\gamma < 0$ and the left-hand side of the distribution will have the longer tail; if the mean is to the right of the mode, $\gamma > 0$ and the right-hand side of the distribution will have the longer tail. For example, $\gamma = 0$ for the normal distribution, $\gamma = 2$ for the exponential distribution, and for the Erlang distribution with m phases we have $\gamma = 2/\sqrt{m}$. The Weibull pdf's shown in Figure 1.12 have skewness coefficients of 3.9 and 0.63, respectively, for the left-hand and right-hand graphs. Thus, the value of γ can help complete the intuitive understanding of a particular distribution.

▶ *Suggestion: Do Exercises 1.14–1.18.*

1.5 MULTIVARIATE DISTRIBUTIONS

The analysis of physical phenomena usually involves many distinct random variables. In this section we discuss the concepts involved when two random variables are defined. The extension to more than two is left to the imagination of the reader and the numerous textbooks that have been written on the subject.

DEFINITION 1.16 *The function F is called the* JOINT CUMULATIVE DISTRIBUTION FUNCTION *for X_1 and X_2 if*

$$F(a, b) = \Pr\{X_1 \le a, X_2 \le b\}$$

for a and b any two real numbers.

In a probability statement such as that shown on the right-hand side of the preceding equation, the comma means intersection of events and is read as "The probability that X_1 is less than or equal to *a and* X_2 is less than or equal to *b*." The initial understanding of joint probabilities is easiest with discrete random variables.

DEFINITION 1.17 *The function f is a* JOINT PMF *for the discrete random variables X_1 and X_2 if*

$$f(a, b) = \Pr\{X_1 = a, X_2 = b\}$$

for each (a, b) in the range of (X_1, X_2).

For the single-variable pmf, the height of the pmf at a specific value gives the probability that the random variable will equal that value. It is the same for the joint pmf except that the graph is in three dimensions. Thus, the height of the pmf evaluated at a specified ordered pair gives the probability that the random variables will equal those specified values (Figure 1.14).

It is sometimes necessary to obtain from the joint pmf the probability of one random variable without regard to the value of the second random variable.

DEFINITION 1.18 *The* MARGINAL PMF*'s for X_1 and X_2, denoted by f_1 and f_2, respectively, are*

$$f_1(a) = \Pr\{X_1 = a\} = \sum_k f(a, k)$$

for a in the range of X_1, and

$$f_2(b) = \Pr\{X_2 = b\} = \sum_k f(k, b)$$

for b in the range of X_2.

EXAMPLE 1.12 We return again to Example 1.1 to illustrate these concepts. The random variable R will indicate whether a randomly chosen box contains radio phones or plain phones; namely, if the box contains radio phones then we set $R = 1$ and if plain phones then $R = 0$. Also the random variable N will denote the number of defective phones in the box. Thus, according to the

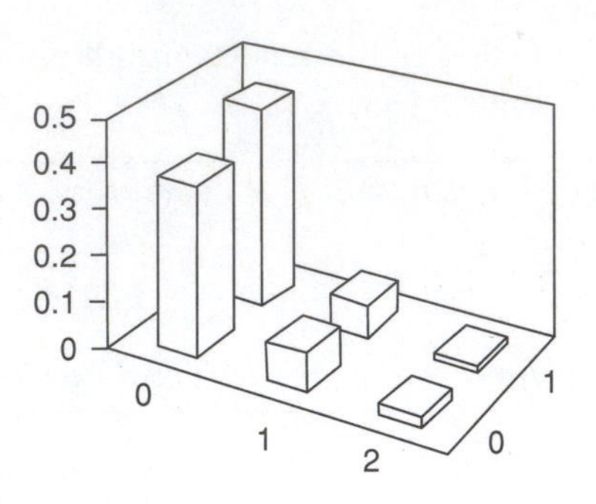

FIGURE 1.14 Probability mass function for two discrete random variables from Example 1.12.

probabilities defined in Example 1.1, the joint pmf,

$$f(a, b) = \Pr\{R = a, N = b\},$$

is defined by

	0	1	2
0	0.37	0.08	0.02
1	0.45	0.07	0.01

By summing in the "margins," we obtain the marginal pmf for R and N separately; namely,

	0	1	2	$f_1(a)$
0	0.37	0.08	0.02	0.47
1	0.45	0.07	0.01	0.53
$f_2(b)$	0.82	0.15	0.03	

Thus we see, for example, that the probability of choosing a box with radio phones (i.e., $\Pr\{R = 1\}$) is 53%, the probability of choosing a box of radio phones that has one defective phone (i.e., $\Pr\{R = 1, N = 1\}$) is 7%, and the probability that both phones in a randomly chosen box (i.e., $\Pr\{N = 2\}$) are defective is 3%. ■

Continuous random variables are treated in an analogous manner to the discrete case. The major difference in moving from one continuous random variable to two is that probabilities are given in terms of a volume under a surface instead of an area under a curve (see Figure 1.15 for representation of a joint pdf).

DEFINITION 1.19 *The functions g, g_1, and g_2 are the* JOINT PDF *for X_1 and X_2, the* MARGINAL PDF *for X_1, and the* MARGINAL PDF *for X_2, respectively, as the following hold:*

$$\Pr\{a_1 \leq X_1 \leq b_1,\ a_2 \leq X_2 \leq b_2\} = \int_{a_2}^{b_2} \int_{a_1}^{b_1} g(s_1, s_2)\, ds_1\, ds_2,$$

$$g_1(a) = \int_{-\infty}^{\infty} g(a, s)\, ds,$$

$$g_2(b) = \int_{-\infty}^{\infty} g(s, b)\, ds,$$

where

$$\Pr\{a \leq X_1 \leq b\} = \int_a^b g_1(s)\, ds,$$

$$\Pr\{a \leq X_2 \leq b\} = \int_a^b g_2(s)\, ds.$$

We return now to the concept of conditional probabilities (Definition 1.3). The situation often arises in which the experimentalist has knowledge regarding one random variable and would like to use that knowledge in predicting the value of the other (unknown) random variable. Such predictions are possible through conditional probability functions

---

DEFINITION 1.20 *Let f be a joint pmf for the discrete random variables X_1 and X_2 with f_2 the marginal pmf for X_2. Then the* CONDITIONAL PMF *for X_1,* GIVEN *that $X_2 = b$, is defined, if $\Pr\{X_2 = b\} \neq 0$, to be*

$$f_{1|b}(a) = \frac{f(a, b)}{f_2(b)},$$

where

$$\Pr\{X_1 = a \mid X_2 = b\} = f_{1|b}(a).$$

FIGURE 1.15 Probability density function for two continuous random variables from Example 1.13.

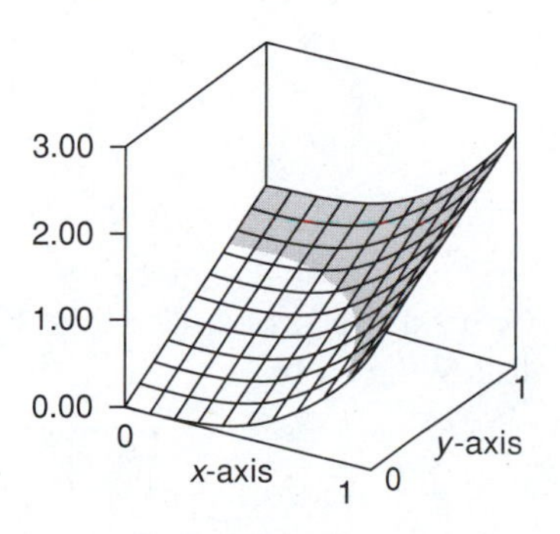

DEFINITION 1.21 *Let g be a joint pdf for continuous random variables X_1 and X_2 with g_2 the marginal pdf for X_2. Then the* CONDITIONAL PDF *for X_1,* GIVEN *that $X_2 = b$, is defined to be*

$$g_{1|b}(a) = \frac{g(a, b)}{g_2(b)},$$

where

$$\Pr\{a_1 \leq X_1 \leq a_2 \mid X_2 = b\} = \int_{a_1}^{a_2} g_{1|b}(s)\, ds.$$

The conditional statements for X_2 given a value for X_1 are made similar to those of Definitions 1.20 and 1.21 with the subscripts reversed. These conditional statements can be illustrated by using Example 1.12. We have already determined that the probability of having a box full of defective phones is 3%; however, let us assume that it is already known that we have picked a box of radio phones. Now, given a box of radio phones, the probability of both phones being defective is

$$f_{2|a=1}(2) = \frac{f(1, 2)}{f_1(1)} = \frac{0.01}{0.53} = 0.0189;$$

thus, knowledge that the box consisted of radio phones enabled a more accurate prediction of the probabilities that both phones were defective. Or to consider a different situation, assume that we know the box has both phones defective. The probability that the box contains plain phones is

$$f_{1|b=2}(0) = \frac{f(0, 2)}{f_2(2)} = \frac{0.02}{0.03} = 0.6667.$$

EXAMPLE 1.13 Let X and Y be two continuous random variables with joint pdf given by

$$f(x, y) = \frac{4}{3}(x^3 + y) \quad \text{for } 0 \leq x \leq 1, 0 \leq y \leq 1.$$

Utilizing Definition 1.19, we obtain

$$f_1(x) = \frac{4}{3}(x^3 + 0.5) \quad \text{for } 0 \leq x \leq 1,$$

$$f_2(y) = \frac{4}{3}(y + 0.25) \quad \text{for } 0 \leq y \leq 1.$$

To find the probability that Y is less than or equal to 0.5, we perform the following steps:

$$\Pr\{Y \leq 0.5\} = \int_0^{0.5} f_2(y)\, dy = \frac{4}{3}\int_0^{0.5} (y + 0.25)\, dy = \frac{1}{3}.$$

To find the probability that Y is less than or equal to 0.5 given that we know $X = 0.1$, we perform

$$\begin{aligned}\Pr\{Y \le 0.5 \mid X = 0.1\} &= \int_0^{0.5} f_{2|0.1}(y)\,dy \\ &= \int_0^{0.5} \frac{0.1^3 + y}{0.1^3 + 0.5}\,dy \\ &= \frac{0.1255}{0.501} \approx \frac{1}{4}.\end{aligned}$$

■

■ **EXAMPLE 1.14** Let U and V be two continuous random variables with a joint pdf given by

$$g(u, v) = 8u^3 v \quad \text{for } 0 \le u \le 1, 0 \le v \le 1.$$

The marginal pdf's are

$$\begin{aligned}g_1(u) &= 4u^3 \quad \text{for } 0 \le u \le 1, \\ g_2(v) &= 2v \quad \text{for } 0 \le v \le 1.\end{aligned}$$

The following two statements are easily verified.

$$\Pr\{0.1 \le V \le 0.5\} = \int_{0.1}^{0.5} 2v\,dv = 0.24,$$

$$\Pr\{0.1 \le V \le 0.5 \mid U = 0.1\} = 0.24.$$

■

The preceding example illustrates independence. Notice in the example that knowledge of the value of U did not change the probabilities regarding the probability statement of V.

DEFINITION 1.22 *Let f be the joint probability distribution (pmf if discrete and pdf if continuous) of two random variables X_1 and X_2. Furthermore, let f_1 and f_2 be the marginals for X_1 and X_2, respectively. If*

$$f(a, b) = f_1(a) f_2(b)$$

for all a and b, then X_1 and X_2 are called INDEPENDENT.

Independent random variables are much easier to work with because of their separability. However, in the use of Definition 1.22, it is important to test the property for *all* values of a and b. It would be easy to make a mistake by stopping after the equality was shown to hold for only one particular pair of a, b values. Once independence has been shown, the following property is very useful.

Property 1.23 Let X_1 and X_2 be independent random variables. Then

$$E[X_1 X_2] = E[X_1]E[X_2]$$

and

$$\mathrm{var}(X_1 + X_2) = \mathrm{var}(X_1) + \mathrm{var}(X_2).$$

EXAMPLE 1.15 Consider again the random variables R and N defined in Example 1.12. We see from the marginal pmf's given in that example that $E[R] = 0.53$ and $E[N] = 0.21$. We also have

$$\begin{aligned} E[R \cdot N] &= 0 \times 0 \times 0.37 + 0 \times 1 \times 0.08 + 0 \times 2 \times 0.02 \\ &\quad + 1 \times 0 \times 0.45 + 1 \times 1 \times 0.07 + 1 \times 2 \times 0.01 = 0.09. \end{aligned}$$

Thus, it is possible to say that the random variables R and N are not independent since $0.53 \times 0.21 \neq 0.09$. If, however, the expected value of the product of two random variables equals the product of the two individual expected values, the claim of independence is *not* proven. ■

We close this chapter by giving two final measures that are used to express the relationship between two dependent random variables. The first measure is called the *covariance* and the second measure is called the *correlation coefficient*.

DEFINITION 1.24 *The* COVARIANCE *of two random variables, X_1 and X_2, is defined by*

$$\mathrm{cov}(X_1, X_2) = E[(X_1 - E[X_1])(X_2 - E[X_2])].$$

Property 1.25 The following is often helpful as a computational aid:

$$\mathrm{cov}(X_1, X_2) = E[X_1 X_2] - \mu_1 \mu_2.$$

where μ_1 and μ_2 are the means for X_1 and X_2, respectively.

If you compare Property 1.23 to Property 1.25, it should be clear that random variables that are independent have zero covariance. However, it is possible to obtain random variables with zero covariance that are not independent (see Example 1.17 later on in this section). A principal use of the covariance is in the definition of the correlation coefficient, which is a measure of the linear relationship between two random variables.

DEFINITION 1.26 *Let X_1 be a random variable with mean μ_1 and variance σ_1^2. Let X_2 be a random variable with mean μ_2 and variance σ_2^2. The* CORRELATION COEFFICIENT, *denoted by ρ, of X_1 and X_2 is defined by*

$$\rho = \frac{\mathrm{cov}(X_1, X_2)}{\sqrt{\mathrm{var}(X_1)\,\mathrm{var}(X_2)}} = \frac{E[X_1 X_2] - \mu_1 \mu_2}{\sigma_1 \sigma_2}$$

The correlation coefficient is always between -1 and $+1$. A negative correlation coefficient indicates that if one random variable happens to be large, the other random variable is likely to be small. A positive correlation coefficient indicates that if one random variable happens to be large, the other random variable is also likely to be large. The following examples illustrate this concept.

EXAMPLE 1.16 Let X_1 and X_2 denote two discrete random variables, where X_1 ranges from 1 to 3 and X_2 ranges from 10 to 30. Their joint and marginal pmf's are given in the following table:

	10	20	30	$f_1(\cdot)$
1	0.28	0.08	0.04	0.4
2	0.04	0.12	0.04	0.2
3	0.04	0.08	0.28	0.4
$f_2(\cdot)$	0.36	0.28	0.36	

The following facts should not be difficult to verify: $\mu_1 = 2.0$, $\sigma_1^2 = 0.8$, $\mu_2 = 20.0$, $\sigma_2^2 = 72.0$, and $E[X_1X_2] = 44.8$. Therefore, the correlation coefficient of X_1 and X_2 is given by

$$\rho = \frac{44.8 - 2 \times 20}{\sqrt{0.8 \times 72}} = 0.632.$$

The conditional probabilities will help verify the intuitive concept of a positive correlation coefficient. Figure 1.16 contains a graph illustrating the conditional probabilities of X_2 given various values of X_1; the area of each circle in the figure is proportional to the conditional probability. Thus, the figure gives a visual representation that as X_1 increases, it is likely (but *not* necessary) that X_2 will increase. For example, the top right-hand circle represents $\Pr\{X_2 = 30 \mid X_1 = 3\} = 0.7$, and the middle right-hand circle represents $\Pr\{X_2 = 20 \mid X_1 = 3\} = 0.2$.

As a final example, we switch the top and middle right-hand circles in Figure 1.16 so that the appearance is not so clearly linear. (That is, let $\Pr\{X_1 = 3, X_2 = 20\} = 0.28, \Pr\{X_1 = 3, X_2 = 30\} = 0.08$, and all other

FIGURE 1.16 Graphical representation for conditional probabilities of X_2 given X_1 from Example 1.16, where the correlation coefficient is 0.632.

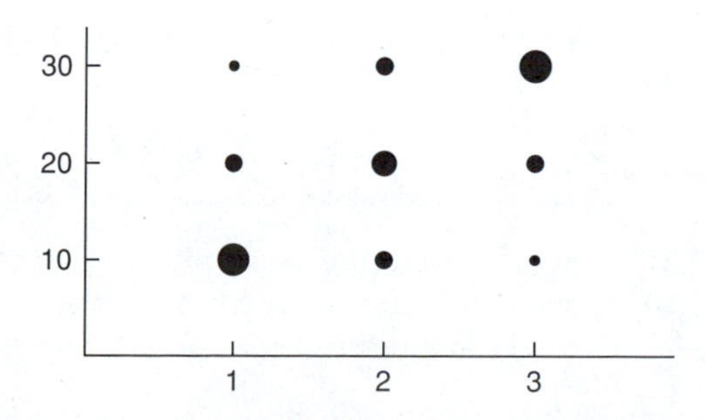

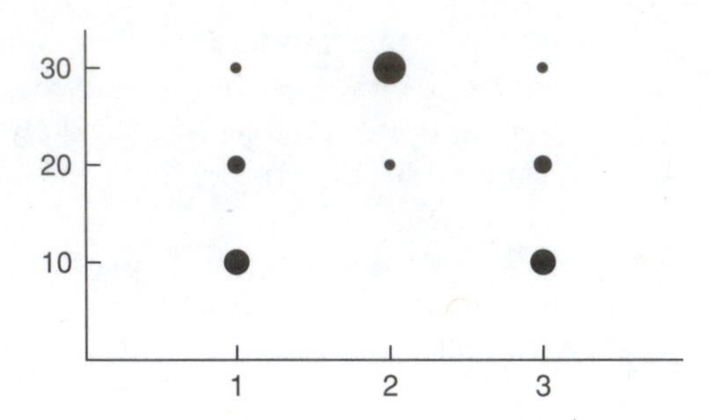

FIGURE 1.17 Graphical representation for conditional probabilities of X_2 given X_1 from Example 1.17, where the correlation coefficient is zero.

probabilities the same.) With this change, μ_1 and σ_1^2 remain unchanged, $\mu_2 = 18$, $\sigma_2^2 = 48.0$, $\text{cov}(X_1, X_2) = 2.8$, and the correlation coefficient is $\rho = 0.452$. Thus, as the linear relationship between X_1 and X_2 weakens, the value of ρ becomes smaller. ■

■ **EXAMPLE 1.17** Let X_1 and X_2 denote two discrete random variables, where X_1 ranges from 1 to 3 and X_2 ranges from 10 to 30. Their joint and marginal pmf's are given in the following table:

	10	20	30	$f_1(\cdot)$
1	0.28	0.08	0.04	0.4
2	0.00	0.02	0.18	0.2
3	0.28	0.08	0.04	0.4
$f_2(\cdot)$	0.56	0.18	0.26	

Again, we give the various measures and allow the reader to verify their accuracy: $\mu_1 = 2, \mu_2 = 17$, and $E[X_1 X_2] = 34$. Therefore, the correlation coefficient of X_1 and X_2 is zero so there is no *linear* relation between X_1 and X_2; however, the two random variables are clearly dependent. If X_1 is either one or three, then the most likely value of X_2 is 10; whereas, if X_1 is 2, then it is impossible for X_2 to have the value of 10; thus, the random variables must be dependent. If you observe the representation of the conditional probabilities in Figure 1.17, then the lack of a linear relationship is obvious. ■

▶ *Suggestion: Do Exercises 1.19–1.25.*

1.6 EXERCISES

1.1 A manufacturing company ships (by truckload) its product to three different distribution centers on a weekly basis. Demands vary from week to week ranging over 0, 1, and 2 truckloads needed at each distribution center.

Conceptualize an experiment in which a week is selected and then the number of truckloads demanded at each of the three centers is recorded.

(a) Describe the sample space, i.e., list all outcomes.

(b) How many possible different events are there?

(c) Write the event that represents "a total of three truckloads are needed for the week."

(d) If each event containing a single outcome has the same probability, what is the probability that a total demand for three truckloads will occur?

1.2 A library has classified its books into fiction and nonfiction. Furthermore, all books can also be described as hardback and paperback. As an experiment, we pick a book at random and record whether it is fiction or nonfiction and whether it is paperback or hardback.

(a) Describe the sample space, i.e., list all outcomes.

(b) Describe the event space, i.e., list all events.

(c) Define a probability measure such that the probability of picking a nonfiction paperback is 0.15, the probability of picking a nonfiction book is 0.30, and the probability of picking a fiction hardback is 0.65.

(d) Using the probabilities from part (c), find the probability of picking a fiction book given that the book chosen is known to be a paperback.

1.3 Let N be a random variable describing the number of defective items in a box from Example 1.1. Draw the graph for the cumulative distribution function of N and give its pmf.

1.4 Let X be a random variable with cumulative distribution function given by

$$G(a) = \begin{cases} 0 & \text{for } a < 0 \\ a^2 & \text{for } 0 \le a < 1 \\ 1 & \text{for } a \ge 1 \end{cases} .$$

(a) Give the pdf for X.

(b) Find $\Pr\{X \ge 0.5\}$.

(c) Find $\Pr\{0.5 < X \le 0.75\}$.

(d) Let X_1 and X_2 be independent random variables with their CDF given by $G(\cdot)$. Find $\Pr\{X_1 + X_2 \le 1\}$.

1.5 Let T be a random variable with pdf given by

$$f(t) = \begin{cases} 0 & \text{for } t < 0.5 \\ ke^{-2(t-0.5)} & \text{for } t \ge 0.5 \end{cases} .$$

(a) Find k.

(b) Find $\Pr\{0.25 \le T \le 1\}$.

(c) Find $\Pr\{T \le 1.5\}$.

(d) Give the cumulative distribution function for T.

(e) Let the independent random variables T_1 and T_2 have their pdf's given by $f(\cdot)$. Find $\Pr\{1 \leq T_1 + T_2 \leq 2\}$.

(f) Let $Y = X + T$, where X is independent of T and is defined by the previous problem. Give the pdf for Y.

1.6 Let U be a random variable with pdf given by

$$h(u) = \begin{cases} 0 & \text{for } u < 0 \\ u & \text{for } 0 \leq u < 1 \\ 2 - u & \text{for } 1 \leq u < 2 \\ 0 & \text{for } u \geq 2 \end{cases}.$$

(a) Find $\Pr\{0.5 < U < 1.5\}$.

(b) Find $\Pr\{0.5 \leq U \leq 1.5\}$.

(c) Find $\Pr\{0 \leq U \leq 1.5,\ 0.5 \leq U \leq 2\}$. (A comma acts as an intersection and is read as "and.")

(d) Give the cumulative distribution function for U and calculate $\Pr\{U \leq 1.5\} - \Pr\{U \leq 0.5\}$.

1.7 An independent roofing contractor has determined that the number of jobs obtained for the month of September varies. From previous experience, the probabilities of obtaining 0, 1, 2, or 3 jobs have been determined to be 0.1, 0.35, 0.30, and 0.25, respectively. The profit obtained from each job is $300. What is the expected profit for September?

1.8 There are three investment plans for your consideration. Each plan calls for an investment of $25,000 and the return will be one year later. Plan A will return $27,500. Plan B will return $27,000 or $28,000 with probabilities 0.4 and 0.6, respectively. Plan C will return $24,000, $27,000, or $33,000 with probabilities 0.2, 0.5, and 0.3, respectively. If your objective is to maximize the expected return, which plan should you choose? Are there considerations that might be relevant other than simply the expected values?

1.9 Let the random variables A, B, C denote the returns from investment plans A, B, and C, respectively, from Problem 1.8. What are the mean and standard deviations of the three random variables?

1.10 Let N be a random variable with cumulative distribution function given by

$$F(x) = \begin{cases} 0 & \text{for } x < 1 \\ 0.2 & \text{for } 1 \leq x < 2 \\ 0.5 & \text{for } 2 \leq x < 3 \\ 0.8 & \text{for } 3 \leq x < 4 \\ 1 & \text{for } x \geq 4 \end{cases}.$$

Find the mean and standard deviation of N.

1.11 Prove that the $E[(X - \mu)^2] = E[X^2] - \mu^2$ for any random variable X whose mean is μ.

1.12 Find the mean and standard deviation for X as defined in Problem 1.4.

1.13 Find the mean and standard deviation for U as defined in Problem 1.6. Also, find the mean and standard deviation using the last two properties mentioned in Property 1.14.

Use the appropriate distribution from Section 1.4 to answer the questions in Problems 1.14 through 1.18.

1.14 A manufacturing company produces parts of which 97% are within specifications and 3% are defective (outside specifications). There is apparently no pattern to the production of defective parts; thus, we assume that whether or not a part is defective is independent of other parts.

(a) What is the probability that there will be no defective parts in a box of five?

(b) What is the probability that there will be exactly two defective parts in a box of five?

(c) What is the probability that there will be two or more defective parts in a box of five?

(d) Use the Poisson distribution to approximate the probability that there will be 4 or more defective parts in a box of 40.

(e) Use the normal distribution to approximate the probability that there will be 20 or more defective parts in a box of 400.

1.15 A store sells two types of tables: plain and deluxe. When an order for a table arrives, there is an 80% chance that the plain table will be desired.

(a) Out of five orders, what is the probability that no deluxe tables will be desired?

(b) Assume that each day five orders arrive and that today (Monday) an order came for a deluxe table. What is the probability that the first day in which one or more deluxe tables are again ordered will be in three more days (Thursday)? What is the expected number of days until a deluxe table is desired?

(c) Actually, the number of orders each day is a Poisson random variable with a mean of 5. What is the probability that exactly five orders will arrive on a given day?

1.16 A vision system is designed to measure the angle at which the arm of a robot deviates from the vertical; however, the vision system is not totally accurate. The result from observations is a continuous random variable with a uniform distribution. If the measurement indicates that the range of the angle is between 9.7 and 10.5 degrees, what is the probability that the actual angle is between 9.9 and 10.1 degrees?

1.17 The dispatcher at a central fire station has observed that the time between calls is an exponential random variable with a mean of 32 minutes.

(a) A call has just arrived. What is the probability that the next call will arrive within the next half hour?

(b) What is the probability that there will be exactly two calls during the next hour?

1.18 In an automated soldering operation, the location at which the solder is placed is very important. The deviation from the center of the board is a normally distributed random variable with a mean of 0 inches and a standard deviation of 0.01 inches. (A positive deviation indicates a deviation to the right of the center and a negative deviation indicates a deviation to the left of the center.)

(a) What is the probability that on a given board the actual location of the solder deviated by less than 0.005 inches (in absolute value) from the center?

(b) What is the probability that on a given board the actual location of the solder deviated by more than 0.02 inches (in absolute value) from the center?

1.19 The purpose of this problem is to illustrate the dangers of statistics, especially with respect to categorical data. In this example, the data can be used to support contradicting claims, depending on the inclinations of the person doing the reporting! The population in which we are interested is made up of males and females, those who are sick and not sick, and those who received treatment prior to becoming sick and those who did not receive prior treatment. (In the following questions, assume that the treatment has no adverse side effects.) The population numbers are as follows:

	MALES	
	Sick	*Not sick*
Treated	200	300
Not treated	50	50

	FEMALES	
	Sick	*Not sick*
Treated	50	100
Not treated	200	370

(a) What is the conditional probability of being sick given that the treatment was received and the patient is a male?

(b) Considering only the population of males, should the treatment be recommended?

(c) Considering only the population of females, should the treatment be recommended?

(d) Considering the entire population, should the treatment be recommended?

1.20 Let X and Y be two discrete random variables where their joint pmf

$$f(a, b) = \Pr\{X = a,\ Y = b\}$$

is defined by

	0	1	2
10	0.01	0.06	0.03
11	0.02	0.12	0.06
12	0.02	0.18	0.10
13	0.07	0.24	0.09

with the possible values for X being 10 through 13 and the possible values for Y being 0 through 2.

(a) Find the marginal pmf's for X and Y and then find the $\Pr\{X = 11\}$ and $E[X]$.

(b) Find the conditional pmf for X given that $Y = 1$ and then find the $\Pr\{X = 11 \mid Y = 1\}$ and find the $E[X \mid Y = 1]$.

(c) Are X and Y independent? Why or why not?

(d) Find $\Pr\{X = 13,\ Y = 2\}, \Pr\{X = 13\}$, and $\Pr\{Y = 2\}$. [Now make sure your answer to part (c) was correct.]

1.21 Let S and T be two continuous random variables with a joint pdf given by

$$f(s, t) = kst^2 \quad \text{for } 0 \le s \le 1,\ 0 \le t \le 1,$$

and zero elsewhere.

(a) Find the value of k.

(b) Find the marginal pdf's for S and T and then find the $\Pr\{S \le 0.5\}$ and $E[S]$.

(c) Find the conditional pdf for S given that $T = 0.1$ and then find the $\Pr\{S \le 0.5 \mid T = 0.1\}$ and find the $E[S \mid T = 0.1]$.

(d) Are S and T independent? Why or why not?

1.22 Let U and V be two continuous random variables with joint pdf given by

$$g(u, v) = e^{-u-v} \quad \text{for } u \ge 0, v \ge 0,$$

and zero elsewhere.

(a) Find the marginal pdf's for U and V and then find the $\Pr\{U \le 0.5\}$ and $E[U]$.

(b) Find the conditional pdf for U given that $V = 0.1$ and then find the $\Pr\{U \le 0.5 \mid V = 0.1\}$ and find the $E[U \mid V = 0.1]$.

(c) Are U and V independent? Why or why not?

1.23 This problem is to consider the importance of keeping track of history when discussing the reliability of a machine. Let T be a random variable that indicates the time until failure for the machine. Assume that T has a uniform distribution from zero to two years and answer the question, "What is the probability that the machine will continue to work for at least three more months?"

(a) Assume the machine is new.

(b) Assume the machine is one year old and has not yet failed.

(c) Now assume that T has an exponential distribution with mean one year, and answer parts (a) and (b) again.

(d) Is it important to know how old the machine is in order to answer the question, "What is the probability that the machine will continue to work for at least three more months?"

1.24 Determine the correlation coefficient for the random variables X and Y from Example 1.13.

1.25 A shipment containing 1000 steel rods has just arrived. Two measurements are of interest: the cross-sectional area and the force that each rod can support. We conceptualize two random variables: A and B. The random variable A is the cross-sectional area, in square centimeters, of the chosen rod, and B is the force, in kilonewtons, that causes the rod to break. Both random variables can be approximated by a normal distribution. (A generalization of the normal distribution to two random variables is called a *bivariate normal distribution.*) The random variable A has a mean of 6.05 cm^2 and a standard deviation of 0.1 cm^2. The random variable B has a mean of 132 kN and a standard deviation of 10 kN. The correlation coefficient for A and B is 0.8.

To answer the following questions, use these facts: If X_1 and X_2 are bivariate normal random variables with means μ_1 and μ_2, respectively, variances σ_1^2 and σ_2^2, respectively, and a correlation coefficient ρ, then

- The marginal distribution of X_1 is normal,
- The conditional distribution of X_2 given X_1 is normal,
- $E[X_2 \mid X_1 = x] = \mu_2 + \rho(\sigma_2/\sigma_1)(x - \mu_1)$,
- $\text{var}(X_2 \mid X_1 = x) = \sigma_2^2(1 - \rho^2)$.

(a) Specifications call for the rods to have a cross-sectional area of between 5.9 and 6.1 cm^2. What is the expected number of rods that will have to be discarded because of size problems?

(b) The rods must support a force of 31 kN, and the engineer in charge has decided to use a safety factor of 4; therefore, design specifications call for each rod to support a force of at least 124 kN. What is the expected number of rods that will have to be discarded because of strength problems?

(c) A rod has been selected, and its cross-sectional area measures 5.94 cm^2. What is the probability that it will not support the force required in the specifications?

(d) A rod has been selected, and its cross-sectional area measures 6.08 cm^2. What is the probability that it will not support the force required in the specifications?

2

Markov Chains

Many decisions must be made within the context of randomness. Random failures of equipment, fluctuating production rates, and unknown demands are all part of normal decision-making processes. In an effort to quantify, understand, and predict the effects of randomness, the mathematical theory of probability and stochastic processes has been developed.

This chapter introduces one type of process called a *Markov chain*. The key feature of Markov chains is their lack of memory (the next section gives the specific definitions). In particular, a Markov chain has the property that the future is independent of the past given the present. These processes are named after the probabilist A. A. Markov, who published a series of papers starting in 1907 that laid the theoretical foundations for finite state Markov chains.[1] The foundations for the infinite state problems were developed by A. N. Kolmogorov in the mid-1930s.

An interesting example from the second half of the nineteenth century is the so-called Galton-Watson[2] process. (Of course, since this was before Markov's time, Galton and Watson did not use Markov chains in their analyses, but the process they studied is a Markov chain and serves as an interesting example of one.) Galton, a British scientist and cousin of Charles Darwin, and Watson, a clergyman and mathematician, were interested in answering the question of when and with what probability would a given family name become extinct. In the nineteenth century, the propagation or extinction of aristocratic family names was important, since land and titles stayed with the name. The process they investigated was as follows: At generation zero, the process starts with a single ancestor. Generation one consists of all the sons

[1] E. Çinlar, *Introduction to Stochastic Processes* (Englewood Cliffs, N.J.: Prentice-Hall, Inc., 1975) p. 143.

[2] F. Galton and H. W. Watson "On the Probability of the Extinction of Families," *J. Anthropol. Soc. London* **4** (1875), pp. 138–144.

of the ancestor (sons were modeled since it was the male that carried the family name). The next generation consists of all the sons of each son from the first generation (i.e., grandsons to the ancestor), generations continuing *ad infinitum* or until extinction. The assumption is that for each individual in a generation, the probability of having zero, one, two, etc., sons is given by some specified (and unchanging) probability mass function, and that mass function is identical for all individuals at any generation. Such a process might continue to expand or it might become extinct, and Galton and Watson were able to address the questions of whether or not extinction occurred and, if extinction did occur, how many generations it would take. The distinction that makes a Galton-Watson process a Markov chain is the fact that, at any generation, the number of individuals in the next generation is completely independent of the number of individuals in previous generations, as long as the number of individuals in the current generation is known. It is processes with this feature (the future being independent of the past given the present) that are studied next. They are interesting, not only because of their mathematical elegance, but also because of their practical utilization in probabilistic modeling.

2.1 BASIC DEFINITIONS

As stated in the previous chapter, a random variable is simply a function that assigns a real number to each possible outcome in the sample space. However, we are usually interested in more than just a single random variable. For example, the daily demand for a particular product during a year would form a sequence of 365 random variables, all of which may influence a single decision. For this reason we define a stochastic process.

DEFINITION 2.1 *A* STOCHASTIC PROCESS *is a sequence of random variables.*

It is possible for a stochastic process to consist of a countable number of random variables, such as the sequence of daily temperatures, in which case it is called a *discrete parameter process*. It is also possible that a stochastic process consists of an uncountable number of random variables, in which case it is called a *continuous parameter process*. An example of a continuous parameter process is the continuous monitoring of the working condition of a machine. That is, at every point in time during the day a random variable is defined that designates the machine's working condition by letting the random variable equal one if the machine is working and zero if the machine is not working. For example, if the machine was working 55.78 minutes after the day had started, then $X_{55.78} = 1$.

DEFINITION 2.2 *The set of real numbers containing the ranges of all the random variables in a stochastic process is called the* STATE SPACE *for the stochastic process.*

The state space of a process may be either discrete or continuous. Most of the processes considered in this chapter have a discrete state space. Often, for ease of presentation, finite state spaces (i.e., the number of elements in the state space is finite) are assumed.

The easiest stochastic process to consider is one made up of independent and identically distributed (iid) random variables. For example, suppose that we decided to play a game with a fair, unbiased coin. We each start with five pennies and repeatedly toss the coin. If it turns up heads, then you give me a penny; if tails, I give you a penny. We continue until one of us has none and the other ten pennies. The sequence of heads and tails from the successive tosses of the coin would form an iid stochastic process.

One step more interesting to analyze than an iid process is a Markov chain. Intuitively, *a Markov chain is a discrete parameter process in which the future is independent of the past given the present*. The previously mentioned game where we started with five pennies each illustrates a Markov chain. Consider the stochastic process formed by the number of pennies you have immediately after each toss of the coin. The sequence would start at five and increase or decrease by one at each step until eventually the process would cease with a value of either zero or ten. Assume, after several tosses, you currently have three pennies. The probability that after the next toss you will have four pennies is 0.5 and knowledge of the past (i.e., how many pennies you had one or two tosses ago) does not help in calculating the probability of 0.5. Thus, the future (how many pennies you will have after the next toss) is independent of the past (how many pennies you had several tosses ago) given the present (you currently have three pennies).

Another example of the Markov property comes from Mendelian genetics. Mendel demonstrated that the seed color of peas was a genetically controlled trait. Thus, knowledge about the gene pool of the current generation of peas is sufficient information to predict the seed color for the next generation. In fact, if full information about the current generation's genes are known, then knowing about previous generations does not help in predicting the future; thus, we would say that the future is independent of the past given the present.

DEFINITION 2.3 *The stochastic process* $X = \{X_n; n = 0, 1, \cdots\}$ *with discrete state space* E *is a* MARKOV CHAIN *if the following holds for each* $j \in E$ *and* $n = 0, 1, \cdots$

$$\Pr\{X_{n+1} = j \mid X_0 = i_0, X_1 = i_1, \cdots, X_n = i_n\} = \Pr\{X_{n+1} = j \mid X_n = i_n\},$$

for any set of states $i_0, \cdots, i_n$ *in the state space. Furthermore, the Markov chain is said to have* STATIONARY *transition probabilities if*

$$\Pr\{X_1 = j \mid X_0 = i\} = \Pr\{X_{n+1} = j \mid X_n = i\}.$$

The first equation in Definition 2.3 is a mathematical statement of the Markov property. To interpret the equation, think of time n as the present. The left-hand side is the probability of going to state j next, given the history of all past states. The right-hand side is the probability of going to state j next, given only the present state. Because they are equal, we have that the past history of states provides no additional information helpful in predicting the future if the present state is known. The second equation (i.e., the stationary property) simply indicates that the probability of a one-step transition does not change as time increases (in other words, the probabilities are the same in the winter and the summer).

In this chapter we always assume that we are working with stationary transition probabilities. Because the probabilities are stationary, the only information needed to describe the process is the initial conditions (a probability mass function for X_0) and the one-step transition probabilities. A square matrix is used for the transition probabilities and is often denoted by the capital letter $\boldsymbol{P}$, where

$$P(i,j) = \Pr\{X_1 = j \mid X_0 = i\}. \qquad \textbf{(2.1)}$$

Since the matrix $\boldsymbol{P}$ contains probabilities, it is always nonnegative (a matrix is nonnegative if every element of it is nonnegative) and the sum of the elements within each row equals one. In fact, any nonnegative square matrix with row sums equal to one is called a *Markov matrix*.

EXAMPLE 2.1 Consider a farmer using an old tractor. The tractor is often in the repair shop but it always takes only one day to get it running again. The first day out of the shop it always works but on any given day thereafter, independent of its previous history, there is a 10% chance of it not working and thus being sent back to the shop. Let $X_0, X_1, \cdots$ be random variables denoting the daily condition of the tractor, where a one denotes the working condition and a zero denotes the failed condition. In other words, $X_n = 1$ denotes that the tractor is working on day n and $X_n = 0$ denotes it being in the repair shop on day n. Thus, $X_0, X_1, \cdots$ is a Markov chain with state space $E = \{0, 1\}$ and with Markov matrix

$$\boldsymbol{P} = \begin{matrix} 0 \\ 1 \end{matrix} \begin{bmatrix} 0 & 1 \\ 0.1 & 0.9 \end{bmatrix}.$$

To develop a "mental image" of the Markov chain, it is very helpful to draw a state diagram (Fig. 2.1) of the Markov matrix. In the diagram, each state is represented by a circle and the transitions with positive probabilities are represented by an arrow. Until the student is very familiar with Markov chains, we recommend that state diagrams be drawn for any chain being discussed.

FIGURE 2.1 State diagram for the Markov chain of Example 2.1.

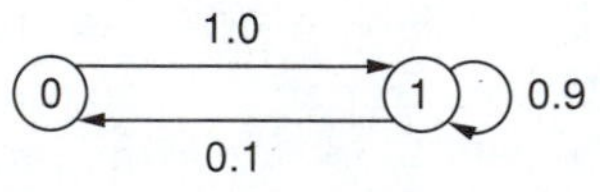

1.0
0 1 0.9
0.1

EXAMPLE 2.2 A salesman lives in town a and is responsible for towns a, b, and c. Each week he is required to visit a different town. When he is in his home town, it makes no difference which town he visits next so he flips a coin and if it is heads he goes to b and if tails he goes to c. However, after spending a week away from home he has a slight preference for going home so when he is in either towns b or c he flips two coins. If two heads occur, then he goes to the other town; otherwise he goes to a. The successive towns that he visits form a Markov chain with state space $E = \{a, b, c\}$ where the random variable X_n equals a, b, or c according to his location during week n. The state diagram for this system is given in Fig. 2.2 and the associated Markov matrix is

$$\boldsymbol{P} = \begin{array}{c} a \\ b \\ c \end{array}\begin{bmatrix} 0 & 0.50 & 0.50 \\ 0.75 & 0 & 0.25 \\ 0.75 & 0.25 & 0 \end{bmatrix}.$$

EXAMPLE 2.3 Let $X = \{X_n; n = 0, 1, \cdots\}$ be a Markov chain with state space $E = \{1, 2, 3, 4\}$ and transition probabilities given by

$$\boldsymbol{P} = \begin{array}{c} 1 \\ 2 \\ 3 \\ 4 \end{array}\begin{bmatrix} 1 & 0 & 0 & 0 \\ 0 & 0.3 & 0.7 & 0 \\ 0 & 0.5 & 0.5 & 0 \\ 0.2 & 0 & 0.1 & 0.7 \end{bmatrix}.$$

The chain in this example (Figure 2.3) is structurally different than the other two examples in that you might start in state 4, then go to state 3, and never get to state 1; or you might start in state 4 and go to state 1 and stay there forever. The other two examples, however, involved transitions in which it was always possible to reach every state eventually from every other state.

EXAMPLE 2.4 The final example in this section is taken from Parzen[3] and illustrates that the parameter n need not refer to time. (It is often true that the "steps"

FIGURE 2.2 State diagram for the Markov chain of Example 2.2.

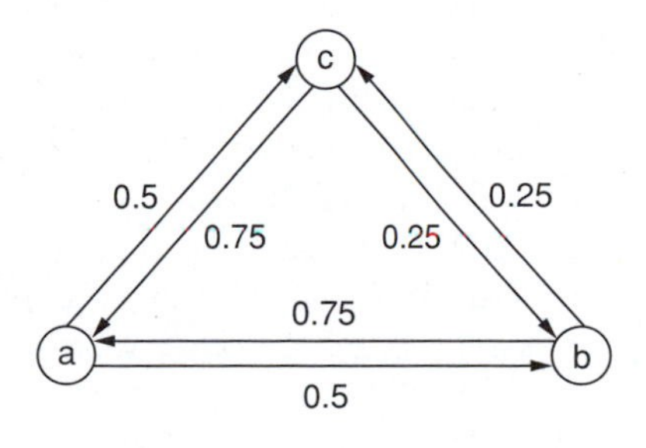

[3]E. Parzen, *Stochastic Processes* (Oakland, Calif.: Holden-Day, Inc., 1962), p. 191.

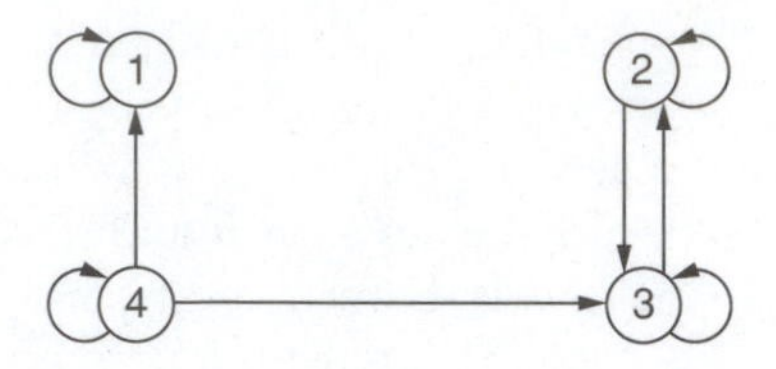

FIGURE 2.3 State diagram for the Markov chain of Example 2.3.

of a Markov chain refer to days, weeks, or months, but that need not be the case.) Consider a page of text and represent vowels by zeros and consonants by ones. Thus the page becomes a string of zeros and ones. It has been indicated[4] that the sequence of vowels and consonants in the Samoan language forms a Markov chain, where a vowel always follows a consonant and a vowel follows another vowel with a probability of 0.51. Thus, a sequence of ones and zeros on a page of Samoan text would evolve according to the Markov matrix

$$\boldsymbol{P} = \begin{matrix} 0 \\ 1 \end{matrix} \begin{bmatrix} 0.51 & 0.49 \\ 1 & 0 \end{bmatrix}. \qquad \blacksquare$$

After a Markov chain has been formulated, there are many questions that might be of interest. For example: In what state will the Markov chain be five steps from now? What percent of time is spent in a given state? Starting from one state, will the chain ever reach another fixed state? If a profit is realized for each visit to a particular state, what is the long run average profit per unit of time? The remainder of this chapter is devoted to answering questions of this nature.

▶ *Suggestion: Do part (a) of Exercises 2.1–2.3 and 2.6–2.9.*

2.2 MULTISTEP TRANSITIONS

The Markov matrix provides direct information about one-step transition probabilities and it can also be used in the calculation of probabilities for transitions involving more than one step. Consider the salesman of Example 2.2 starting in town b. The Markov matrix indicates that the probability of being in state a after one step (in one week) is 0.75, but what is the probability that he will be in state a after two steps? Figure 2.4 illustrates the paths that go from b to a in two steps (some of the paths shown have probability zero). Thus, to compute the probability, we need to sum over all possible routes. In other words we would have the following calculations:

[4] G. A. Miller, "Finite Markov Processes in Psychology," *Psychometrika* **17** (1952), pp. 149–167, quotes a result by Newman.

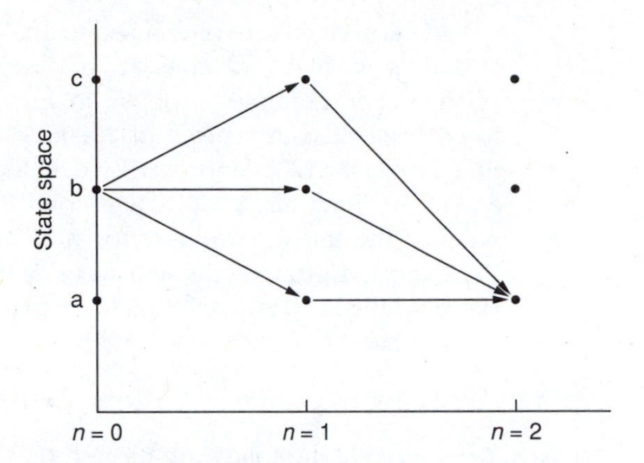

FIGURE 2.4 Possible paths of a two-step transition from state b to state a for a three-state Markov chain.

$$\begin{aligned}\Pr\{X_2 = a \mid X_0 = b\} &= \Pr\{X_1 = a \mid X_0 = b\} \times \Pr\{X_2 = a \mid X_1 = a\} \\ &\quad + \Pr\{X_1 = b \mid X_0 = b\} \times \Pr\{X_2 = a \mid X_1 = b\} \\ &\quad + \Pr\{X_1 = c \mid X_0 = b\} \times \Pr\{X_2 = a \mid X_1 = c\} \\ &= P(b,a)P(a,a) + P(b,b)P(b,a) + P(b,c)P(c,a)\end{aligned}$$

The final equation should be recognized as the definition of matrix multiplication; thus,

$$\Pr\{X_2 = a \mid X_0 = b\} = P^2(b,a).$$

This result is easily generalized into the following property.

Property 2.4 Let $X = \{X_n; n = 0, 1, \cdots\}$ be a Markov chain with state space E and Markov matrix $\boldsymbol{P}$, then for $i, j \in E$, and $n = 1, 2, \cdots$,

$$\Pr\{X_n = j \mid X_0 = i\} = P^n(i,j),$$

where the right-hand side represents the $i - j$ element of the matrix $\boldsymbol{P}^n$.

Property 2.4 indicates that $P^n(i,j)$ is interpreted to be the probability of going from state i to state j in n steps. It is important to remember that the notation $P^n(i,j)$ means that the matrix is *first* raised to the nth power and then the $i - j$ element of the resulting matrix is taken for the answer.

Returning to Example 2.2, the squared matrix is

$$\boldsymbol{P}^2 = \begin{bmatrix} 0.75 & 0.125 & 0.125 \\ 0.1875 & 0.4375 & 0.375 \\ 0.1875 & 0.375 & 0.4375 \end{bmatrix}.$$

The $b - a$ element of $\boldsymbol{P}^2$ is 0.1875 and thus there is a 0.1875 probability of being in town a two weeks after being in town b.

Markov chains are often used to analyze the cost or profit of an operation and thus we need to consider a cost or profit function imposed on the process. For example, suppose in Example 2.2 that every week spent in town a resulted in a profit of \$1000, every week spent in town b resulted in a profit of \$1200, and every week spent in town c resulted in a profit of \$1250. We then might ask what would be the expected profit after the first week if the initial town was town a? Or, more generally, what would be the expected profit of the nth week if the initial town was a? It should not be too difficult to convince yourself that the following calculation would be appropriate:

$$E[\text{Profit for week } n] = P^n(a,a) \times 1000 + P^n(a,b) \times 1200 + P^n(a,c) \times 1250.$$

We thus have the following property.

Property 2.5 Let $X = \{X_n; n = 0, 1, \cdots\}$ be a Markov chain with state space E, Markov matrix $\boldsymbol{P}$, and profit function $\boldsymbol{f}$ [i.e., each time the chain visits state i, a profit of $f(i)$ is obtained]. The expected profit at the nth step is given by

$$E[f(X_n) \mid X_0 = i] = \boldsymbol{P}^n\boldsymbol{f}(i).$$

Note again that in the right-hand side of the equation, the matrix $\boldsymbol{P}^n$ is multiplied by the vector $\boldsymbol{f}$ first and then the ith component of the resulting vector is taken as the answer. Thus, in Example 2.2, we have that the expected profit during the second week given that the initial town was a is

$$\begin{aligned} E[f(X_2) \mid X_0 = a] &= 0.75 \times 1000 + 0.125 \times 1200 + 0.125 \times 1250 \\ &= 1056.25 \end{aligned}$$

Up until now we have always assumed that the initial state was known. However, that is not always the situation. The manager of the traveling salesman might not know for sure the location of the salesman; instead, all that is known is a probability mass function describing his initial location. Again, using Example 2.2, suppose that we do not know for sure the salesman's initial location but know that there is a 50% chance he is in town a, a 30% chance he is in town b, and a 20% chance he is in town c. We now ask, what is the probability that the salesman will be in town a next week? The calculations for that are

$$\Pr\{X_1 = a\} = 0.50 \times 0 + 0.30 \times 0.75 + 0.20 \times 0.75 = 0.375.$$

These calculations generalize to the following.

Property 2.6 Let $X = \{X_n; n = 0, 1, \cdots\}$ be a Markov chain with state space E, Markov matrix $\boldsymbol{P}$, and initial probability vector $\boldsymbol{\mu}$ [i.e., $\mu(i) = \Pr\{X_0 = i\}$]. Then

$$\Pr_{\mu}\{X_n = j\} = \boldsymbol{\mu}\boldsymbol{P}^n(j).$$

Note that μ is a subscript to the probability statement on the left-hand side of the equation. The purpose for the subscript is to ensure that there is no confusion over the given conditions. Again, when interpreting the right-hand side of the equation, the vector $\boldsymbol{\mu}$ is multiplied by the matrix $\boldsymbol{P}^n$ first and then the jth element is taken from the resulting vector.

The last two properties can be combined, when necessary, into one statement.

Property 2.7 Let $X = \{X_n; n = 0, 1, \cdots\}$ be a Markov chain with state space E, Markov matrix $\boldsymbol{P}$, initial probability vector $\boldsymbol{\mu}$, and profit function $\boldsymbol{f}$. The expected profit at the nth step is given by

$$E_\mu[f(X_n)] = \boldsymbol{\mu}\boldsymbol{P}^n\boldsymbol{f}.$$

Returning to Example 2.2 and using the initial probabilities and profit function just given, the expected profit in the second week is calculated to be

$$\begin{aligned}\boldsymbol{\mu}\boldsymbol{P}^2\boldsymbol{f} &= (0.50, 0.30, 0.20)\begin{bmatrix}0.75 & 0.125 & 0.125\\ 0.1875 & 0.4375 & 0.375\\ 0.1875 & 0.375 & 0.4375\end{bmatrix}\begin{pmatrix}1000\\1200\\1250\end{pmatrix} \\ &= 1119.375\end{aligned}$$

EXAMPLE 2.5 A market analysis concerning consumer behavior in auto purchases has been conducted. Body styles traded in and purchased have been recorded by a particular dealer with the following results:

Number of Customers	Trade
275	Sedan for sedan
180	Sedan for station wagon
45	Sedan for convertible
80	Station wagon for sedan
120	Station wagon for station wagon
150	Convertible for sedan
50	Convertible for convertible

These data are believed to be representative of average consumer behavior, and it is assumed that the Markov assumptions are appropriate.

We develop a Markov chain to describe the changing body styles over the life of a customer. (Notice, for purposes of instruction, we are simplifying the process by assuming that the age of the customer does not affect the customer's choice of a body style.) We define the Markov chain to be the body style of the automobile that the customer has immediately after a trade-in, where the state space is $E = \{s, w, c\}$ with s for the sedan, w for the wagon, and c for the convertible. Of 500 customers who have a sedan,

275 will stay with the sedan during the next trade; therefore, the s-s element of the transition probability matrix will be $\frac{275}{500}$. Thus, the Markov matrix for the "body style" Markov chain is

$$\boldsymbol{P} = \begin{matrix} s \\ w \\ c \end{matrix} \begin{bmatrix} \frac{275}{500} & \frac{180}{500} & \frac{45}{500} \\ \frac{80}{200} & \frac{120}{200} & \frac{0}{200} \\ \frac{150}{200} & \frac{0}{200} & \frac{50}{200} \end{bmatrix} = \begin{bmatrix} 0.55 & 0.36 & 0.09 \\ 0.40 & 0.60 & 0.0 \\ 0.75 & 0.0 & 0.25 \end{bmatrix}.$$

Let us assume we have a customer whose behavior is described by this model. Furthermore, the customer always buys a new car in January of every year. It is now January 1997, and the customer enters the dealership with a sedan (i.e., $X_{1996} = s$). From the preceding matrix, $P\{X_{1997} = s \mid X_{1996} = s\} = 0.55$, or the probability that the customer will leave with another sedan is 55%. We would now like to predict what body style the customer will have for trade-in during January 2000. Notice that the question involves three transitions of the Markov chain; therefore, the cubed matrix must be determined, which is

$$\boldsymbol{P}^3 = \begin{bmatrix} 0.5023 & 0.4334 & 0.0643 \\ 0.4816 & 0.4680 & 0.0504 \\ 0.5355 & 0.3780 & 0.0865 \end{bmatrix}.$$

Thus, there is approximately a 50% chance the customer will enter in the year 2000 with a sedan, a 43% chance with a station wagon, and almost a 7% chance of having a convertible. The probability that the customer will leave this year (1997) with a convertible and then leave next year (1998) with a sedan is given by $P(s, c) \times P(c, s) = 0.09 \times 0.75 = 0.0675$. Now determine the probability that the customer who enters the dealership now (1997) with a sedan will leave with a sedan and also leave with a sedan in the year 2000. The mathematical statement is

$$\Pr\{X_{1997} = s, X_{2000} = s \mid X_{1996} = s\} = P(s, s) \times P^3(s, s) \approx 0.28.$$

Notice in this probability statement that no mention is made of the body style for the intervening years; thus, the customer may or may not switch in the years 1998 and 1999.

Now to illustrate profits. Assume that a sedan yields a profit of \$1200, a station wagon yields \$1500, and a convertible yields \$2500. The expected profit this year from the customer entering with a sedan is \$1425, and the expected profit in the year 1999 from a customer who enters the dealership with a sedan in 1997 is approximately \$1414. Or, to state this mathematically,

$$E[\boldsymbol{f}(X_{1999}) \mid X_{1996} = s] \approx 1414,$$

where $\boldsymbol{f} = (1200, 1500, 2500)^T$. ■

▶ *Suggestion: Do Exercises 2.10a–c, 2.11a–f, and 2.12.*

2.3 CLASSIFICATION OF STATES

There are two types of states possible in a Markov chain, and before most questions can be answered, the individual states for a particular chain must be classified into one of these two types. As we begin our study of Markov chains, it is important to cover a significant amount of new notation. The student will discover that the time spent learning the new terminology will be rewarded with a fuller understanding of Markov chains. Furthermore, a good understanding of the dynamics involved with Markov chains will greatly aid in the understanding of the entire area of stochastic processes.

Two random variables that will be extremely important denote "first passage times" and "number of visits to a fixed state." To describe these random variables, consider a fixed state, call it state j, in the state space for a Markov chain. The first passage time is a random variable, denoted by T^j, that equals the time (i.e., number of steps) it takes to reach the fixed state *for the first time*. Mathematically, the first passage time random variable is defined by

$$T^j = \min\{n \geq 1 : X_n = j\}, \tag{2.2}$$

where the minimum of the empty set is taken to be $+\infty$. For example, in Example 2.3, if $X_0 = 1$ then $T^2 = \infty$, i.e., if the chain starts in state 1 then it will never reach state 2.

The number of visits to a state is a random variable, denoted by N^j, that equals the total number of visits (including time zero) that the Markov chain makes to the fixed state throughout the life of the chain. Mathematically, the "number of visits" random variable is defined by

$$N^j = \sum_{n=0}^{\infty} I(X_n, j), \tag{2.3}$$

where $\boldsymbol{I}$ is the identity matrix. The identity matrix is used simply as a counter. Because the identity has ones on the diagonal and zeros off the diagonal, it follows that $I(X_n, j) = 1$ if $X_n = j$; otherwise, it is zero. Note that the summation in Eq. (2.3) starts at $n = 0$; thus, if $X_0 = j$ then N^j must be at least one.

EXAMPLE 2.6 Let us consider a realization for the Markov chain of Example 2.3. (By realization, we mean conceptually that an experiment is conducted and we record the random outcomes of the chain.) Assume that the first part of the realization (Figure 2.5) is $X_0 = 4$, $X_1 = 4$, $X_2 = 4$, $X_3 = 3$, $X_4 = 2$, $X_5 = 3$, $X_6 = 2$, $\cdots$. The first passage random variables for this realization are $T^1 = \infty$, $T^2 = 4$, $T^3 = 3$, and $T^4 = 1$. To see why $T^1 = \infty$, it is easiest to refer to the state diagram describing the Markov chain (Figure 2.3). By inspecting the diagram it becomes obvious that once the chain is in either states 2 or 3 that it will never get to state 1 and thus $T^1 = \infty$.

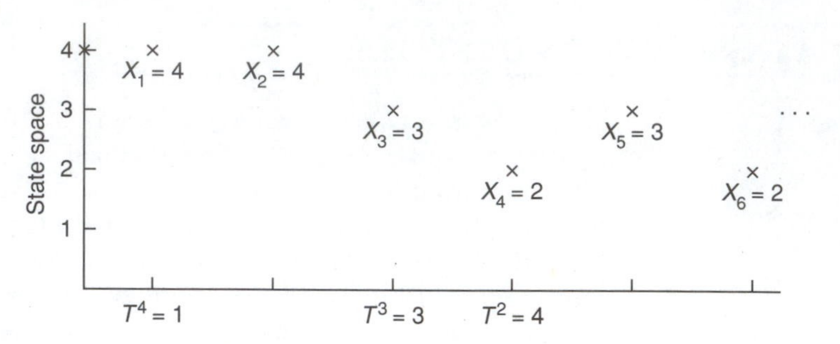

FIGURE 2.5 One possible realization for the Markov chain of Example 2.3.

Using the same realization, the number of visits random variables are $N^1 = 0$, $N^2 = \infty$, $N^3 = \infty$, and $N^4 = 3$. Again, the values for N^1, N^2, and N^3 are obtained by inspecting the state diagram and observing that if the chain ever gets to states 2 or 3 it will stay in those two states forever and thus will visit them an infinite number of times.

Let us perform the experiment one more time. Assume that our second realization results in the values $X_0 = 4$, $X_1 = 4$, $X_2 = 1$, $X_3 = 1$, $X_4 = 1$, $\cdots$. The new outcomes for the first passage random variables for this second realization are $T^1 = 2$, $T^2 = \infty$, $T^3 = \infty$, and $T^4 = 1$. Furthermore, $N^1 = \infty$, $N^2 = 0$, $N^3 = 0$, and $N^4 = 2$. Thus, you should understand from this example that we may not be able to say what the value of T^j will be before an experiment, but we should be able to describe its probability mass function. ■

The primary quantity of interest regarding the first passage times is the so-called first passage probabilities. The major question of interest is whether or not it is possible to reach a particular state from a given initial state. To answer this question, we determine the first passage probability, denoted by $F(i, j)$, which is the probability of eventually reaching state j at least once given that the initial state was i. The probability $F(i, j)$ (F for *first* passage) is defined by

$$F(i, j) = \Pr\{T^j < \infty \mid X_0 = i\}. \quad \textbf{(2.4)}$$

By inspecting the state diagram in Figure 2.3, it should be obvious that the first passage probabilities for the chain of Example 2.3 are given by

$$\boldsymbol{F} = \begin{bmatrix} 1 & 0 & 0 & 0 \\ 0 & 1 & 1 & 0 \\ 0 & 1 & 1 & 0 \\ <1 & <1 & <1 & <1 \end{bmatrix}.$$

The primary quantity of interest for the number of visits random variable is its expected value. The expected number of visits to state j given that the initial state was i is denoted by $R(i, j)$ (R for *returns*) and is defined by

$$R(i, j) = E[N^j \mid X_0 = i]. \quad \textbf{(2.5)}$$

Again the state diagram of Figure 2.3 allows the determination of some of the values of $\boldsymbol{R}$ as follows:

$$\boldsymbol{R} = \begin{bmatrix} \infty & 0 & 0 & 0 \\ 0 & \infty & \infty & 0 \\ 0 & \infty & \infty & 0 \\ \infty & \infty & \infty & < \infty \end{bmatrix}.$$

This matrix may appear unusual because of the occurrence of ∞ for elements of the matrix. In Section 2.5, numerical methods will be given to calculate the values of the $\boldsymbol{R}$ and $\boldsymbol{F}$ matrices; however, the values of ∞ are obtained by an understanding of the structures of the Markov chain and not through specific formulas. For now, our goal is to develop an intuitive understanding of these processes; therefore, the major concern for the $\boldsymbol{F}$ matrix is whether an element is zero or one, and the concern for the $\boldsymbol{R}$ matrix is whether an element is zero or infinity. As might be expected, there is a close relation between $R(i,j)$ and $F(i,j)$ as is shown in the following property.

Property 2.8 Let $R(i,j)$ and $F(i,j)$ be as defined in Eqs. (2.5) and (2.4), respectively. Then

$$R(i,j) = \begin{cases} 1/[1 - F(j,j)] & \text{for } i = j \\ F(i,j)/[1 - F(j,j)] & \text{for } i \neq j \end{cases},$$

where the convention 0/0=0 is used.

The discussion utilizing Example 2.3 should help to point out that there is a basic difference between states 1, 2, or 3 and state 4 of the example. Consider that if the chain ever gets to state 1 or to state 2 or to state 3 then that state will continually reoccur. However, even if the chain starts in state 4, it will only stay there a finite number of times and will eventually leave state 4 never to return. These ideas give rise to the terminology of recurrent and transient states. Intuitively, a state is called recurrent if, starting in that state, it will continuously reoccur; and a state is called transient if, starting in that state, it will eventually leave that state never to return. Or equivalently, a state is recurrent if, starting in that state, the chain must (i.e., with probability one) eventually return to the state at least once; and a state is transient if, starting in that state, there is a chance (i.e., with probability greater than zero) that the chain will leave the state and never return. The following mathematical definitions are based on these notions.

DEFINITION 2.9 *A state j is called* TRANSIENT *if $F(j,j) < 1$. Equivalently, state j is* TRANSIENT *if $R(j,j) < \infty$.*

DEFINITION 2.10 *A state j is called* RECURRENT *if $F(j,j) = 1$. Equivalently, state j is* RECURRENT *if $R(j,j) = \infty$.*

From these two definitions, a state must either be transient or recurrent. A dictionary[5] definition for the word *transient* is "passing esp. quickly into and out of existence." Thus, the use of the word is justified since transient states will only occur for a finite period of time. For a transient state, there will be a time after which the transient state will never again be visited. A dictionary[5] definition for *recurrent* is "returning or happening time after time." So again the mathematical concept of a recurrent state parallels the common English usage: Recurrent states are recognized by those states that are continually revisited.

Recurrent states might also be periodic, as with the Markov matrix

$$\boldsymbol{P} = \begin{bmatrix} 0 & 1 \\ 1 & 0 \end{bmatrix},$$

but discussions regarding periodic chains are left to other texts.

By drawing the state diagrams for the Markov chains of Examples 2.1 and 2.2, it should become clear that all states in those two cases are recurrent. The only transient state so far illustrated is state 4 of Example 2.3.

EXAMPLE 2.7 Let $X = \{X_n; n = 0, 1, \cdots\}$ be a Markov chain with state space $E = \{1, 2, 3, 4\}$ and transition probabilities given by

$$\boldsymbol{P} = \begin{matrix} 1 \\ 2 \\ 3 \\ 4 \end{matrix} \begin{bmatrix} 1 & 0 & 0 & 0 \\ 0.01 & 0.29 & 0.7 & 0 \\ 0 & 0.5 & 0.5 & 0 \\ 0.2 & 0 & 0.1 & 0.7 \end{bmatrix}.$$

Notice the similarity between this Markov chain and the chain in Example 2.3. Although the numerical difference between the two Markov matrices is slight, there is a radical difference between the structures of the two chains. (The difference between a zero term and a nonzero term, no matter how small, can be very significant.) The chain of this example has one recurrent state and three transient states.

By inspecting the state diagram for this Markov chain (Figure 2.6) the following matrices are obtained (through observation, not through calculation):

$$\boldsymbol{F} = \begin{bmatrix} 1 & 0 & 0 & 0 \\ 1 & <1 & <1 & 0 \\ 1 & 1 & <1 & 0 \\ 1 & <1 & <1 & <1 \end{bmatrix}.$$

$$\boldsymbol{R} = \begin{bmatrix} \infty & 0 & 0 & 0 \\ \infty & <\infty & <\infty & 0 \\ \infty & <\infty & <\infty & 0 \\ \infty & <\infty & <\infty & <\infty \end{bmatrix}.$$

[5] *Webster's Ninth New Collegiate Dictionary* (Springfield, Mass.: Merriam-Webster, Inc., 1989).

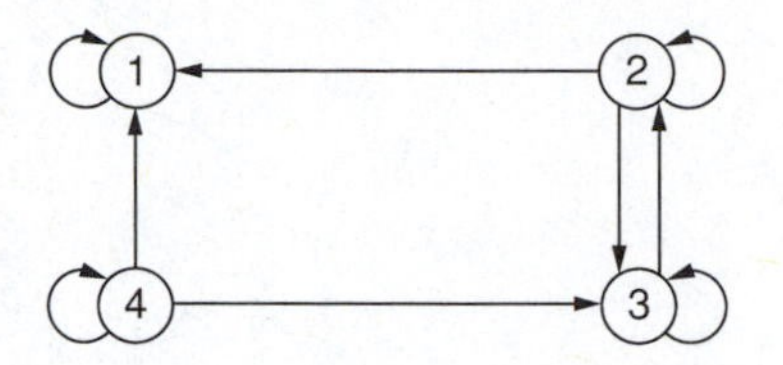

FIGURE 2.6 State diagram for the Markov chain of Example 2.7.

Observe that $F(3,2) = 1$ even though both state 2 and state 3 are transient. It is only the diagonal elements of the $\boldsymbol{F}$ and $\boldsymbol{R}$ matrices that determine whether a state is transient or recurrent. ■

■ **EXAMPLE 2.8** A manufacturing process consists of two processing steps in sequence. After step 1, 20% of the parts must be reworked, 10% of the parts are scrapped, and 70% proceed to the next step. After step 2, 5% of the parts must be returned to the first step, 10% must be reworked, and 5% are scrapped; the remainder are sold. The diagram of Figure 2.7 illustrates the dynamics of the manufacturing process.

The Markov matrix associated with this manufacturing process is given by

$$\boldsymbol{P} = \begin{matrix} 1 \\ 2 \\ s \\ g \end{matrix} \begin{bmatrix} 0.2 & 0.7 & 0.1 & 0.0 \\ 0.05 & 0.1 & 0.05 & 0.8 \\ 0.0 & 0.0 & 1.0 & 0.0 \\ 0.0 & 0.0 & 0.0 & 1.0 \end{bmatrix}. \tag{2.6}$$

Consider the dynamics of the chain. A part to be manufactured will begin the process by entering state 1. After possibly cycling for awhile, the part will end the process by entering either the g state or the s state. Therefore, states 1 and 2 are transient, and states g and s are recurrent. ■

Along with classifying states, we also need to be able to classify sets of states. We first define a closed set, which is a set that, once the Markov chain has entered the set, cannot leave the set.

DEFINITION 2.11 *Let $X = \{X_n; n = 0, 1, \cdots\}$ be a Markov chain with Markov matrix* $\boldsymbol{P}$ *and let C be a set of states contained in its state space. Then C is* CLOSED *if*

$$\sum_{j \in C} P(i,j) = 1 \text{ for all } i \in C.$$

To illustrate the concept of closed sets, refer again to Example 2.3. The sets $\{2,3,4\}$, $\{2\}$, and $\{3,4\}$ are *not* closed sets. The set $\{1,2,3,4\}$ is obviously closed, but it can be reduced to a smaller closed set. The set $\{1,2,3\}$ is also closed, but again it can be further reduced. Both sets $\{1\}$ and $\{2,3\}$ are closed and cannot be reduced further. This idea of taking a closed set and trying to reduce it is extremely important and leads to the definition of an irreducible set.

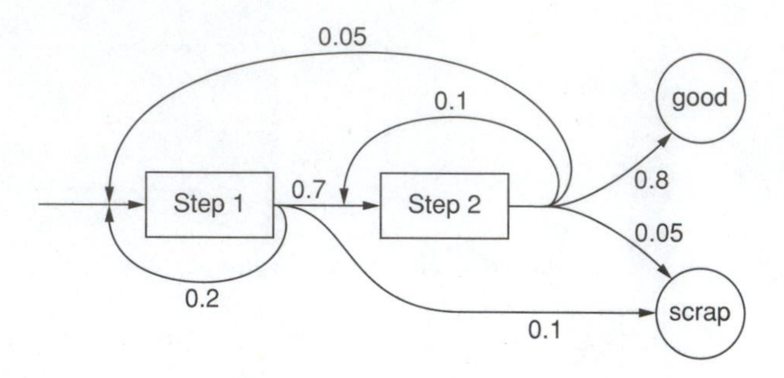

FIGURE 2.7 Two-step manufacturing process of Example 2.8.

DEFINITION 2.12 *A closed set of states that contains no proper subset that is also closed is called* IRREDUCIBLE. *A state that forms an irreducible set by itself is called an* ABSORBING *state.*

The Markov chain of Example 2.3 has two irreducible sets: the set $\{1\}$ and the set $\{2, 3\}$. Because the first irreducible set constains only one state, that state is an absorbing state.

EXAMPLE 2.9 Let X be a Markov chain with state space $E = \{a, \cdots, g\}$ and Markov matrix

$$\boldsymbol{P} = \begin{matrix} a \\ b \\ c \\ d \\ e \\ f \\ g \end{matrix} \begin{bmatrix} 0.3 & 0.7 & 0 & 0 & 0 & 0 & 0 \\ 0.5 & 0.5 & 0 & 0 & 0 & 0 & 0 \\ 0 & 0.2 & 0.4 & 0.4 & 0 & 0 & 0 \\ 0 & 0 & 0.5 & 0.5 & 0 & 0 & 0 \\ 0 & 0 & 0 & 0.8 & 0.1 & 0.1 & 0 \\ 0 & 0 & 0 & 0 & 0.7 & 0.3 & 0 \\ 0 & 0 & 0 & 0 & 0 & 0.4 & 0.6 \end{bmatrix}.$$

By drawing a state diagram (Figure 2.8), you should observe that states a and b are recurrent and all others are transient. (If that is not obvious, notice that once the chain reaches either a or b, it will stay in those two states, there are no paths leading away from the set $\{a, b\}$; furthermore, all states will

FIGURE 2.8 State diagram for the Markov chain of Example 2.9.

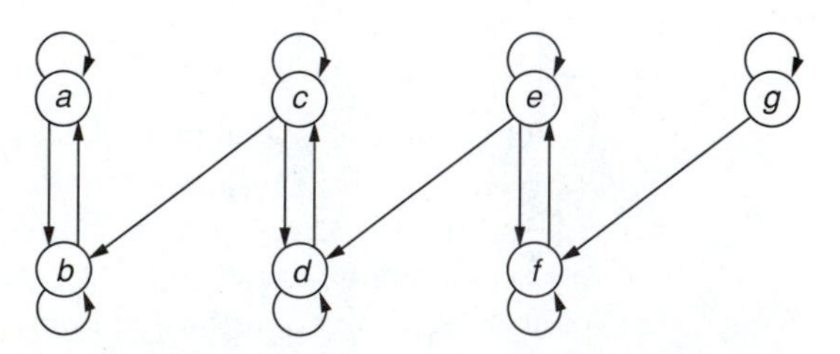

eventually reach state b.) Obviously, the entire state space is closed. Since no state goes to g, we can exclude g and still have a closed set; namely, the set $\{a, b, c, d, e, f\}$ is closed. However, the set $\{a, b, c, d, e, f\}$ can be reduced to a smaller closed set, so it is *not* irreducible. Because there is a path going from e to f, the set $\{a, b, c, d, e\}$ is *not* closed. By excluding the state e, a closed set is again obtained; namely, $\{a, b, c, d\}$ is also closed since there are no paths out of the set $\{a, b, c, d\}$ to another state. Again, however, it can also be reduced so the set $\{a, b, c, d\}$ is *not* irreducible. Finally, consider the set $\{a, b\}$. This two-state set is closed and cannot be further reduced to a smaller closed set; therefore, the set $\{a, b\}$ is an irreducible set. ■

Definitions 2.11 and 2.12 are important because of the following property.

Property 2.13 All states within an irreducible set are of the same classification.

The significance of this property is that if you can identify one state within an irreducible set as being transient, then all states within the set are transient, and if one state is recurrent, then all states within the set are recurrent. We are helped even more by recognizing that it is impossible to have an irreducible set of transient states if the set contains only a finite number of states; thus we have the following property.

Property 2.14 Let C be an irreducible set of states such that the number of states within C is finite. Then each state within C is recurrent.

One final concept will help us to identify irreducible sets; namely, communication between states. Communication between states is like communication between people; there is communication only if messages can go both ways. In other words, two states, i and j, *communicate* if and only if it is possible to eventually reach j from i *and* it is possible to eventually reach i from j. In Example 2.3, states 2 and 3 communicate, but states 4 and 2 do not communicate. Although state 2 can be reached from 4, it does not go both ways because state 4 cannot be reached from 2. The communication must be both ways but it does not have to be in one step. For example, in the Markov chain with state space $\{a, b, c, d\}$ and with Markov matrix

$$\boldsymbol{P} = \begin{bmatrix} 0.5 & 0.5 & 0 & 0 \\ 0.2 & 0.4 & 0.4 & 0 \\ 0 & 0.1 & 0.8 & 0.1 \\ 0 & 0 & 0.3 & 0.7 \end{bmatrix}, \qquad \textbf{(2.7)}$$

all states communicate with every other state. In particular, state a communicates with state d even though they cannot reach each other in one step (see Figure 2.9).

The notion of communication is often the concept used to identify irreducible sets as is given in the following property.

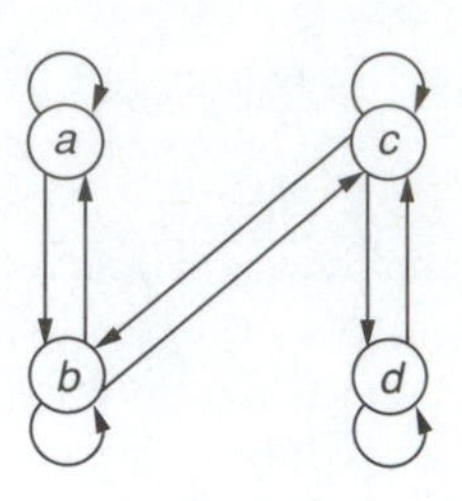

FIGURE 2.9 State diagram for the Markov chain of Eq. (2.7).

Property 2.15 The closed set of states C is irreducible if and only if every state within C communicates with every other state within C.

The procedure for identifying irreducible sets of states for a Markov chain with a finite state space is first to draw the state diagram, then pick a state and identify all states that communicate with it. If the set made up of all those states that communicate with it is closed, then the set is an irreducible, recurrent set of states. If it is not closed, then the originally chosen state and all the states that communicate with it are transient states.

EXAMPLE 2.10 Let X be a Markov chain with state space $E = \{a, \cdots, g\}$ and Markov matrix

$$\mathbf{P} = \begin{matrix} a \\ b \\ c \\ d \\ e \\ f \\ g \end{matrix} \begin{bmatrix} 0.3 & 0.1 & 0.2 & 0.2 & 0.1 & 0.1 & 0 \\ 0 & 0.5 & 0 & 0 & 0 & 0 & 0.5 \\ 0 & 0 & 0.4 & 0.6 & 0 & 0 & 0 \\ 0 & 0 & 0.3 & 0.2 & 0 & 0.5 & 0 \\ 0 & 0 & 0.2 & 0.3 & 0.4 & 0.1 & 0 \\ 0 & 0 & 1 & 0 & 0 & 0 & 0 \\ 0 & 0.8 & 0 & 0 & 0 & 0 & 0.2 \end{bmatrix}.$$

By drawing a state diagram (Figure 2.10), we see that there are two irreducible recurrent sets of states, $\{b, g\}$ and $\{c, d, f\}$. States a and e are transient. ■

▶ *Suggestion: Do Exercises 2.13a–c and 2.14a–d.*

2.4 STEADY-STATE BEHAVIOR

We have put much effort into classifying states and sets of states because the analysis of the limiting behavior of a chain is dependent on the type of state under consideration. To illustrate long-run behavior, we again return to Example 2.2 and focus attention on state a. Figure 2.11 shows a graph that gives the probabilities of being in state a at time n given that at time zero the chain was in state a. [In other words, the graph gives the values of $P^n(a, a)$ as a function of n.] Although the graph varies dramatically for small n, it

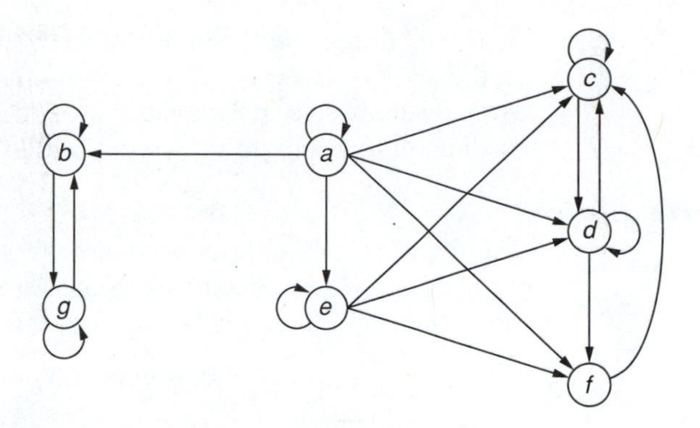

FIGURE 2.10 State diagram for the Markov chain of Example 2.10.

reaches an essentially constant value as n becomes large. It is seen from the graph that

$$\lim_{n\to\infty} \Pr\{X_n = a \mid X_0 = a\} = 0.42857.$$

In fact, if you spent the time to graph the probabilities of being in state a starting from state b instead of state a, you would discover the same limiting value, namely,

$$\lim_{n\to\infty} \Pr\{X_n = a \mid X_0 = b\} = 0.42857.$$

When discussing steady-state (or limiting) conditions, this is what is meant: not that the chain stops changing—it is dynamic by definition—but that enough time has elapsed so that the probabilities do not change with respect to time. It is often stated that steady-state results are independent of initial conditions, and this is true for the chain in Example 2.2; however, it is not always true. In Example 2.3, it is clear that the steady-state conditions are radically different when the chain starts in state 1 as compared to starting in state 2. The important consideration in steady-state analysis is the identification of irreducible, recurrent sets.

FIGURE 2.11 Probabilities (from Example 2.2) of being in state a as a function of time.

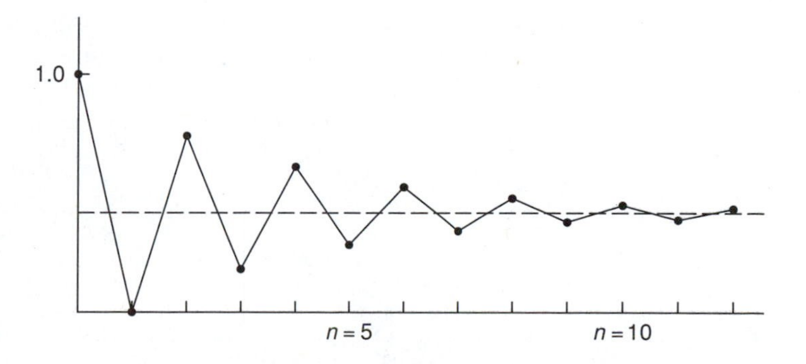

If the entire state space of a Markov chain forms an irreducible recurrent set, the Markov chain is called an *irreducible recurrent Markov chain*. In this case the steady-state probabilities are independent of the initial state and are not difficult to compute as is seen in the following property.

Property 2.16 Let $X = \{X_n; n = 0, 1, \cdots\}$ be a Markov chain with finite state space E and Markov matrix $\boldsymbol{P}$. Furthermore, assume that the entire state space forms an irreducible set (and thus all states are recurrent). Let the steady-state probabilities be denoted by the vector $\boldsymbol{\pi}$; that is,

$$\pi(j) = \lim_{n\to\infty} \Pr\{X_n = j \mid X_0 = i\}.$$

The vector $\boldsymbol{\pi}$ is the solution to the following system

$$\boldsymbol{\pi P} = \boldsymbol{\pi},$$
$$\sum_{i\in E} \pi(i) = 1.$$

To illustrate the determination of $\boldsymbol{\pi}$, observe that the Markov chain of Example 2.2 is irreducible, recurrent. Applying Property 2.16, we obtain

$$\begin{aligned}
0.75\pi_b + 0.75\pi_c &= \pi_a \\
0.5\pi_a \qquad\qquad + 0.25\pi_c &= \pi_b \\
0.5\pi_a + 0.25\pi_b \qquad\qquad &= \pi_c \\
\pi_a + \quad \pi_b + \quad \pi_c &= 1.
\end{aligned}$$

There are four equations and only three variables, so normally there would not be a unique solution; however, for an irreducible Markov matrix there is always exactly one redundant equation from the system formed by $\boldsymbol{\pi P} = \boldsymbol{\pi}$. Thus, to solve the preceding system, arbitrarily choose one of the first three equations to discard and solve the remaining 3×3 system (*never* discard the final or norming equation), which yields

$$\pi_a = \frac{3}{7}, \pi_b = \frac{2}{7}, \pi_c = \frac{2}{7}.$$

Property 2.16 cannot be directly applied to the chain of Example 2.3 because the state space is not irreducible. All irreducible, recurrent sets must be identified and grouped together and then the Markov matrix is rearranged so that the irreducible sets are together and transient states are last. In such a manner, the Markov matrix for a chain can always be rewritten in the form

$$\boldsymbol{P} = \begin{bmatrix} \boldsymbol{P_1} & & & & \\ & \boldsymbol{P_2} & & & \\ & & \boldsymbol{P_3} & & \\ & & & \ddots & \\ \boldsymbol{B_1} & \boldsymbol{B_2} & \boldsymbol{B_3} & \cdots & \boldsymbol{Q} \end{bmatrix}. \tag{2.8}$$

After a chain is in this form, each submatrix $\boldsymbol{P}_i$ is a Markov matrix and can be considered as an independent Markov chain for which Property 2.16 is applied.

The Markov matrix of Example 2.3 is already in the form of Eq. (2.8). Because state 1 is absorbing (i.e., an irreducible set of one state), its associated steady-state probability is easy; namely, $\pi_1 = 1$. States 2 and 3 form an irreducible set so Property 2.16 can be applied to the submatrix from those states, resulting in the following system:

$$\begin{aligned} 0.3\pi_2 + 0.5\pi_3 &= \pi_2 \\ 0.7\pi_2 + 0.5\pi_3 &= \pi_3 \\ \pi_2 + \quad \pi_3 &= 1, \end{aligned}$$

which yields (after discarding one of the first two equations)

$$\pi_2 = \frac{5}{12} \text{ and } \pi_3 = \frac{7}{12}.$$

The values π_2 and π_3 are interpreted to mean that if a snapshot of the chain is taken a long time after it started and if it started in states 2 or 3, then there is a $\frac{5}{12}$ probability that the picture will show the chain in state 2 and a $\frac{7}{12}$ probability that it will be in state 3. Another interpretation of the steady-state probabilities is that if we recorded the time spent in states 2 and 3, then over the long run the fraction of time spent in state 2 would equal $\frac{5}{12}$ and the fraction of time spent in state 3 would equal $\frac{7}{12}$. These steady-state results are summarized in the following property.

Property 2.17 Let $X = \{X_n; n = 0, 1, \cdots\}$ be a Markov chain with finite state space with k distinct irreducible sets. Let the lth irreducible set be denoted by C_l, and let $\boldsymbol{P}_l$ be the Markov matrix restricted to the lth irreducible set [as in Eq. (2.8)]. Finally, let $\boldsymbol{F}$ be the matrix of first passage probabilities.

- If state j is transient,

$$\lim_{n\to\infty} \Pr\{X_n = j \mid X_0 = i\} = 0.$$

- If states i and j both belong to the lth irreducible set,

$$\lim_{n\to\infty} \Pr\{X_n = j \mid X_0 = i\} = \pi(j)$$

where

$$\boldsymbol{\pi P}_l = \boldsymbol{\pi} \quad \text{and} \quad \sum_{i\in C_l} \pi(i) = 1.$$

- If state j is recurrent and i is not in its irreducible set,

$$\lim_{n\to\infty} \Pr\{X_n = j \mid X_0 = i\} = F(i,j)\pi(j),$$

where $\boldsymbol{\pi}$ is determined as above.

- If state j is recurrent and X_0 is in the same irreducible set as j,

$$\lim_{n\to\infty} \frac{1}{n} \sum_{m=0}^{n-1} I(X_m, j) = \pi(j),$$

where $\boldsymbol{I}$ is the identity matrix.

- If state j is recurrent,

$$E[T^j \mid X_0 = j] = \frac{1}{\pi(j)}.$$

The intuitive idea of the second to last item in the preceding property is obtained by considering the role that the identity matrix plays in the left-hand side of the equation. As mentioned previously, the identity matrix acts as a counter so that the summation on the left-hand side of the equation is the total number of visits that the chain makes to state j. Thus, the equality indicates that the fraction of time spent in state j is equal to the steady-state probability of being in state j. This property is called the *ergodic property*. The last property indicates that the reciprocal of the long-run probabilities equals the expected number of steps to return to the state. Intuitively, this is as one would expect it, because the higher the probability, the quicker the return.

EXAMPLE 2.11 We consider again the Markov chain of Example 2.10. To determine its steady-state behavior, the first step is to rearrange the matrix as in Eq. (2.8). The irreducible sets were identified in Example 2.10 and on that basis we order the state space as b, g, c, d, f, a, e. The Markov matrix thus becomes

$$\boldsymbol{P} = \begin{array}{c} b \\ g \\ \hline c \\ d \\ f \\ \hline a \\ e \end{array} \left[\begin{array}{cc|ccc|cc} 0.5 & 0.5 & & & & & \\ 0.8 & 0.2 & & & & & \\ \hline & & 0.4 & 0.6 & 0 & & \\ & & 0.3 & 0.2 & 0.5 & & \\ & & 1 & 0 & 0 & & \\ \hline 0.1 & 0 & 0.2 & 0.2 & 0.1 & 0.3 & 0.1 \\ 0 & 0 & 0.2 & 0.3 & 0.1 & 0 & 0.4 \end{array} \right]$$

(Blank blocks in a matrix are always interpreted to be zeros.) The steady-state probabilities of being in states b or g are found by solving

$$\begin{aligned} 0.5\pi_b + 0.8\pi_g &= \pi_b \\ 0.5\pi_b + 0.2\pi_g &= \pi_g \\ \pi_b + \quad \pi_g &= 1. \end{aligned}$$

Thus, after deleting one of the first two equations, we solve the 2×2 system and obtain $\pi_b = \frac{8}{13}$ and $\pi_g = \frac{5}{13}$. Of course, if the chain starts in state c, d, or f the long-run probability of being in state b or g is zero, because it is impossible to reach the irreducible set $\{b, g\}$ from any state in the irreducible set $\{c, d, f\}$. The steady-state probabilities of being in state $c, d,$ or f starting in that set are given by the solution to

$$\begin{aligned} 0.4\pi_c + 0.3\pi_d + \pi_f &= \pi_c \\ 0.6\pi_c + 0.2\pi_d \qquad &= \pi_d \\ 0.5\pi_d \qquad &= \pi_f \\ \pi_c + \quad \pi_d + \pi_f &= 1, \end{aligned}$$

which yields $\pi_c = \frac{8}{17}$, $\pi_d = \frac{6}{17}$, and $\pi_f = \frac{3}{17}$. (Note that the reasonable person would delete the first equation before solving the above system.)

As a final point, assume the chain is in state c, and we wish to know the expected number of steps until the chain returns to state c. Since $\pi_c = \frac{8}{17}$, it follows that the expected number of steps until the first return is $\frac{17}{8}$. ■

■ **EXAMPLE 2.12** The market analysis discussed in Example 2.5 gave switching probabilities for body styles and associated profits with each body style. The long-term behavior for a customer can be estimated by calculating the long-run probabilities. The following system of equations

$$\begin{aligned} 0.36\pi_s + 0.60\pi_w \qquad\qquad &= \pi_w \\ 0.09\pi_s \qquad\qquad + 0.25\pi_c &= \pi_c \\ \pi_s + \qquad \pi_w + \qquad \pi_c &= 1 \end{aligned}$$

is solved to obtain $\boldsymbol{\pi} = (0.495, 0.446, 0.059)$. Therefore, the long-run expected profit per customer trade-in is

$$0.495 \times 1200 + 0.446 \times 1500 + 0.059 \times 2500 = \$1410.50.$$ ■

▶ *Suggestion: Do Exercises 2.10d, 2.11g–i, and 2.15.*

2.5 COMPUTATIONS

The determination of $\boldsymbol{R}$ [Eq. (2.5)] is straightforward for recurrent states. If states i and j are in the same irreducible set, then $R(i,j) = \infty$. If i is recurrent and j is either transient or in a different irreducible set than i, then $R(i,j) = F(i,j) = 0$. For i transient and j recurrent, then $R(i,j) = \infty$ if $F(i,j) > 0$ and $R(i,j) = 0$ if $F(i,j) = 0$. In the case where i and j are transient, we have to do a little work as is given in the following property.

Property 2.18 Let $X = \{X_n; n = 0, 1, \cdots\}$ be a Markov chain and let A denote the (finite) set of all transient states. Let $\boldsymbol{Q}$ be the matrix of transition probabilities restricted to the set A. Then, for $i, j \in A$

$$R(i,j) = (\boldsymbol{I} - \boldsymbol{Q})^{-1}(i,j).$$

Continuing with Example 2.11, we first note that $\boldsymbol{Q}$ is

$$\boldsymbol{Q} = \begin{bmatrix} 0.3 & 0.1 \\ 0 & 0.4 \end{bmatrix},$$

and evaluating $(\boldsymbol{I} - \boldsymbol{Q})^{-1}$ we have

$$\boldsymbol{R} = \left[\begin{array}{cc|ccc|cc} \infty & \infty & 0 & 0 & 0 & 0 & 0 \\ \infty & \infty & 0 & 0 & 0 & 0 & 0 \\ \hline 0 & 0 & \infty & \infty & \infty & 0 & 0 \\ 0 & 0 & \infty & \infty & \infty & 0 & 0 \\ 0 & 0 & \infty & \infty & \infty & 0 & 0 \\ \hline \infty & \infty & \infty & \infty & \infty & \frac{10}{7} & \frac{10}{42} \\ 0 & 0 & \infty & \infty & \infty & 0 & \frac{10}{6} \end{array}\right]$$

The calculations for the matrix $\boldsymbol{F}$ are slightly more complicated than for $\boldsymbol{R}$. The matrix $\boldsymbol{P}$ must be rewritten so that each irreducible, recurrent set is treated as a single "super" state. Once the Markov chain gets into an irreducible, recurrent set, it will remain in that set forever and all states within the set will be visited infinitely often. Therefore, in determining the probability of reaching a recurrent state from a transient state, it is only necessary to find the probability of reaching the appropriate irreducible set. The transition matrix in which a single irreducible set is treated as a single state is denoted by $\hat{\boldsymbol{P}}$. The matrix $\hat{\boldsymbol{P}}$ has the form

$$\hat{\boldsymbol{P}} = \begin{bmatrix} 1 & & & & \\ & 1 & & & \\ & & 1 & & \\ & & & \ddots & \\ \boldsymbol{b}_1 & \boldsymbol{b}_2 & \boldsymbol{b}_3 & \cdots & \boldsymbol{Q} \end{bmatrix}, \tag{2.9}$$

where $\boldsymbol{b}_l$ is a vector giving the one-step probability of going from transient state i to irreducible set l, that is,

$$b_l(i) = \sum_{j \in C_l} P(i, j)$$

for i transient and C_l denoting the lth irreducible set.

Property 2.19 Let $\{X_n; n = 0, 1, \cdots\}$ be a Markov chain with a finite state space ordered so that its Markov matrix can be reduced to the form in Eq. (2.9). Then for a transient state i and a recurrent state j, we have

$$F(i, j) = ((\boldsymbol{I} - \boldsymbol{Q})^{-1}\boldsymbol{b}_l)(i)$$

for each state j in the lth irreducible set.

We again return to Example 2.11 to illustrate the calculations for $\boldsymbol{F}$. We will also take advantage of Property 2.8 to determine the values in the lower right-hand portion of $\boldsymbol{F}$. Note that for i and j both transient, Property 2.8 can be rewritten as

$$F(i,j) = \begin{cases} 1 - 1/R(j,j) & \text{for } i = j \\ R(i,j)/R(j,j) & \text{for } i \neq j \end{cases}. \tag{2.10}$$

Using Property 2.19, the following holds:

$$\hat{\boldsymbol{P}} = \begin{bmatrix} 1 & 0 & 0 & 0 \\ 0 & 1 & 0 & 0 \\ 0.1 & 0.5 & 0.3 & 0.1 \\ 0 & 0.6 & 0 & 0.4 \end{bmatrix},$$

$$(\boldsymbol{I} - \boldsymbol{Q})^{-1}\boldsymbol{b}_1 = \begin{bmatrix} \frac{1}{7} \\ 0 \end{bmatrix},$$

$$(\boldsymbol{I} - \boldsymbol{Q})^{-1}\boldsymbol{b}_2 = \begin{bmatrix} \frac{6}{7} \\ 1 \end{bmatrix};$$

thus using Property 2.19 and Eq. (2.10), we have

$$\boldsymbol{F} = \left[\begin{array}{cc|ccc|cc} 1 & 1 & 0 & 0 & 0 & 0 & 0 \\ 1 & 1 & 0 & 0 & 0 & 0 & 0 \\ \hline 0 & 0 & 1 & 1 & 1 & 0 & 0 \\ 0 & 0 & 1 & 1 & 1 & 0 & 0 \\ 0 & 0 & 1 & 1 & 1 & 0 & 0 \\ \hline \frac{1}{7} & \frac{1}{7} & \frac{6}{7} & \frac{6}{7} & \frac{6}{7} & \frac{3}{10} & \frac{1}{7} \\ 0 & 0 & 1 & 1 & 1 & 0 & \frac{4}{10} \end{array}\right].$$

We can now use Property 2.17 and the previously computed steady-state probabilities to finish the limiting probability matrix for this example; namely,

$$\lim_{n\to\infty} \boldsymbol{P}^n = \left[\begin{array}{cc|ccc|cc} \frac{8}{13} & \frac{5}{13} & 0 & 0 & 0 & 0 & 0 \\ \frac{8}{13} & \frac{5}{13} & 0 & 0 & 0 & 0 & 0 \\ \hline 0 & 0 & \frac{8}{17} & \frac{6}{17} & \frac{3}{17} & 0 & 0 \\ 0 & 0 & \frac{8}{17} & \frac{6}{17} & \frac{3}{17} & 0 & 0 \\ 0 & 0 & \frac{8}{17} & \frac{6}{17} & \frac{3}{17} & 0 & 0 \\ \hline \frac{8}{91} & \frac{5}{91} & \frac{48}{119} & \frac{36}{119} & \frac{18}{119} & 0 & 0 \\ 0 & 0 & \frac{8}{17} & \frac{6}{17} & \frac{3}{17} & 0 & 0 \end{array}\right].$$

The goal of many modeling projects is the determination of revenues or costs. The ergodic property (the last item in Property 2.17) of an irreducible recurrent Markov chain gives an easy formula for the long-run average return for the process. In Example 2.1, assume that each day the tractor is running, a profit of \$100 is realized; however, each day it is in the repair shop, a cost of \$25 is incurred. The Markov matrix of Example 2.1 yields steady-state results of $\pi_0 = \frac{1}{11}$ and $\pi_1 = \frac{10}{11}$; thus, the daily average return in the long run is

$$-25 \times \frac{1}{11} + 100 \times \frac{10}{11} = 88.64.$$

This intuitive result is given in the following property.

Property 2.20 Let $X = \{X_n; n = 0, 1, \cdots\}$ be an irreducible Markov chain with finite state space E and with steady-state probabilities given by the vector $\boldsymbol{\pi}$. Let the vector $\boldsymbol{f}$ be a profit function [i.e., $f(i)$ is the profit received for each visit to state i]. Then (with probability one) the long-run average profit per unit of time is

$$\lim_{n\to\infty} \frac{1}{n} \sum_{k=0}^{n-1} f(X_k) = \sum_{j\in E} \pi(j) f(j).$$

EXAMPLE 2.13 We return to the simplified manufacturing system of Example 2.8. Refer again to the diagram in Figure 2.7 and the Markov matrix of Eq. (2.6). The cost structure for the process is as follows: The cost of the raw material going into step 1 is \$150; each time a part is processed through step 1 a cost of \$200 is incurred; and every time a part is processed through step 2 a cost of \$300 is incurred. (Thus if a part is sold that was reworked once in step 1 but was not reworked in step 2, that part would have \$850 of costs associated with it.) Because the raw material is toxic, there is also a disposal cost of \$50 per part sent to scrap.

Each day, we start with enough raw material to make 100 parts so that at the end of each day, 100 parts are finished: some good, some scrapped. To establish a reasonable strategy for setting the price of the parts to be sold, we first must determine the cost that should be attributed to the parts. To answer the relevant questions, we will first need the $\boldsymbol{F}$ and $\boldsymbol{R}$ matrices, which are

$$\boldsymbol{F} = \begin{matrix} 1 \\ 2 \\ s \\ g \end{matrix} \begin{bmatrix} 0.239 & 0.875 & 0.183 & 0.817 \\ 0.056 & 0.144 & 0.066 & 0.934 \\ 0.0 & 0.0 & 1.0 & 0.0 \\ 0.0 & 0.0 & 0.0 & 1.0 \end{bmatrix}$$

$$\boldsymbol{R} = \begin{matrix} 1 \\ 2 \\ s \\ g \end{matrix} \begin{bmatrix} 1.31 & 1.02 & \infty & \infty \\ 0.07 & 1.17 & \infty & \infty \\ 0.0 & 0.0 & \infty & 0.0 \\ 0.0 & 0.0 & 0.0 & \infty \end{bmatrix}.$$

An underlying assumption for this model is that each part that begins the manufacturing process is an independent entity, so that the actual number of finished "good" parts is a random variable following a binomial distribution. As you recall, the binomial distribution needs two parameters, the number of trials and the probability of success. The number of trials is given in the problem statement as 100; the probability of a success is the probability that a part which starts in state 1 ends in state g. Therefore, the expected number of good parts at the end of the day is

$$100 \times F(1, g) = 81.7.$$

The cost per part started is given by

$$150 + 200 \times R(1, 1) + 300 \times R(1, 2) + 50 \times F(1, s) = 727.15.$$

Therefore, the cost that should be associated to each part sold is

$$(100 \times 727.15)/81.7 = \$890/\text{part sold}.$$

A rush order for 100 parts from a very important customer has just been received, and there are no parts in inventory. Therefore, we wish to start tomorrow's production with enough raw material to be 95% confident that there will be at least 100 good parts at the end of the day. How many parts should we plan on starting tomorrow morning? Let the random variable N_n denote the number of finished parts that are good given that the day started with enough raw material for n parts. From the preceding discussion, the random variable N_n has a binomial distribution where n is the number of trials and $F(1, g)$ is the probability of success. Therefore, the question of interest is to find n such that $\Pr\{N_n \geq 100\} = 0.95$. Hopefully, you also recall that the binomial distribution can be approximated by the normal distribution; therefore, define X to be a normally distributed random variable with mean $nF(1, g)$ and variance $nF(1, g)F(1, s)$. We now have the following equation:

$$\begin{aligned}\Pr\{N_n \geq 100\} &\approx \Pr\{X > 99.5\}\\ &= \Pr\{Z > (99.5 - 0.817n)/\sqrt{0.1495n}\} = 0.95,\end{aligned}$$

where Z is normally distributed with mean 0 and variance 1. The tables in Appendix C yield

$$\frac{99.5 - 0.817n}{\sqrt{0.1495n}} = -1.645.$$

This equation is solved for n. Since it becomes a quadratic equation, there are two roots: $n_1 = 113.5$ and $n_2 = 130.7$. We must take the second root (why?) and round up; thus, the day must start with enough raw material for 131 parts. ■

■ **EXAMPLE 2.14** A missile is launched and, as it is tracked, a sequence of course correction signals is sent to it. Suppose that the system has four states that are labeled as follows.

State 0: on-course, no further correction necessary

State 1: minor deviation

State 2: major deviation

State 3: abort, off-course so badly a self-destruct signal is sent.

Let X_n represent the state of the system after the nth course correction and assume that the behavior of X can be modeled by a Markov chain with the following probability transition matrix:

$$\boldsymbol{P} = \begin{array}{c} 0 \\ 1 \\ 2 \\ 3 \end{array} \begin{bmatrix} 1.0 & 0.0 & 0.0 & 0.0 \\ 0.5 & 0.25 & 0.25 & 0.0 \\ 0.0 & 0.5 & 0.25 & 0.25 \\ 0.0 & 0.0 & 0.0 & 1.0 \end{bmatrix}$$

As always, the first step is to classify the states. Therefore, observe that states 0 and 3 are absorbing, and states 1 and 2 are transient. After a little work, you should be able to obtain the following matrices:

$$\boldsymbol{F} = \begin{array}{c} 0 \\ 1 \\ 2 \\ 3 \end{array} \begin{bmatrix} 1.0 & 0.0 & 0.0 & 0.0 \\ 0.857 & 0.417 & 0.333 & 0.143 \\ 0.571 & 0.667 & 0.417 & 0.429 \\ 0.0 & 0.0 & 0.0 & 1.0 \end{bmatrix}$$

$$\boldsymbol{R} = \begin{array}{c} 0 \\ 1 \\ 2 \\ 3 \end{array} \begin{bmatrix} \infty & 0.0 & 0.0 & 0.0 \\ \infty & 1.714 & 0.571 & \infty \\ \infty & 1.143 & 1.714 & \infty \\ 0.0 & 0.0 & 0.0 & \infty \end{bmatrix}.$$

Suppose that upon launch, the missile starts in state 2. The probability that it will eventually get on-course is 57.1% [namely, F(2,0)]; whereas, the probability that it will eventually have to be destroyed is 42.9%. When a missile is launched, 50,000 pounds of fuel are used. Every time a minor correction is made, 1000 pounds of fuel are used; and every time a major correction is made, 5000 pounds of fuel are used. Assuming that the missile started in state 2, we wish to determine the expected fuel usage for the mission. This calculation is

$$50000 + 1000 \times R(2,1) + 5000 \times R(2,2) = 59713.$$

■

The decision maker sometimes uses a discounted cost criterion instead of average costs over an infinite planning horizon. To determine the expected total discounted return for a Markov chain, we have the following property.

Property 2.21 Let $X = \{X_n; n = 0, 1, \cdots\}$ be a Markov chain with Markov matrix $\boldsymbol{P}$. Let $\boldsymbol{f}$ be a return function and let $\alpha (0 < \alpha < 1)$ be a discount factor. Then the expected total discounted cost is given by

$$E\left[\sum_{n=0}^{\infty} \alpha^n f(X_n) \mid X_0 = i\right] = \left[(\boldsymbol{I} - \alpha\boldsymbol{P})^{-1}\boldsymbol{f}\right](i).$$

We finish Example 2.11 by considering a profit function associated with it. Let the state space for the example be $E = \{1, 2, 3, 4, 5, 6, 7\}$; that is, state 1 is state b, state 2 is state g, etc. Let $\boldsymbol{f}$ be the profit function with $\boldsymbol{f} = (500, 450, 400, 350, 300, 350, 200)^T$; that is, each visit to state 1 produces \$500, each visit to state 2 produces \$450, etc. If the chain starts in states 1 or 2, then the long-run average profit per period is

$$500 \times \frac{8}{13} + 450 \times \frac{5}{13} = \$480.77.$$

If the chain starts in states 3, 4, or 5, then the long-run average profit per period is

$$400 \times \frac{8}{17} + 350 \times \frac{6}{17} + 300 \times \frac{3}{17} = \$364.71.$$

If the chain starts in state 6, then the expected value of the long-run average would be a weighted average of the two ergodic results, namely,

$$480.77 \times \frac{1}{7} + 364.71 \times \frac{6}{7} = \$381.29,$$

and starting in state 7 gives the same long-run result as starting in states 3, 4, or 5.

Now assume that a discount factor should be used. In other words, our criterion is total discounted profit instead of long-run average profit per period. Let the monetary rate of return be 20% per period, which gives $\alpha = \frac{1}{1.20} = \frac{5}{6}$. Then

$$(\boldsymbol{I} - \alpha\boldsymbol{P})^{-1} = \left[\begin{array}{cc|ccc|cc} 4.0 & 2.0 & 0 & 0 & 0 & 0 & 0 \\ 3.2 & 2.8 & 0 & 0 & 0 & 0 & 0 \\ \hline 0 & 0 & 3.24 & 1.95 & 0.81 & 0 & 0 \\ 0 & 0 & 2.32 & 2.59 & 1.08 & 0 & 0 \\ 0 & 0 & 2.70 & 1.62 & 1.68 & 0 & 0 \\ \hline 0.44 & 0.22 & 1.76 & 1.37 & 0.70 & 1.33 & 0.17 \\ 0 & 0 & 2.02 & 1.66 & 0.82 & 0 & 1.50 \end{array}\right].$$

Denote the total discounted return when the chain starts in state i by $h(i)$. Then, multiplying the preceding matrix by the profit function $\boldsymbol{f}$ yields

$$\boldsymbol{h}^T = (2900, 2860, 2221.5, 2158.5, 2151, 2079, 1935).$$

In other words, if the chain started in state 1, the present worth of all future profits would be \$2900 where the present worth of one dollar one period from now is 83.33 cents. If the initial state is unknown and a distribution of probabilities for the starting state is given by $\boldsymbol{\mu}$, the total discounted profit is given by $\boldsymbol{\mu h}$.

▶ *Suggestion: Do Exercises 2.13d–f, 2.14e–g, and finish 2.1–2.9.*

2.6 EXERCISES

The exercises given here are designed to help in your understanding of Markov chains and to indicate some of the potential uses for Markov chain analysis. All numbers are fictitious.

2.1 Joe and Pete each have two cents in their pockets. They have decided to match pennies; that is, they will each take one of their own pennies and flip them. If the pennies match (two heads or two tails), Joe gets Pete's penny; if the pennies do not match, Pete gets Joe's penny. They will keep repeating the game until one of them has four cents, and the other one is broke. Although they do not realize it, all four pennies are biased. The probability of tossing a head is 0.6, and the probability of a tail is 0.4. Let X be a Markov chain where X_n denotes the amount that Joe has after the nth play of the game.

(a) Give the Markov matrix for X.

(b) What is the probability that Joe will have four pennies after the second toss?

(c) What is the probability that Pete will be broke after three tosses?

(d) What is the probability that the game will be over before the third toss?

(e) What is the expected amount of money Pete will have after two tosses?

(f) What is the probability that Pete will end up broke?

(g) What is the expected number of tosses until the game is over?

2.2 At the start of each week, the condition of a machine is determined by measuring the amount of electrical current it uses. According to its amperage reading, the machine is categorized as being in one of the following four states: low, medium, high, failed. A machine in the low state has a probability of 0.05, 0.03, and 0.02 of being in the medium, high, or failed state, respectively, at the start of the next week. A machine in the medium state has a probability of 0.09 and 0.06 of being in the high or failed state, respectively, at the start of the next week (it cannot, by itself, go to the low state). And, a machine in the high state has a probability of 0.1 of being in the failed state at the start of the next week (it cannot, by itself, go to the low or medium state). If a machine is in the failed state at the start of a week, repair is immediately begun on the machine so that it will (with probability 1) be in the low state at the start of the following week. Let X be a Markov chain where X_n is the state of the machine at the start of week n.

(a) Give the Markov matrix for X.

(b) A new machine always starts in the low state. What is the probability that the machine is in the failed state three weeks after it is new?

(c) What is the probability that a machine has at least one failure three weeks after it is new?

(d) What is the expected number of weeks after a new machine is installed until the first failure occurs?

(e) On average, how many weeks per year is the machine working?

(f) Each week that the machine is in the low state, a profit of \$1000 is realized; each week that the machine is in the medium state, a profit of \$500 is realized; each week that the machine is in the high state, a profit of \$400 is realized; and the week in which a failure is fixed, a cost of \$700 is incurred. What is the long-run average profit per week realized by the machine?

(g) A suggestion has been made to change the maintenance policy for the machine. If at the start of a week the machine is in the high state, the machine will be taken out of service and repaired so that at the start of the next week it will again be in the low state. When a repair is made due to the machine being in the high state instead of a failed state, a cost of \$600 is incurred. Is this new policy worthwhile?

2.3 We are interested in the movement of patients within a hospital. For purposes of our analysis, we consider the hospital to have three different types of rooms: general care, special care, and intensive care. Based on past data, 60% of arriving patients are initially admitted into the general care category, 30% in the special care category, and 10% in intensive care. A "general care" patient has a 55% chance of being released healthy the following day, a 30% chance of remaining in the general care room, and a 15% chance of being moved to the special care facility. A "special care" patient has a 10% chance of being released the following day, a 20% chance of being moved to general care, a 10% chance of being upgraded to intensive care, and a 5% chance of dying during the day. An "intensive care" patient is never released from the hospital directly from the intensive care unit (ICU), but is always moved to another facility first. The probabilities that the patient is moved to general care, special care, or remains in intensive care are 5%, 30%, or 55%, respectively. Let X be a Markov chain where X_0 is the type of room that an admitted patient initially uses, and X_n is the room category of that patient at the end of day n.

(a) Give the Markov matrix for X.

(b) What is the probability that a patient admitted into the intensive care room eventually leaves the hospital healthy?

(c) What is the expected number of days that a patient, admitted into intensive care, will spend in the ICU?

(d) What is the expected length of stay for a patient admitted into the hospital as a general care patient?

(e) During a typical day, 100 patients are admitted into the hospital. What is the average number of patients in the ICU?

2.4 Consider again the manufacturing process of Example 2.8. New production plans call for an expected production level of 2000 good parts per month. (In other words, enough raw material must be used so that the expected number of good parts produced each month is 2000.) For a capital investment and an increase in operating costs, all rework and scrap can be eliminated. The sum of the capital investment and operating cost increase is equivalent to an annual cost of $5 million. Is it worthwhile to increase the annual cost by $5 million in order to eliminate the scrap and rework? (Refer to Example 2.13 for costs.)

2.5 Assume the description of the manufacturing process of Example 2.8 is for the process at branch A of the manufacturing company. Branch B of the same company has been closed. Branch B had an identical process and they had 1000 items in stock that had been through step 1 of the process when they were shut down. Because the process was identical, these items can be fed into branch A's process at the start of step 2. (However, since the process was identical there is still a 5% chance that after finishing step 2 the item will have to be reworked at step 1, a 10% chance the item will have to be reworked at step 2, and a 5% chance that the item will have to be scrapped.) Branch A purchases these (partially completed) items for a total of $300,000, and they will start processing at step 2 of branch A's system. After branch A finishes processing this batch of items, they must determine the cost of these items so they will know how much to charge customers in order to recover the cost. (They may want to give a discount.) Your task is to determine the cost that would be attributed to each item shipped. (Refer to Example 2.13 for costs.)

2.6 The manufacture of a certain type of electronic board consists of four steps: tinning, forming, insertion, and solder. After the forming step, 5% of the parts must be retinned; after the insertion step, 20% of the parts are bad and must be scrapped; and after the solder step, 30% of the parts must be returned to insertion and 10% must be scrapped. (We assume that when a part is returned to a processing step, it is treated like any other part entering that step.) Figure 2.12 gives a schematic showing the flow of a job through the manufacturing steps.

(a) Model this process using Markov chains and give its Markov matrix.

(b) If a batch of 100 boards begins this manufacturing process, what is the expected number that will end up scrapped?

(c) How many boards should we start with if the goal is to have the expected number of boards that finish in the good category equal to 100?

(d) How many boards should we start with if we want to be 90% sure that we end up with a batch of 100 boards? (*Hint:* The final status of each board can be considered a Bernoulli random variable, the sum of independent

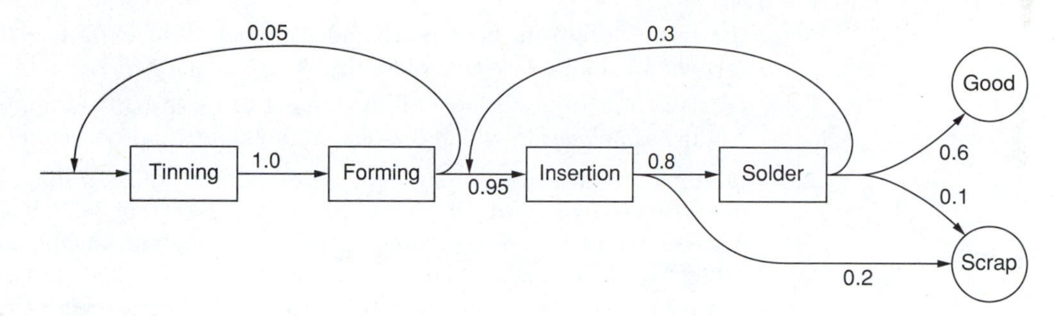

FIGURE 2.12 Manufacturing structure for board processing steps.

Bernoullis is binomial, and the binomial can be approximated by a normal.)

(e) Each time a board goes through a processing step, direct labor and material costs are $10 for tinning, $15 for forming, $25 for insertion, and $20 for solder. The raw material costs $8, and a scrapped board returns $2. The average overhead rate is $1,000,000 per year, which includes values for capital recovery. The average processing rate is 5000 board starts per week. We would like to set a price per board so that expected revenues are 25% higher than expected costs. At what value would you set the price per board?

2.7 The government has done a study of the flow of money among three types of financial institutions: banks, savings associations, and credit unions. We assume that the Markov assumptions are valid and model the monthly movement of money among the various institutions. Some recent data are:

June Amounts	*Units in Billions of Dollars May*				*June Totals*
	Bank	*Savings*	*Credit*	*Other*	
Bank	10	4.5	1.5	3.6	19.6
Savings	4	6	0.5	1.2	11.7
Credit	2	3	6	2.4	13.4
Other	4	1.5	2	4.8	12.3
May totals	20	15	10	12	57.0

For example, of the 20 billion dollars in banks during the month of May, half of it remained in banks for the month of June, and 2 out of the 20 billion left banks to be invested in credit unions during June.

(a) Use the data given to estimate a Markov matrix. What would be the problems with such a model? (That is, be critical of the Markov assumption of the model.)

(b) In the long run, how much money would you expect to be in credit unions during any given month?

(c) How much money would you expect to be in banks during August of the same year in which the data in the table were collected?

2.8 Within a certain market area, there are two brands of soap that most people use: "super soap" and "cheap soap," with the current market split evenly between the two brands. A company is considering introducing a third brand called "extra clean soap," and they have done some initial studies of market conditions. Their estimates of weekly shopping patterns are as follows: If a customer buys super soap this week, there is a 75% chance that next week the super soap will be used again, a 10% chance that extra clean will be used, and a 15% chance that the cheap soap will be used. If a customer buys the extra clean this week, there is a fifty-fifty chance the customer will switch, and if a switch is made it will always be to super soap. If a customer buys cheap soap this week, it is equally likely that next week the customer will buy any of the three brands.

(a) Assuming that the Markov assumptions are good, use the data given to estimate a Markov matrix. What would be the problems with such a model?

(b) What is the long-run market share for the new soap?

(c) What will be the market share of the new soap two weeks after it is introduced?

(d) The market consists of approximately one million customers each week. Each purchase of super soap yields a profit of 15 cents; a purchase of cheap soap yields a profit of 10 cents; and a purchase of extra clean will yield a profit of 25 cents. Assume that the market was at steady state with the even split between the two products. The initial advertising campaign to introduce the new brand was $100,000. How many weeks will it be until the $100,000 is recovered from the added revenue of the new product?

(e) The company feels that with these three brands, an advertising campaign of $30,000 per week will increase the weekly total market by a quarter of a million customers. Is the campaign worthwhile? (Use a long-term average criterion.)

2.9 A small company sells high-quality laser printers and they use a simple periodic inventory ordering policy. If there are two or fewer printers in inventory at the end of the day on Friday, the company will order enough printers so that there will be five printers in stock at the start of Monday. (It only takes the weekend for printers to be delivered from the wholesaler.) If there are more than two printers at the end of the week, no order is placed. Weekly demand data have been analyzed yielding the probability mass function for weekly demand as $\Pr\{D = 0\} = 0.05$, $\Pr\{D = 1\} = 0.1$, $\Pr\{D = 2\} = 0.2$, $\Pr\{D = 3\} = 0.4$, $\Pr\{D = 4\} = 0.1$, $\Pr\{D = 5\} = 0.1$, and

$\Pr\{D = 6\} = 0.05$. Let X be a Markov chain where X_n is the inventory at the end of week n. (*Note:* Backorders are not accepted.)

(a) Give the Markov matrix for X.

(b) If at the end of week 1 there were five items in inventory, what is the probability that there will be five items in inventory at the end of week 2?

(c) If at the end of week 1 there were five items in inventory, what is the probability that there will be five items in inventory at the end of week 3?

(d) What is the expected number of times each year that an order is placed?

(e) Each printer sells for $1800. Each time an order is placed, it costs $500 plus $1000 times the number of items ordered. At the end of each week, each unsold printer costs $25 (in terms of keeping them clean, money tied up, and other such inventory-type expenses). Whenever a customer wants a printer not in stock, the company buys it retail and sends it by airfreight to the customer; thus the customer spends the $1800 sales price but it costs the company $1900. To reduce the lost sales, the company is considering raising the reorder point to three, but still keeping the order-up-to quantity at five. Would you recommend the change in the reorder point?

2.10 Let X be a Markov chain with state space $\{a, b, c\}$ and transition probabilities given by

$$\boldsymbol{P} = \begin{bmatrix} 0.3 & 0.4 & 0.3 \\ 1.0 & 0.0 & 0.0 \\ 0.0 & 0.3 & 0.7 \end{bmatrix}.$$

(a) What is $\Pr\{X_2 = a \mid X_1 = b\}$?

(b) What is $\Pr\{X_2 = a \mid X_1 = b, X_0 = c\}$?

(c) What is $\Pr\{X_{35} = a \mid X_{33} = a\}$?

(d) What is $\Pr\{X_{200} = a \mid X_0 = b\}$? (Use steady-state probability to answer this.)

2.11 Let X be a Markov chain with state space $\{a, b, c\}$ and transition probabilities given by

$$\boldsymbol{P} = \begin{bmatrix} 0.3 & 0.7 & 0.0 \\ 0.0 & 0.6 & 0.4 \\ 0.4 & 0.1 & 0.5 \end{bmatrix}.$$

Let the initial probabilities be given by the vector $(0.1, 0.3, 0.6)$ with a profit function given by $(10, 20, 30)^T$, which means, for example, each visit to state a yields a profit of $10. Find the following:

(a) $\Pr\{X_2 = b \mid X_1 = c\}$.

(b) $\Pr\{X_3 = b \mid X_1 = c\}$.

(c) $\Pr\{X_3 = b \mid X_1 = c, X_0 = c\}$.

(d) $\Pr\{X_2 = b\}$.

(e) $\Pr\{X_1 = b, X_2 = c \mid X_0 = c\}$.

(f) $E[f(X_2) \mid X_1 = c]$.

(g) $\lim_{n\to\infty} \Pr\{X_n = a \mid X_0 = a\}$.

(h) $\lim_{n\to\infty} \Pr\{X_n = b \mid X_{10} = c\}$.

(i) $\lim_{n\to\infty} E[f(X_n) \mid X_0 = c]$.

2.12 Let X be a Markov chain with state space $\{a, b, c, d\}$ and transition probabilities given by

$$\boldsymbol{P} = \begin{bmatrix} 0.1 & 0.3 & 0.6 & 0.0 \\ 0.0 & 0.2 & 0.5 & 0.3 \\ 0.5 & 0.0 & 0.0 & 0.5 \\ 0.0 & 1.0 & 0.0 & 0.0 \end{bmatrix}.$$

Each time the chain is in state a, a profit of \$20 is made; each visit to state b yields a \$5 profit; each visit to state c yields \$15 profit; and each visit to state d costs \$10. Find the following:

(a) $E[f(X_5) \mid X_3 = c, X_4 = d]$.

(b) $E[f(X_1) \mid X_0 = b]$.

(c) $E[f(X_1)^2 \mid X_0 = b]$.

(d) $\text{var}[f(X_1) \mid X_0 = b]$.

(e) $\text{var}[f(X_1)]$ given that $P\{X_0 = a\} = 0.2, P\{X_0 = b\} = 0.4, P\{X_0 = c\} = 0.3$, and $P\{X_0 = d\} = 0.1$.

2.13 Consider the following Markov matrix representing a Markov chain with state space $\{a, b, c, d, e\}$.

$$\boldsymbol{P} = \begin{bmatrix} 0.3 & 0.0 & 0.0 & 0.7 & 0.0 \\ 0.0 & 1.0 & 0.0 & 0.0 & 0.0 \\ 1.0 & 0.0 & 0.0 & 0.0 & 0.0 \\ 0.0 & 0.0 & 0.5 & 0.5 & 0.0 \\ 0.0 & 0.2 & 0.4 & 0.0 & 0.4 \end{bmatrix}.$$

(a) Draw the state diagram.

(b) List the transient states.

(c) List the irreducible set(s).

(d) Let $F(i, j)$ denote the first passage probabilities of reaching (or returning to) state j given that $X_0 = i$. Calculate the $\boldsymbol{F}$ matrix.

(e) Let $R(i, j)$ denote the expected number of visits to state j given that $X_0 = i$. Calculate the $\boldsymbol{R}$ matrix.

(f) Calculate the $\lim_{n\to\infty} \boldsymbol{P}^n$ matrix.

2.14 Let X be a Markov chain with state space $\{a, b, c, d, e, f\}$ and transition probabilities given by

$$P = \begin{bmatrix} 0.3 & 0.5 & 0.0 & 0.0 & 0.0 & 0.2 \\ 0.0 & 0.5 & 0.0 & 0.5 & 0.0 & 0.0 \\ 0.0 & 0.0 & 1.0 & 0.0 & 0.0 & 0.0 \\ 0.0 & 0.3 & 0.0 & 0.0 & 0.0 & 0.7 \\ 0.1 & 0.0 & 0.1 & 0.0 & 0.8 & 0.0 \\ 0.0 & 1.0 & 0.0 & 0.0 & 0.0 & 0.0 \end{bmatrix}.$$

(a) Draw the state diagram.

(b) List the recurrent states.

(c) List the irreducible set(s).

(d) List the transient states.

(e) Calculate the $\boldsymbol{F}$ matrix.

(f) Calculate the $\boldsymbol{R}$ matrix.

(g) Calculate the $\lim_{n\to\infty} \boldsymbol{P}^n$ matrix.

2.15 Let X be a Markov chain with state space $\{a, b, c, d\}$ and transition probabilities given by

$$P = \begin{bmatrix} 0.3 & 0.7 & 0.0 & 0.0 \\ 0.5 & 0.5 & 0.0 & 0.0 \\ 0.0 & 0.0 & 1.0 & 0.0 \\ 0.1 & 0.1 & 0.2 & 0.6 \end{bmatrix}.$$

(a) Find $\lim_{n\to\infty} \Pr\{X_n = a \mid X_0 = a\}$.

(b) Find $\lim_{n\to\infty} \Pr\{X_n = a \mid X_0 = b\}$.

(c) Find $\lim_{n\to\infty} \Pr\{X_n = a \mid X_0 = c\}$.

(d) Find $\lim_{n\to\infty} \Pr\{X_n = a \mid X_0 = d\}$.

3 Simulation

Simulation is one of the most widely used probabilistic modeling tools in industry. It is used for the analysis of existing systems and for the selection of hypothetical systems. For example, suppose a bank has been receiving complaints from customers regarding the length of time customers spend waiting in line at the drive-in window. Management has decided to add some extra windows; they now need to decide how many to add. Simulation models can be used to help management determine the number of windows to add. Even though the main focus of this textbook is on building analytical (as opposed to simulation) models, there will be times when the physical system is too complicated for analytical modeling; in such a case, simulation would be an appropriate tool. The idea behind simulation, applied to this banking problem, is that a computer program would be written to generate randomly arriving customers, and then process each customer through the drive-in facility. In such a manner, the effect of having different windows could be determined before the expense of building them is incurred. Or, to continue this example, it may be possible (after covering Chapter 5) to build an analytical model of the banking problem; however, the analyst may want to furnish additional evidence for the validity of the analytical model. (When a model is to be used for decision making that involves large capital expenditures, validation efforts are always time-consuming and essential.) A simulation could be built to model the same system as the analytical model describes. Then, if the two models agree, the analyst would have confidence in their use.

A final reason for simulation is that it can be used to increase understanding of the process being modeled. In fact, this is why Chapter 3 is devoted to simulation. Now that the theory of Markov chains has been taught, simulation can be used to help the intuitive understanding of the dynamics of the process. (We could have used it similarly after Chapter 1, but we felt that it is more interesting to begin work on stochastic processes as soon as possible.) It is impossible to build a simulation of something not understood, so just the process of developing a simulation of a Markov chain will force an understanding of Markov chains.

In this chapter, only a cursory introduction to simulation is given.[1] Our goal is to give enough of an introduction to enable the student to simulate simple probabilistic events, Markov chains, and Markov processes. A more detailed treatment is postponed until Chapter 6. In particular, the use of future event lists and the statistical analyses of simulation output are discussed in the second simulation chapter.

One important topic that is not discussed in this book is the use of a special simulation language. To facilitate the building of simulations, many special-purpose simulation languages have been developed. Some of these languages are very easy to learn, and we recommend that students interested in simulation investigate the various simulation languages.

3.1 EXAMPLES

In this section, we introduce the basic concepts of a simulation through two simple examples. The examples use simulation to determine properties of simple probabilistic events. Although the examples can be solved using analytical equations, they should serve to illustrate the concept of simulation.

EXAMPLE 3.1 Consider a microwave oven salesperson who works in a busy retail store. Often people come into the department merely to browse, while others are interested in purchasing an oven. Of all the people who take up the salesperson's time, 50% end up buying one of the three models available and 50% do not make a purchase. Of those customers who actually buy an oven, 25% purchase the plain model, 50% purchase the standard model, and 25% purchase the deluxe model. The plain model yields a profit of $30, the standard model yields a profit of $60, and the deluxe model yields a profit of $75.

The salesperson wishes to determine the average profit per customer. (We assume that the salesperson does not have a mathematical background and therefore is unable to calculate the exact expected profit per customer. Hence, an alternative to a mathematical computation for estimating the profit must be used.) One approach for estimating the average profit would be to keep records of all the customers who talk to the salesperson and, based on the data, calculate an estimate for the expected profit per customer. However, there is an easier and less time-consuming method: The process can be simulated.

[1]Some of the material presented in Chapters 3 and 6 is taken from G. L. Curry, B. L. Deuermeyer, and R. M. Feldman, *Discrete Simulation: Fundamentals and Microcomputer Support* (Oakland, Calif.: Holden-Day, Inc., 1989).

The basic concept in simulation is to generate random outcomes (for example, we might toss a coin or roll dice to generate outcomes) and then associate appropriate physical behavior with the resultant (random) outcomes. To simulate the microwave oven buying decision, a (fair) coin is tossed with a head representing a customer and a tail representing a browser; thus, there is a fifty-fifty chance that an individual entering the department is a customer. To simulate the type of microwave that is bought by an interested customer, two fair coins are tossed. Two tails (which occur 25% of the time) represent buying the plain model, a head and a tail (which occur 50% of the time) represent buying the standard model, and two heads (which occur 25% of the time) represent buying the deluxe model. Table 3.1 summarizes the relationship between the random coin tossing results and the physical outcomes being modeled. Table 3.2 shows the results of repeating this process 20 times to simulate 20 customers and their buying decisions.

If we take the final cumulative profit and divide it by the number of customers, the estimate for the expected profit per customer is calculated to be $34.50. Two facts are immediately obvious. First, the number of interested customers was 12 out of 20, but because there is a fifty-fifty chance that any given customer will be interested, we expect only 10 out of 20. Furthermore, utilizing some basic probability rules, a person knowledgeable in probability would determine that the theoretical expected profit per customer is only $28.125. Thus, it is seen that the simulation does not provide the exact theoretical values sought. ■

The simulation is, in fact, just a statistical experiment. This point cannot be overemphasized. *The results of a simulation involving random numbers must be interpreted statistically.* In Chapter 6, some basic statistical concepts will be given to aid in the proper analysis of simulation results. For now it is important to realize that the simulation is simply a statistical experiment performed so that the expense and time needed to perform and/or to observe the actual process can be avoided. For instance, this experimental evaluation might require less than an hour to accomplish, whereas obtaining an estimate of the theoretical profit by observing the actual process for 20 customers might take days.

TABLE 3.1 Procedure for determining the interest of a potential customer and the microwave model an interested customer buys.

Random Value	*Simulated Outcome*
Head	Customer
Tail	Browser

Random Value	*Simulated Outcome*
Tail-tail	Buy plain
Tail-head	Buy standard
Head-tail	Buy standard
Head-head	Buy deluxe

TABLE 3.2 Simulated behavior of 20 customers.

Customer Number	*Coin Toss*	*Interested?*	*Two-Coin Toss*	*Profit*	*Cumulative Profit*	*Average Profit*
1	T	No	—	0	0	0.00
2	H	Yes	TH	60	60	30.00
3	H	Yes	HH	75	135	45.00
4	H	Yes	TH	60	195	48.75
5	T	No	—	0	195	39.00
6	T	No	—	0	195	32.50
7	H	Yes	TT	30	225	32.14
8	T	No	—	0	225	28.13
9	H	Yes	HT	60	285	31.67
10	H	Yes	HH	75	360	36.00
11	H	Yes	TT	30	390	35.45
12	H	Yes	TT	30	420	35.00
13	T	No	—	0	420	32.31
14	T	No	—	0	420	30.00
15	T	No	—	0	420	28.00
16	H	Yes	TH	60	480	30.00
17	H	Yes	HT	60	540	31.76
18	T	No	—	0	540	30.00
19	H	Yes	HH	75	615	32.37
20	H	Yes	HH	75	690	34.50

EXAMPLE 3.2 Let us consider a game taken from a popular television game show. The contestant is shown three doors, labeled A, B, and C, and told that behind one of the three doors is a pot of gold worth $120,000; there is nothing behind the other two doors. The contestant is to pick a door. After the door is selected, the host will pick one of the other two doors, one with an empty room. The contestant will then be given the choice of switching doors or keeping the originally selected door. After the contestant decides whether or not to switch, the game is over and the contestant gets whatever is behind the door finally chosen. The question of interest is "Should the contestant switch when given the option?" We simulate here the "no-switch" option and refer to Exercise 3.2 for the "switch" option.

To simulate the "no-switch" policy, it is necessary to simulate the random event of placing the gold behind a door, and simulate the event of selecting a door. Both events will be simulated by rolling a single die and interpreting the outcome according to Table 3.3. Since a switch is never made, it is not necessary to simulate the host selecting a door after the initial selection.

The results from 20 simulated games are shown in Table 3.4. Based on the simulation run of Table 3.4, the estimate for the expected yield is $36,000 per game. (Of course, it is easy to calculate that the actual expected yield is $40,000 per game, which illustrates that simulations only yield statistical estimates.) As can be seen from these two examples, a difficulty in simulation is deciding what information to maintain, and then keeping track

TABLE 3.3 Random events for the "pot-of-gold" game.

Random Value	*Simulated Outcome*
1 or 2	*A*
3 or 4	*B*
5 or 6	*C*

of that information. Once the decision is made as to what information is relevant, the simulation itself is straightforward. ■

▶ *Suggestion: Do Exercises 3.1 and 3.2.*

3.2 GENERATION OF RANDOM VARIATES

System simulation depends heavily on random number generation to model the stochastic nature of the systems being studied. The examples of the previous section simulated random outcomes by having the modeler physically toss a coin or a die. A computer, by contrast, must simulate random

TABLE 3.4 Simulated results from the "pot-of-gold" game using a no-switch policy. Yield is in terms of thousands of dollars.

Game Number	*Die Toss*	*Door for Gold*	*Die Toss*	*Door Chosen*	*Yield (thousands)*	*Cumulative Yield*	*Average Yield*
1	5	*C*	2	*A*	0	0	0.00
2	3	*B*	5	*C*	0	0	0.00
3	5	*C*	4	*B*	0	0	0.00
4	3	*B*	2	*A*	0	0	0.00
5	2	*A*	1	*A*	120	120	24.00
6	1	*A*	6	*C*	0	120	20.00
7	6	*C*	4	*B*	0	120	17.14
8	4	*B*	4	*B*	120	240	30.00
9	5	*C*	5	*C*	120	360	40.00
10	5	*C*	3	*B*	0	360	36.00
11	4	*B*	2	*A*	0	360	32.73
12	3	*B*	4	*B*	120	480	40.00
13	2	*A*	4	*B*	0	480	36.92
14	1	*A*	3	*B*	0	480	34.29
15	1	*A*	3	*B*	0	480	32.00
16	4	*B*	6	*C*	0	480	30.00
17	5	*C*	6	*C*	120	600	35.29
18	4	*B*	4	*B*	120	720	40.00
19	1	*A*	6	*C*	0	720	37.94
20	2	*A*	4	*B*	0	720	36.00

outcomes by generating numbers to give the appearance of randomness. This section deals with the problems involved in programming a computer to simulate randomness. The goal is to introduce the tools necessary for statistically describing a physical situation for modeling purposes.

3.2.1 Uniform Random Numbers

Most mathematical table books contain several pages of random numbers. Such *random numbers* have three properties: (1) The numbers are between zero and one, (2) the probability of selecting a number in the interval (a, b) is equal to $(b - a)$, where $0 \leq a < b \leq 1$, and (3) the numbers are statistically independent. The concept of statistical independence simply means that if several numbers are chosen, knowledge of the value of one number does not provide information that will help in predicting the value of another number. Independence, therefore, rules out the possibility of trends or cyclic patterns occurring within a sequence of random numbers. In terms of probability, these random numbers are sequences of independent, identically distributed random variates having a continuous uniform distribution between zero and one.

Technically, random numbers can refer to the observations of random variables having any arbitrary probability distribution; however, usually the term *random numbers* refers to the observations of uniform zero-one random variables. For observations of random variables governed by a distribution other than the uniform zero-one, the more general term *random variates* is used. Thus, the term *random numbers* as used in this text will always refer to numbers that statistically reproduce observations of zero-one uniform random variables, and the term *random variates* will refer to numbers that statistically reproduce observations of arbitrarily distributed random variables.

As shown later, random numbers are used to generate random variates. Thus, the first task we must be able to do with a computer is generate random numbers by way of a simple and fast algorithm. Because a computer can follow only very specific steps, random numbers generated by a computer are not truly random but are more properly called "pseudo-random numbers" (although the "pseudo" prefix is usually dropped). Numbers generated by a computer program or subprogram are, therefore, called "random" if statistical tests cannot determine the difference between computer-generated and truly random sequences of numbers.

All methods of random number generation in computers begin with an initial value called the *initial random number seed*. Each time a random number is desired, the random number seed is used to produce the next random number and the seed is transformed into another number that becomes the new seed. Thus, the random number seed is continually changed as the random numbers are generated. A simulation program must always furnish an initial random number seed. If the same initial random number seed is used whenever the simulation program is run, the exact same string of random numbers will be generated. If a different string of random numbers is desired, then a different initial random number seed must be furnished.

One of the most popular methods of generating random numbers by a computer is the *congruential method*. In this method the random number seed is always a positive integer, and the random number associated with the seed is the value of the seed divided by the largest possible integer that can be stored in the computer. To obtain the new seed, the old seed is multiplied by a large constant, and a second constant is added to the product. The resulting integer might then be larger than the computer word size (i.e., larger than the largest possible integer for the computer size). If it is too large, it is divided by the largest possible integer, and the remainder is the new seed. If it is not too large for the computer word size, it becomes the new seed. To express this mathematically, let a and b be two fixed integers, and let L denote the largest possible (signed) integer that the computer can store. Let S be a random seed and let S_{next} be the next seed to be determined. The random number associated with the seed is

$$R = \frac{S}{L},$$

and the next seed is

$$S_{\text{next}} = (aS + b)\,\mathbf{mod}\,L.$$

For example, for a 16-bit microcomputer, $L = 32767 = (2^{15} - 1)$ and we might set $a = 1217$, set $b = 0$, and let the initial random number seed be $S_0 = 23$. For this situation, the random number sequence is generated by the following calculations:

$$\begin{aligned}
S_1 &= (1217 \times 23)\,\mathbf{mod}\,32767 = 27991 \\
R_1 &= \frac{27991}{32767} = 0.85424 \qquad \text{(first random number)}, \\
S_2 &= (1217 \times 27991)\,\mathbf{mod}\,32767 \\
&= 34065047\,\mathbf{mod}\,32767 = 20134 \\
R_2 &= \frac{20134}{32767} = 0.61446 \qquad \text{(second random number)}, \\
&\vdots
\end{aligned}$$

Several rules of thumb are used to determine values for the constants a and b that will produce good pseudo-random number strings. However, most high-level languages contain their own random number generators; although you may not need to be concerned with programming[2] these methods, a conceptual understanding of random number generation is beneficial. It is

[2]See Section B.3 of Appendix B if you need a code for random number generation.

especially important to remember the role of the initial random number seed. If the initial seed is the same for every run of a given simulation program, the output will be the same.

3.2.2 Discrete Random Variates

Let us assume that we want to write a computer simulation model of the microwave salesperson situation of Example 3.1. An immediate problem arises because the example involves two random variables, neither of which is a continuous uniform zero-one random variable. The first random variable represents the decision as to whether or not a customer will end up buying a microwave. Mathematically, we can denote the customer's interest by the random variable C and give its probability mass function as

$$\Pr\{C = 0\} = 0.5,$$
$$\Pr\{C = 1\} = 0.5,$$

where the random variable being zero ($C = 0$) represents a customer who is not interested in purchasing a microwave, and the random variable being one ($C = 1$) represents the customer who purchases a microwave.

The cumulative probability distribution function (whose values are between zero and one) permits an easy transformation of a random number into a random variate arising from another probability law. The transformation is obtained simply by letting the random number occur along the ordinate (y-axis) and using the cumulative function to form the inverse mapping back to the abscissa (x-axis). For the "customer interest" random variable C, the following rule is obtained: If the random number is less than or equal to 0.5, then $C = 0$, i.e., the customer is not interested in a purchase; otherwise, if the random number is greater than 0.5, then $C = 1$, i.e., the customer is interested. This transformation is represented schematically in Figure 3.1 for one possible value of the random number.

FIGURE 3.1 Transformation using the cumulative distribution function to obtain the random variate C.

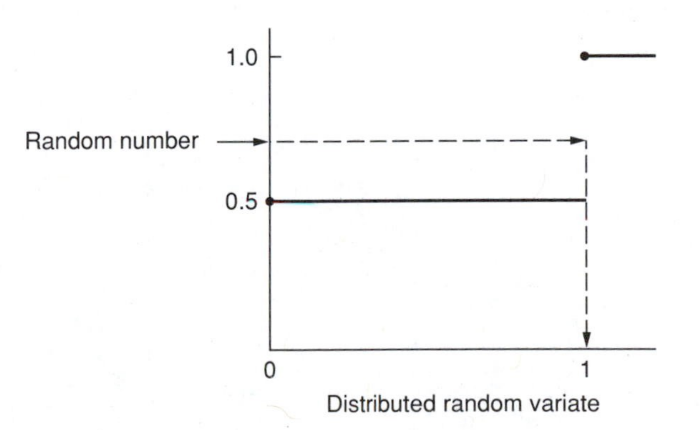

The second random variable that must be simulated in the microwave salesperson example is the profit resulting from the sale of a microwave. Let the "model" random variable be denoted by M with probability mass function given by

$$\Pr\{M = 30\} = 0.25,$$
$$\Pr\{M = 60\} = 0.50,$$
$$\Pr\{M = 75\} = 0.25.$$

In this case, $M = 30$ is the profit from the sale of a basic model, $M = 60$ the profit from a standard model, and $M = 75$ the profit from a deluxe model. Again, it is easiest to simulate this random variable if its cumulative probability distribution function is first determined. The jump points of the cumulative function are given by

$$\Pr\{M \le 30\} = 0.25,$$
$$\Pr\{M \le 60\} = 0.75,$$
$$\Pr\{M \le 75\} = 1.0.$$

The inverse mapping that gives the transformation from the random number to the random variate representing the profit obtained from the sale of a microwave is illustrated in Figure 3.2. The results are as follows: If the random number is less than or equal to 0.25, the plain model is sold; if it is greater than 0.25 and less than or equal to 0.75, the standard model is sold; and if it is greater than 0.75, the deluxe model is sold.

The rather intuitive approach for obtaining the random variates C and M as shown in Figures 3.1 and 3.2 needs to be formalized for use with continuous random variables. Specifically, Figures 3.1 and 3.2 indicate that the random variates were generated by using the inverse of the cumulative probability distribution function (CDF). A mathematical justification for this is now given.

FIGURE 3.2 Transformation using the cumulative distribution function to obtain the random variate M.

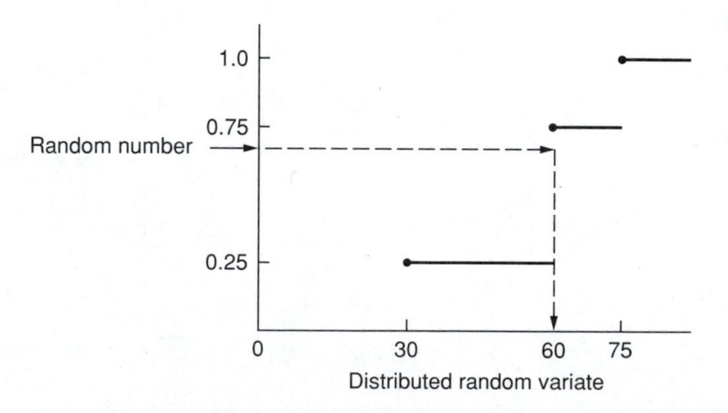

The following property (proven in most introductory mathematical probability and statistics books) leads to the so-called *inverse transformation method* of generating arbitrarily distributed random variates.

Property 3.1 Let R be a random variable with a continuous uniform distribution between zero and one, and let F be an arbitrary cumulative distribution function. If the inverse of the function F exists, denote it by F^{-1}; otherwise, let $F^{-1}(a) = \min\{t \mid F(t) \geq a\}$. Then the random variable X defined by

$$X = F^{-1}(R)$$

has a distribution function given by F; that is,

$$P\{X \leq a\} = F(a) \quad \text{for } -\infty < a < \infty.$$

From Property 3.1, we can see that the function that relates the continuous uniform zero-one random variable with another random variable is simply the inverse of the cumulative probability distribution function. We utilize this property in the next section. First, however, we give some examples simulating discrete random variables.

EXAMPLE 3.3 A manufacturing process produces items with a defect rate of 15%; in other words, the probability of an individual item being defective is 0.15 and the probability that any one item is defective is independent of whether or not any other item is defective. The items are packaged in boxes of four. An inspector tests each box by randomly selecting one item in that box. If the selected item is good, the box is passed; if the item is defective, the box is rejected. The inspector has just started for the day. What is the expected number of boxes that the inspector will look at before the first rejection?

Let the random variable M denote the number of defective items per box. The distribution for M is binomial [Eq. (1.10)]; therefore, its pmf is given by

$$\begin{aligned} \Pr\{M = 0\} &= 0.5220, \\ \Pr\{M = 1\} &= 0.3685, \\ \Pr\{M = 2\} &= 0.0975, \\ \Pr\{M = 3\} &= 0.0115, \\ \Pr\{M = 4\} &= 0.0005. \end{aligned}$$

The transformation from random numbers to random variates is based on the CDF, so the above probabilities are summed to obtain Table 3.5. (There are more efficient techniques than the one demonstrated here for generating some of the common random variates;[3] however, our purpose here is to

[3]An excellent reference is A. M. Law and W. D. Kelton, *Simulation Modeling and Analysis* (New York: McGraw-Hill, Inc., 1991).

TABLE 3.5 Procedure for determining the number of defective items per box.

Random Number Range	*Number of Defective Items*
(0.0000, 0.5220]	0
(0.5220, 0.8905]	1
(0.8905, 0.9880]	2
(0.9880, 0.9995]	3
(0.9995, 1.0000]	4

demonstrate the concept of simulation, not to derive the most efficient procedures for large-scale programs.) Notice that the ranges for the random numbers contained in the table are open on the left and closed on the right. That choice is arbitrary; it is only important to be consistent. In other words, if the random number was equal to 0.8905, the random variate representing the number of defective items in the box would equal 1.

The simulation proceeds as follows: First, a box containing a random number of defective items is generated according to the binomial distribution (Table 3.5). Then, an item is selected using a Bernoulli random variable, where the probability of selecting a defective item equals the number of defective items divided by four (i.e., the total number of items in a box). Table 3.6 shows the results of the simulation. Notice that column 4 contains a critical number, which equals the number of defective items divided by four. If the next random number generated is less than the critical number, the box is rejected; otherwise, the box is accepted. If the critical number is 0, no random number is generated since the box will always be accepted no matter which item the inspector chooses. The simulation is over when the first box containing defective items is selected. Since the question is "How many boxes will be accepted?" it is necessary to make several simulation runs. The final estimate would then be the average length of the runs. To generate random numbers, we use Table C.6 and start in (the randomly selected) row 16, picking numbers sequentially going across the row.

Obviously, only three trials is not enough to draw any reasonable conclusion, but the idea should be sufficiently illustrated. Based on the results of Table 3.6, we would estimate that the inspector samples an average of 6.67 boxes until finding a box to be rejected. ■

Before leaving discrete random variables, a word should be said about generating bivariate random variables. One procedure is to use conditional distributions. For example, if the two dependent random variables X and Y are to be generated, we first generate X according to its marginal distribution, then generate Y according to its conditional distribution given the value of X.

TABLE 3.6 Three simulation runs involving boxes containing defective items.

Box Number	Random Number	Number of Defects	Critical Number	Random Number	Box Accepted?
			Trial #1		
1	0.2358	0	0	—	Yes
2	0.5907	1	0.25	0.3483	Yes
3	0.6489	1	0.25	0.9204	Yes
4	0.6083	1	0.25	0.2709	Yes
5	0.7610	1	0.25	0.5374	Yes
6	0.1730	0	0	—	Yes
7	0.5044	0	0	—	Yes
8	0.6206	1	0.25	0.0474	No
			Trial #2		
1	0.8403	1	0.25	0.9076	Yes
2	0.3143	0	0	—	Yes
3	0.0383	0	0	—	Yes
4	0.0513	0	0	—	Yes
5	0.9537	2	0.50	0.6614	Yes
6	0.1637	0	0	—	Yes
7	0.4939	0	0	—	Yes
8	0.6086	1	0.25	0.1542	No
			Trial #3		
1	0.2330	0	0	—	Yes
2	0.8310	1	0.25	0.8647	Yes
3	0.2143	0	0	—	Yes
4	0.5414	1	0.25	0.0214	No

EXAMPLE 3.4 **Bivariate Discrete Random Variates.** Consider the boxes of phones discussed in Chapter 1. Each box contains two phones, both phones being radio phones or plain phones. The joint pmf describing the probabilities of the phone type and the number of defective phones is given in Example 1.12. Our goal is to estimate, through simulation, the expected number of defective phones per box. (Again, the question we are trying to answer is trivial, but we need to simulate easy systems before attempting complex systems.) Although we could use the joint pmf to generate the bivariate random variables directly, it is often easier to use conditional probabilities when dealing with more complex probability laws.

We first determine whether the box contains plain or radio phones, and then determine the number of defective phones in the box. Table 3.7 shows the transformations needed to obtain the boxes of phones; namely, the first transformation is from the CDF of the random variable indicating phone type, and the second two transformations come from the *conditional* CDFs for the number of defective phones in a box given the type of phone. For example, the conditional probability that there are no defective phones given that the

TABLE 3.7 Transformations for determining the contents of phone boxes.

TYPE OF PHONE	
Random Number Range	*Type Phone*
(0.00, 0.47]	Plain
(0.47, 1.0]	Radio

FOR PLAIN PHONES	
Random Number Range	*Number Defective*
(0.000, 0.787]	0
(0.787, 0.957]	1
(0.957, 1.0]	2

FOR RADIO PHONES	
Random Number Range	*Number Defective*
(0.000, 0.849]	0
(0.849, 0.981]	1
(0.981, 1.0]	2

box contains plain phones is $0.37/0.47 = 0.787$ (see Example 1.12); thus, if a box contains plain phones, a random number in the interval $(0, 0.787]$ will result in no defective phones for that box.

The results obtained after simulating ten boxes is contained in Table 3.8. The random numbers begin in (the randomly selected) row 3 of Table C.6. The second random number used for each box (column 4 of Table 3.8) must be interpreted based on the result of the first random number (column 3 of Table 3.8). In particular, notice the results for box 4. The first random number was 0.1454, yielding a box of plain phones. The second random number was 0.9683, yielding two defective phones. However, if the box had been radio phones, the random number of 0.9683 would have yielded only one defective phone. Based on the results of Table 3.8, the expected number of defective phones per box is estimated to be 0.5. ■

■ **EXAMPLE 3.5** **Markov Chains.** Our final example for discrete random variables will be illustrating the simulation of Markov chains, which again use conditional dis-

TABLE 3.8 Simulation for boxes of phones.

Box Number	*Random Number*	*Phone Type*	*Random Number*	*Number Defective*	*Cumulative Defectives*
1	0.6594	Radio	0.7259	0	0
2	0.6301	Radio	0.1797	0	0
3	0.3775	Plain	0.5157	0	0
4	0.1454	Plain	0.9683	2	2
5	0.7156	Radio	0.5140	0	2
6	0.9734	Radio	0.9375	1	3
7	0.7269	Radio	0.2228	0	3
8	0.1020	Plain	0.9991	2	5
9	0.9537	Radio	0.6292	0	5
10	0.4340	Plain	0.2416	0	5

tributions for the random number to state space transformation. Consider a Markov chain with state space $\{a, b, c\}$ and with the following transition matrix:

$$P = \begin{bmatrix} 0.8 & 0.1 & 0.1 \\ 0.4 & 0.6 & 0.0 \\ 0.0 & 0.0 & 1.0 \end{bmatrix}.$$

From the previous chapter, we know that the expected number of steps taken until the chain is absorbed into state c given that it starts in state a is $R(a, a) + R(a, b) = 12.5$. However, let us assume that as a new student of Markov chains, our confidence level is low in asserting that the mean absorption time starting from state a is 12.5. (Or equivalently, $E[T^c \mid X_0 = a] = 12.5$.) One of the uses of simulation is as a confidence builder; therefore, we wish to simulate this process to help give some evidence that our calculations are correct. In other words, our goal is to simulate this Markov chain starting from state a and recording the number of steps taken until reaching state c.

In this simulation, we use the random numbers listed in Appendix C (Table C.6) to produce the necessary randomness. Therefore, the first step is to record the transformations from the random numbers to the distributions defined by the matrix $\boldsymbol{P}$. These transformations are given in Table 3.9.

We begin the simulation by rolling two dice and coming up with the number 7; therefore, the first random number will be at the start of row 7 in Table C.6. (We choose the starting point randomly so that all the simulations will not be identical.) Table 3.10 contains the results of the three realizations of the experiment, where each experiment records the steps of the Markov chain until it reaches the absorbing state.

The estimate for the expected number of steps until the chain reaches state c based on the above three trials is $(5 + 7 + 1)/3 = 4\frac{1}{3}$. If we did not know how to determine the theoretical answer and if we were satisfied with only three trials, our simulation would give very misleading results. This should emphasize the extremely important fact that a simulation is a statistical experiment whose output must be interpreted statistically. Therefore, our suggestion is that a minimum of 25 to 30 trials should be used when estimating means of random variables. (Reasonable variations from the 25- to 30-trial minimum will make sense after Chapter 6.) ■

▶ *Suggestion: Do Exercises 3.3–3.9.*

TABLE 3.9 Procedure for determining the transitions for the Markov chain.

From State a	
Random Number Range	*Next State*
(0.0, 0.8]	a
(0.8, 0.9]	b
(0.9, 1.0]	c

From State b	
Random Number Range	*Next State*
(0.0, 0.4]	a
(0.4, 1.0]	b

TABLE 3.10 Three simulation runs for the Markov chain.

Step Number	Current State	Random Number	Next State
Trial # 1			
0	a	0.2419	a
1	a	0.1215	a
2	a	0.6602	a
3	a	0.7957	a
4	a	0.9518	c
5	c		
Trial # 2			
0	a	0.1589	a
1	a	0.5695	a
2	a	0.8823	b
3	b	0.4605	b
4	b	0.5212	b
5	b	0.1180	a
6	a	0.9862	c
7	c		
Trial # 3			
0	a	0.9965	c
1	c		

3.2.3 Continuous Random Variates

Although many physical systems can be modeled using discrete random variables, many systems are more appropriately modeled using continuous random variables. For example, the times between arrivals of customers to a teller window in a bank would be best described by a continuous value instead of a discrete value.

One of the most commonly encountered continuous random variables is the exponentially distributed random variable. Specifically, if T is an exponentially distributed random variable, then its cumulative distribution function is given by

$$F(a) = 1 - e^{-a/\theta} \quad \text{for } a \geq 0, \tag{3.1}$$

where θ is a positive scalar with $E[T] = \theta$.

The *inverse transformation method* (Property 3.1) can be used to generate exponentially distributed random variates from random numbers. (See Figure 3.3 for a graph illustrating the transformation from R to T.) To see this, let R be a random number and T be an exponentially distributed random variate. By Property 3.1, we have the following:

$$R = 1 - e^{-T/\theta}$$
$$e^{-T/\theta} = 1 - R$$

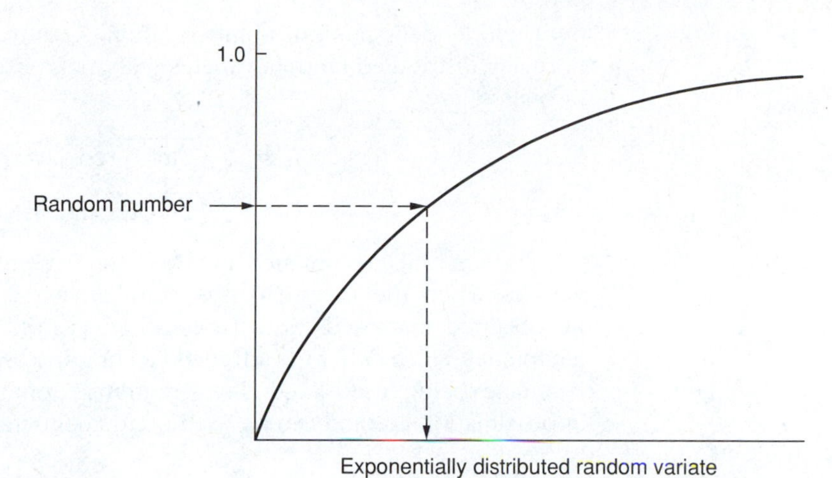

FIGURE 3.3 Inverse transformation for the exponentially distributed random variable T.

$$\ln(e^{-T/\theta}) = \ln(1 - R)$$
$$-\frac{T}{\theta} = \ln(1 - R)$$
$$T = -\theta \ln(1 - R).$$

One final simplification can be made by observing that if R is uniformly distributed between 0 and 1, then $(1 - R)$ must also be; therefore,

$$T = -\theta \ln(R).$$

Since the Erlang distribution [Eq. (1.16)] is the sum of exponentials, the preceding transformation is easily extended for Erlang random variates. Specifically, if X is an Erlang random variate with parameters m and λ, then

$$X = -\frac{1}{m\lambda}\sum_{k=1}^{m} \ln(R_k) = -\frac{1}{m\lambda}\ln\left(\prod_{k=1}^{m} R_k\right),$$

where R_k represents independent random numbers.

The exponential distribution is convenient to use because the expression for its cumulative distribution function is easy to invert. However, many distributions are not so "well behaved." A normally distributed random variable (its probability density function has the familiar bell-shaped curve) is an example of a distribution whose cumulative probability distribution function cannot be written in closed form, much less inverted. An efficient method of generating normally distributed random variables is based on a derivation by Box and Muller.[4] They established that if R_1 and R_2 are

[4]G. E. P. Box and M. E. Muller, "A Note on the Generation of Random Normal Deviates," *Annals of Mathematical Statistics* **29** (1958) pp. 610–611.

two independent random numbers, then Z_1 and Z_2 are two independent normally distributed random variates with mean zero and standard deviation of one, where

$$Z_1 = \sqrt{-2\ln(R_1)}\cos(2\pi R_2)$$

$$Z_2 = \sqrt{-2\ln(R_1)}\sin(2\pi R_2).$$

Several other techniques are used for generating nonuniform random variates when the inverse of the cumulative distribution function cannot be obtained in closed form. However, it is not necessary to know these techniques, since either the advanced techniques are buried in the simulation language being used (and thus are transparent to the user) or else an approximation method can be used quite adequately.

EXAMPLE 3.6 **Compound Poisson Process.** A monitoring device has been placed on a highway to record the number of vehicles using the highway. Actually, what is recorded are axles, so we would like to simulate the number of axles passing a particular point. The time between vehicle arrivals at the monitoring device is exponentially distributed with a mean of 1.2 minutes between vehicles. The probability that a given vehicle has 2, 3, 4, or 5 axles is 0.75, 0.10, 0.07, or 0.08, respectively. We are interested in simulating this process to obtain an estimate for the expected number of axles that pass over the monitoring device during any fixed 5-minute interval. There are two sources of randomness: (1) the interarrival times and (2) the number of axles per vehicle. Table 3.11 shows the mapping from a random number to the number of axles per vehicle; the transformation yielding the random interarrival times is given by

$$T = -1.2\ln(R),$$

where R is a random number and T is the interarrival time.

We now have the necessary information to begin the simulation. The random numbers in Table C.6 will again be used to simulate the random

TABLE 3.11 Procedure for determining the number of axles per vehicle.

Axles per Vehicle	
Random Number Range	*Number of Axles*
(0.0, 0.75]	2
(0.75, 0.85]	3
(0.85, 0.92]	4
(0.92, 1.0]	5

phenomenon for this situation. This time we roll four dice to determine the starting row. The roll gives 19, so we turn to the nineteenth row of Table C.6 and build Table 3.12.

The result of this simulation produces an estimate of $\frac{38}{3} = 12\frac{2}{3}$ axles per 5-minute interval. Since only three 5-minute intervals were simulated, it should be obvious that this estimate should be taken with a great deal of skepticism; however, for purposes of illustration, the 15-minute time span should be sufficient.

Before closing this example, it may be helpful to discuss the type of process that was just simulated. As mentioned briefly in Chapter 1, a Poisson process satisfies the following two conditions: (1) The times between arrivals form a sequence of independent, identically distributed exponential random variables and (2) the arrival times refer to the arrival of single units. Thus, the arrival of vehicles forms a Poisson process. A relaxation of condition (2) yields a compound Poisson process. That is, a compound Poisson process satisfies (1) the times between arrivals form a sequence of independent, identically distributed exponential random variables and (2) the number of units arriving at each arrival time forms a sequence of independent, identically distributed random variables. Thus, the arrival process of axles is a compound Poisson process. ■

EXAMPLE 3.7 **Nonhomogenous Poisson Process.** A bank has decided to increase its drive-in window capacity, and therefore desires to model the facility. A first step in modeling the drive-in bank facility is to simulate the arrival process. The drive-in windows open at 7:30 A.M. during the week. The arrival process is a Poisson process with a varying arrival rate. For the first 15 minutes (from

TABLE 3.12 Simulation of the axle arrival process.

Random Number	*Interarrival Time*	*Arrival Time*	*Random Number*	*Number of Axles*	*Cumulative Number*
0.2330	1.75	1.75	0.8310	3	3
0.8647	0.17	1.92	0.2143	2	5
0.5414	0.74	2.66	0.0214	2	7
0.3664	1.20	3.86	0.6326	2	9
0.7129	0.41	4.27	0.1671	2	11
0.2222	1.81	6.08	0.9869	5	16
0.9278	0.09	6.17	0.1854	2	18
0.3735	1.18	7.35	0.8792	4	22
0.9019	0.12	7.47	0.4779	2	24
0.1472	2.30	9.77	0.3953	2	26
0.5929	0.63	10.40	0.6838	2	28
0.4449	0.97	11.37	0.2235	2	30
0.3689	1.20	12.57	0.5881	2	32
0.9092	0.11	12.68	0.1110	2	34
0.3456	1.27	13.95	0.2000	2	36
0.7063	0.42	14.37	0.3995	2	38
0.2013	1.92	16.29			

7:30 A.M. until 7:45 A.M.), the mean arrival rate is 10 cars per hour; for the next 15 minutes (from 7:45 A.M. until 8:00 A.M.), the mean arrival rate is 15 cars per hour, and then the mean arrival rate returns to 10 cars per hour for the next 30 minutes (from 8:00 A.M. until 8:30 A.M.). As the previous example illustrates, simulating this process involves simulating the exponential interarrival times. During the first interval of time, the interarrival times average 6 minutes; from 7:45 A.M. until 8:00 A.M., the interarrival times average 4 minutes; then starting again at 8:00 A.M., the average returns to 6 minutes.

The main difficulty in simulating this process is handling the interarrival time that crosses the boundary at 7:45 A.M. and at 8:00 A.M.. For example, say the current time is 7:43 A.M., and we generate an interarrival time using an exponential distribution with a mean of 6 minutes. Assume this (randomly generated) time turned out to be 6.3 minutes. This would imply that most of the time interval occurred in the 7:45 to 8:00 time period, in which case an average of 4 minutes should have been used instead of the 6 minutes that was used. To handle an interarrival time that crosses a boundary we do the following: Let λ_1 be the mean arrival rate during the time interval $(0, t_1)$, and let λ_2 be the mean arrival rate during the time interval (t_1, t_2). Assume the current time is in the first time interval and is denoted by t. Furthermore let $T = (-1/\lambda_1)\ln(R)$; that is, T is an exponentially generated interarrival time. Now assume that $t + T > t_1$; that is, although T was generated in the first interval, it causes the time to cross into the second interval. The idea is to take the portion of time that is within the second interval (namely, $t + T - t_1$) and factor it up or down according to the new rate. Thus the new time would be

$$t_{\text{new}} = t_1 + \frac{\lambda_1}{\lambda_2}(t + T - t_1).$$

To continue our illustration using the numbers at the beginning of this paragraph, we first observe that the 6.3 minutes includes 4.3 minutes of time into the second interval. The 4.3 minutes is multiplied by the ratio $\frac{4}{6}$ to obtain 2.9, yielding the next arrival time at 7:47.9 (i.e., 2.9 minutes past 7:45).

Table 3.13 presents a simulation of the arrival process for the first 30 minutes. (Notice that the fourth column of the table gives the time in terms of cumulative amount since startup; whereas the first column gives the time in terms of the "time of day.") The simulation crosses two boundaries, and these points are indicated by the superscript "*" on the interarrival times. In particular notice the calculations at time 7:40.72. The random number 0.2143 was generated, its natural logarithm was taken and multiplied by 6, yielding an interarrival time of 9.24. The current time is 10.72 minutes from the start, and we add 9.24 minutes to that time and obtain 19.96 minutes. Notice that the 19.96 minutes is recorded under the column with a heading of "Potential Arrival Time." Since it crosses a boundary, the time must be modified before it can be recorded as the actual arrival time. In particular, the 9.24 minutes includes 4.96 minutes into the next time interval; therefore, the modified

TABLE 3.13 Simulation of nonhomogenous Poisson process.

Current Time	*Random Number*	*Interarrival Time*	*Potential Arrival Time*	*Amount into Next Interval*	*Arrival Time*
7:30.00 A.M.	0.2330	8.74	8.74	—	8.74
7:38.74	0.8310	1.11	9.85	—	9.85
7:39.85	0.8647	0.87	10.72	—	10.72
7:40.72	0.2143	9.24*	19.96	4.96	18.31
7:48.31	0.5415	2.45	20.76	—	20.76
7:50.76	0.0214	15.38*	36.14	6.14	39.21
8:09.21					

interarrival time will be $4.96 \times \frac{4}{6} = 3.31$ minutes into the second interval yielding the next time being 18.31 minutes after startup, or 7:48.31 A.M.. The second "crossover" occurs when the current time is 7:50.76. In this case, the amount of time into the next interval is 6.14 minutes; therefore, the modified time will be $6.14 \times \frac{6}{4} = 9.21$ minutes into the next time interval, yielding the next arrival time of 8:09.21. ■

■ **EXAMPLE 3.8** **Nonhomogenous Poisson Process—Revisited.** Let us revisit the drive-in bank example with a slightly different arrival process. Assume again that the drive-in windows open at time zero and cars arrive according to a nonhomogenous Poisson process. Whereas in the previous example, we assumed that the mean arrival rate was constant over different intervals; we now assume that the mean arrival rate is described by a continuous function. In particular, assume that for the first hour, the mean arrival rate increases linearly from 10 per hour to 15 per hour. Thus, we have a rate function defined, for $t \leq 60$, by

$$\lambda(t) = 10 + t/12,$$

where t is in minutes and $\lambda(\cdot)$ is in cars per hour.

The concept for simulating this process is to generate arrivals at the maximum rate and then "thin" the process by removing some of the arrivals. In particular, let λ^* denote the maximum arrival rate. If an arrival (generated using λ^*) occurs at time t, then it is accepted with a probability equal to $[\lambda(t)/\lambda^*]$ and rejected with a probability equal to $1 - [\lambda(t)/\lambda^*]$. When an arrival is rejected, the clock is still updated, but an arrival event is not recorded. Table 3.14 shows the results of a simulation that generated arrivals for the first 32.56 minutes. Notice that for each time, two random numbers are generated: a random number that will produce an interarrival time and a random number that will be used to test the interarrival time. Table 3.14 starts at time 0 and the first random number results in a "potential" arrival time of 5.83. At that arrival time, we have $\lambda(5.83) = 10.4858$; therefore, the first arrival is accepted with probability $10.4858/15 = 0.699$. Because the second

TABLE 3.14 Simulation of nonhomogenous Poisson process using a continuous rate function.

Current Time	*Random Number*	*Interarrival Time*	*Arrival Time*	*Critical Number*	*Random Number*	*Event Occur?*
0.00	0.2330	5.83	5.83	0.699	0.8310	No
5.83	0.8647	0.58	6.41	0.702	0.2143	Yes
6.41	0.5415	2.45	8.86	0.716	0.0214	Yes
8.86	0.3664	4.02	12.88	0.738	0.6326	Yes
12.88	0.7129	1.35	14.23	0.746	0.1671	Yes
14.23	0.2222	6.02	20.25	0.779	0.9869	No
20.25	0.9278	0.30	20.55	0.781	0.1854	Yes
20.55	0.3735	3.94	24.49	0.803	0.8792	No
24.49	0.9019	0.41	24.90	0.805	0.4779	Yes
24.90	0.1472	7.66	32.56	0.848	0.3953	Yes
32.56						

random number is greater than 0.699, the arrival event occurring at time 5.83 is rejected. Notice that in the first 32.56 minutes, a total of 10 arrivals were generated, but the simulation only has 7 arrival events actually occurring, with the first interarrival time being 6.41 minutes. ■

3.2.4 Random Variates from Empirical Distributions

Sometimes data are available to approximate a distribution function and so a form of an empirical function is needed. For example, suppose we need to reproduce a continuous random variable we know is always between 1 and 4. Furthermore, some data have been collected, and we know that 25% of the data are between 1 and 1.5, 50% of the data are between 1.5 and 3.5, and 25% are between 3.5 and 4. In such a situation we use a piecewise-linear function containing three segments to approximate the (unknown) cumulative distribution function. The application of the inverse transformation method is illustrated in Figure 3.4. Notice from the figure that the cumulative distribution function is graphed, then a random number is created on the y-axis, and finally the random variate of interest is obtained by an inverse mapping from the y-axis to the x-axis.

As another example, suppose we need to simulate a continuous random variable and all we have are the following ten data points: 5.39, 1.9, 4.62, 2.71, 4.25, 1.11, 2.92, 2.83, 1.88, 2.93. All we know other than the ten points is that the random variable has some (unknown) upper and lower limits that are positive. Our procedure will be the same as that illustrated in the previous example; that is, a cumulative distribution function will be drawn using a piecewise-linear approximation and then the inverse transformation method will be used to go from a random number to the random variate. The distribution function should have the property that 10% of the generated points are centered around 5.39, 10% of the generated points are centered

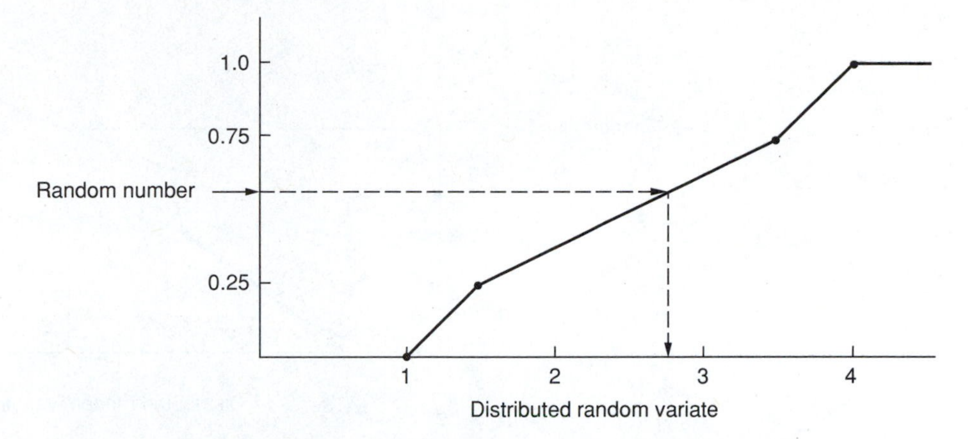

FIGURE 3.4 Mapping used to generate a random variate from a distribution based on experimental data.

around 1.9, etc. To accomplish this, the first step is to order the data, and then calculate the midpoints between adjacent values as shown in the following list.

Data Point	*Midpoint*
1.11	
1.88	1.495
1.90	1.890
2.71	2.305
2.83	2.770
2.92	2.875
2.93	2.925
4.25	3.590
4.62	4.435
5.39	5.005

In developing a distribution function based on these data points, we refer to Figure 3.5. Because there were ten data points, there are ten "cells" with each cell having as its midpoint one of the data points. The major difficulty is the size of the first and last cell; that is, establishing the upper and lower limits. The best way to establish these limits would be through an understanding of the physical limits of whatever it is we are simulating; however, if that cannot be done then one procedure is simply to assume that the lower and upper data points (i.e., 1.11 and 5.39, respectively) are the midpoints of their cells. In other words, the lower and upper limits are established to make this true. Thus, the lower limit is 0.725, and the upper limit is 5.775. [Note that $1.11 = (0.725 + 1.495)/2$ and $5.39 = (5.005 + 5.775)/2$.]

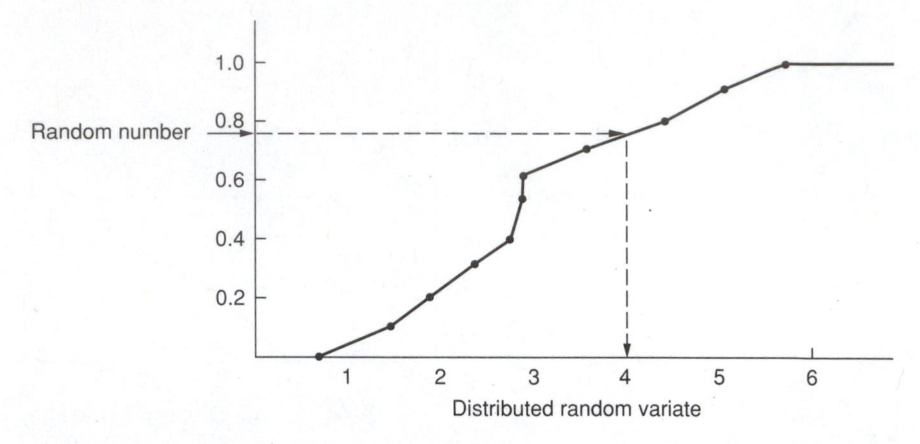

FIGURE 3.5 Mapping used to generate a random variate from a distribution based on ten data points.

The empirically based distribution function increases 0.10 over each cell because there were ten points. In general, if a distribution function is built with n data points, then it would increase $1/n$ over each data cell.

▶ *Suggestion: Do Exercises 3.10–3.14.*

3.3 EXERCISES

3.1 A door-to-door salesman sells pots and pans. He only gets into 50% of the houses that he visits. Of the houses that he enters, $\frac{1}{6}$ of the householders are still not interested in purchasing anything, $\frac{1}{2}$ of them end up placing a \$60 order, and $\frac{1}{3}$ of them end up placing a \$100 order. Estimate the average sales receipts per house visit by simulating 25 house visits using a die. Calculate the theoretical value and compare it with the estimate obtained from your simulation.

3.2 Simulate playing the game of Example 3.2, where the contestant uses the "switch" policy.

(a) Which policy would you recommend?

(b) Can you demonstrate mathematically that your suggested policy is the best policy?

3.3 Given the sequence of three uniform random numbers, 0.15, 0.74, 0.57, generate the corresponding sequences of random variates

(a) from a discrete uniform distribution varying between 1 and 4.

(b) for a discrete random variable, with the probability that it is equal to 1, 2, 3, 4 being 0.3, 0.2, 0.1, 0.4, respectively.

3.4 A manufacturing process has a defect rate of 20% and items are placed in boxes of five. An inspector samples two items from each box. If one or both of the selected items are bad, the box is rejected. Simulate this process to answer the following question: If a customer orders ten boxes, what is the expected number of defective items the customer will receive?

3.5 A binomial random variate with parameters n and p can be obtained by adding n Bernoulli random variates with parameter p. Verify this by writing a computer code that does the following: (1) Generates random numbers (see Appendix B.3), (2) generates four independent Bernoulli random variates with parameter $p = 0.35$ (i.e., if X is Bernoulli, then $X = 1$ if $R \leq p$, and $X = 0$ if $R > p$, where R is a random number), (3) sums the four Bernoulli random variates to obtain a binomial random variate, and (4) estimates the probability mass function for a binomial distribution with $n = 4$ and $p = 0.35$ based on repeated generations of the binomial random variates.

3.6 Consider a Markov chain with state space $\{a, b, c\}$ and with the following transition matrix:

$$\boldsymbol{P} = \begin{bmatrix} 0.35 & 0.27 & 0.38 \\ 0.82 & 0.00 & 0.18 \\ 1.00 & 0.00 & 0.00 \end{bmatrix}.$$

(a) Given that the Markov chain starts in state b, estimate through simulation the expected number of steps until the first return to state b.

(b) The expected return time to a state should equal the reciprocal of the long-run probability of being in that state. Estimate through simulation and analytically the steady-state probability and compare it to your answer to part (a). Explain any differences.

3.7 Use simulation to estimate the answer for part (b) of the problem contained in Exercise 2.6.

3.8 Consider a Markov chain with state space $\{a, b, c\}$ and with the following transition matrix:

$$\boldsymbol{P} = \begin{bmatrix} 0.3 & 0.5 & 0.2 \\ 0.1 & 0.2 & 0.7 \\ 0.8 & 0.0 & 0.2 \end{bmatrix}.$$

Each visit to state a results in a profit of \$5, each visit to state b results in a profit of \$10, and each visit to state c results in a profit of \$12. Write a computer program that will simulate the Markov chain so that an estimate of the expected profit per step can be made. Assume that the chain always starts in state a. The simulation should involve accumulating the profit from each step; then the estimate per simulation run is the cumulative profit divided by the number of steps in the run.

(a) Let the estimate be based on ten replications, where each replication has 25 steps. The estimate is the average value over the ten replicates. Record both the overall average and the range of the averages.

(b) Let the estimate be based on ten replications, where each replication has 1000 steps. Compare the estimates and ranges for parts (a) and this part and explain any differences.

3.9 Consider the problem in Exercise 2.8, part (d). Assume that management wants to recover the initial investment of $100,000 in 3 weeks. Write a computer simulation to estimate the probability that it will take longer than 3 weeks to recover the $100,000.

3.10 Given the sequence of four uniform random numbers, 0.23, 0.74, 0.57, 0.07, generate the corresponding sequences of random variates

(a) from an exponential distribution with a mean of 4.

(b) from a Weibull distribution with scale parameter 0.25 and shape parameter 2.5.

(c) from a normal distribution with a mean of 4 and variance of 4.

(d) from a continuous random variable designed to reproduce the randomness observed by the following data: 4.5, 12.6, 13.8, 6.8, 10.3, 12.5, 8.3, 9.2, 15.3, 11.9, 9.3, 8.1, 16.3, 14.0, 7.3, 6.9, 10.5, 12.3, 9.9, 13.6.

3.11 A mail-order service is open 24 hours per day. Operators receive telephone calls according to a Poisson process. Simulate the process and determine the expected volume (in terms of dollars) of sales during the 15-minute interval from 12:00 noon until 12:15 pm. (Assume each day is probabilistically similar to all other days.)

(a) The Poisson arrival process is homogenous with a mean rate of 50 calls per hour. Furthermore, half of the calls do not result in a sale, 30% of the calls result in a sale of $100, and 20% of the calls result in a sale of $200.

(b) Since more people tend to call on their lunch break, the mean arrival rate slowly increases at the start of the noon hour. For the five-minute interval from 12 noon until 12:05 pm, the mean arrival rate is 40 calls per hour; from 12:05 pm until 12:10 pm, the rate is 45 calls per hour; and from 12:10 pm until 12:50 pm, the rate is 50 calls per hour. The probabilities and amounts of a sale are the same as in part (a).

(c) The arrival rate of calls is approximated by a linear function that equals 40 calls per hour at noon and equals 50 calls per hour at 12:15 pm. The rate is then constant for the next 30 minutes. The probabilities and amounts of a sale are the same as above.

(d) The arrival process is homogenous with a mean rate of 50 calls per hour; however, instead of exponential times, the interarrival times have a type-3 Erlang distribution. The probabilities and amounts of sale are the same as above.

3.12 Write a computer program to produce random variates according to a continuous uniform distribution between 3 and 8. Generate 1000 random variates and compare the sample mean and variance of the generated data with the known mean and variance of the uniform distribution. Do you trust your random variate generator?

3.13 Write a computer program to produce random variates according to a Weibull distribution where the mean and standard deviation are input constants. Generate 1000 random variates and compare the sample mean and standard deviation of the generated data with the input values. Do you trust your random variate generator? (Note, a search routine will have to be written to obtain the scale and shape parameters. See Appendix B.2 for evaluating the gamma function.)

3.14 Write a computer program that will select 100 steel rods from the shipment of rods described in Exercise 1.25. How many rods from your randomly selected sample are outside of the specifications?

4

Markov Processes

Chapter 2 gave a detailed treatment of (discrete-time) Markov chains. The key Markov property was that the future was independent of the past given the present. A Markov process is simply the continuous analogue to a chain, namely, the Markov property is maintained and time is measured continuously. For example, suppose we are keeping track of patients within a large hospital, and the state of our process is the specific department housing the patient. If we record the patient's location every day, the resulting process will be a Markov chain if the Markov property holds. If, however, we record the patient's location continuously throughout the day, then the resulting process will be called a Markov process if the Markov property holds. Thus, instead of transitions only occurring at times 1, 2, 3, etc., transitions may occur at times 2.54 or 7.01.

Our major reason for covering Markov processes is that they play a key role in queueing theory, and queueing theory has applications in many areas. (In particular, any system in which there may be a buildup of waiting lines is a candidate for a queueing model.) Therefore, in this chapter, we give a cursory treatment of Markov processes in preparation for the next chapter dealing with queues.

4.1 BASIC DEFINITIONS

The definition for a Markov process is similar to that for a Markov chain except the Markov property must be shown to hold for all future times instead of just for one step (since "one step" has no meaning in continuous time).

Definition 4.1 *The process $Y = \{Y_t; t \geq 0\}$ with finite state space E is a* Markov Process *if the following holds for all $j \in E$ and $t, s \geq 0$*

$$\Pr\{Y_{t+s} = j \mid Y_u; u \leq t\} = \Pr\{Y_{t+s} = j \mid Y_t\}.$$

Furthermore, the Markov process is said to have STATIONARY *transitions if*

$$\Pr\{Y_{t+s} = j \mid Y_t = i\} = \Pr\{Y_s = j \mid Y_0 = i\}.$$

These equations are completely analogous to those in Definition 2.3. Think of the present time as being time t. The left-hand side of the first equation in Definition 4.1 indicates that a prediction of the future s time units from now is desired given all the past history up to and including the current time t. The right-hand side of the equation indicates that the prediction is the same if the only information available is the state of the system at the present time. (Note that an implication of this definition is that it is a waste of resources to keep historical records of a process if it is Markovian.) The second equation of the definition is the time homogenous condition, which indicates that the probability law governing the process does not change during the life of the process.

EXAMPLE 4.1 A salesman lives in town a and is responsible for towns a, b, and c. The amount of time he spends in each town is random. After some study it has been determined that the amount of consecutive time spent in any one town follows an exponentially distributed random variable [see Eq. (1.15)] with the mean time depending on the town. In his hometown, a mean time of 2 weeks is spent, in town b a mean time of 1 week is spent, and in town c a mean time of 1.5 weeks is spent. When he leaves town a, he flips a coin to determine to which town he goes next; when he leaves either town b or c, he flips two coins so that there is a 75% chance of returning to a and a 25% chance of going to the other town. Let Y_t be a random variable denoting the town that the salesman is in at time t. The process $Y = \{Y_t; t \geq 0\}$ is a Markov process due to the lack of memory inherent in the exponential distribution (see Exercise 1.23). ■

The construction of the Markov process in Example 4.1 is instructive in that it contains the major characterizations for any finite state Markov process. A Markov process remains in each state for an exponential length of time and then when it jumps, it jumps according to a Markov chain (see Figure 4.1). To describe this characterization mathematically, we introduce some additional notation that should be familiar due to its similarity to the previous chapter. We first let $T_0 = 0$ and $X_0 = Y_0$. Then, we denote the successive jump times of the process by $T_1, T_2, \cdots$, and denote the state of the process immediately after the jumps as $X_1, X_2, \cdots$. To be more explicit, for $n = 0, 1, \cdots$,

$$T_{n+1} = \min\{t > T_n : Y_t \neq Y_{T_n}\}, \quad \text{and}$$
$$X_n = Y_{T_n}.$$

The time between jumps (namely, $T_{n+1} - T_n$) is called the *sojourn time*. The Markov property implies that the only information needed for predicting

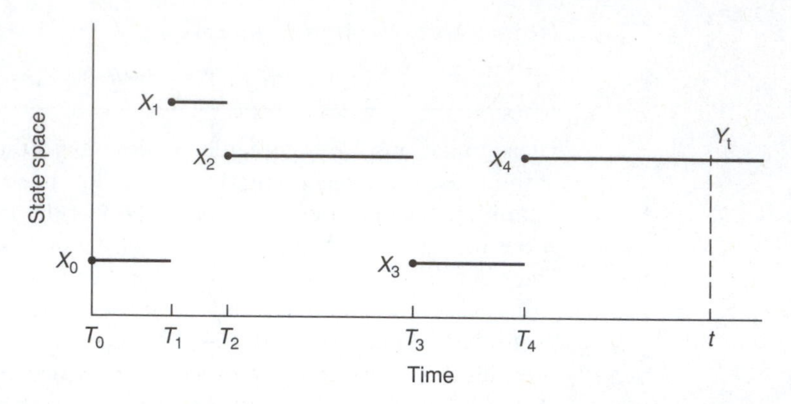

FIGURE 4.1 Typical realization for a Markov process.

the sojourn time is the current state. Or, to say it mathematically,

$$\Pr\{T_{n+1} - T_n \leq t \mid X_0, \cdots, X_n, T_0, \cdots, T_n\} = \Pr\{T_{n+1} - T_n \leq t \mid X_n\}.$$

The Markov property also implies that if we focus our attention only on the process $X_0, X_1, \cdots$, then we would see a Markov chain. This Markov chain is called the *imbedded Markov chain.*

DEFINITION 4.2 *Let $Y = \{Y_t; t \geq 0\}$ be a Markov process with finite state space E and with jump times denoted by $T_0, T_1, \cdots$, and the imbedded process at the jump times denoted by $X_0, X_1, \cdots$. Then there is a collection of scalars, $\lambda(i)$ for $i \in E$, called the* MEAN SOJOURN RATES, *and a Markov matrix,* $\boldsymbol{P}$, *called the* IMBEDDED MARKOV MATRIX, *that satisfy the following:*

$$\Pr\{T_{n+1} - T_n \leq t \mid X_n = i\} = 1 - e^{-\lambda(i)t},$$

$$\Pr\{X_{n+1} = j \mid X_n = i\} = P(i,j),$$

where each $\lambda(i)$ is nonnegative and the diagonal elements of $\boldsymbol{P}$ *are zero.*

EXAMPLE 4.2 Consider again Example 4.1. The imbedded Markov chain described by the above definition is

$$\boldsymbol{P} = \begin{matrix} a \\ b \\ c \end{matrix} \begin{bmatrix} 0 & 0.50 & 0.50 \\ 0.75 & 0 & 0.25 \\ 0.75 & 0.25 & 0 \end{bmatrix},$$

and the mean sojourn rates are $\lambda(a) = \frac{1}{2}$, $\lambda(b) = 1$, and $\lambda(c) = \frac{2}{3}$. (Note that mean rates are always the reciprocal of mean times.) ■

To avoid pathological cases, we shall adopt the policy in this chapter that the mean time spent in each state be finite and nonzero. The notions of state classification can now be carried over from Chapter 2.

Definition 4.3 *Let $Y = \{Y_t; t \geq 0\}$ be a Markov process with finite state space E such that $0 < \lambda(i) < \infty$ for all $i \in E$. A state is called* RECURRENT *or* TRANSIENT *according to whether it is recurrent or transient in the imbedded Markov chain of the process. A set of states is* IRREDUCIBLE *for the process if it is irreducible for the imbedded chain.*

4.2 STEADY-STATE PROPERTIES

There is a close relationship between the steady-state probabilities for the Markov process and the steady-state probabilities of the imbedded chain. Specifically, the Markov process probabilities are obtained by the weighted average of the imbedded Markov chain probabilities according to each state's sojourn times. To show the relationship mathematically, consider $Y = \{Y_t; t \geq 0\}$ to be a Markov process with an irreducible, recurrent state space. Let $X = \{X_n; n = 0, 1, \cdots\}$ be the imbedded Markov chain, with $\boldsymbol{\pi}$ its steady-state probabilities; namely,

$$\pi(j) = \lim_{n \to \infty} \Pr\{X_n = j \mid X_0 = i\}.$$

(In other words, $\boldsymbol{\pi P} = \boldsymbol{\pi}$, where $\boldsymbol{P}$ is the Markov matrix of the imbedded chain.) The steady-state probabilities for the Markov process are denoted by the vector $\boldsymbol{p}$ where

$$p(j) = \lim_{t \to \infty} \Pr\{Y_t = j \mid Y_0 = i\}. \tag{4.1}$$

Note that the limiting values are independent of the initial state since the state space is assumed to be irreducible and recurrent. The relationship between the vectors $\boldsymbol{\pi}$ and $\boldsymbol{p}$ is given by

$$p(j) = \frac{\pi(j)/\lambda(j)}{\sum_{k \in E} \pi(k)/\lambda(k)} \tag{4.2}$$

where E is the state space and $\lambda(j)$ is the mean sojourn rate for state j.

The use of Eq. (4.2) is easily illustrated with Example 4.2 since the steady-state probabilities $\boldsymbol{\pi}$ were calculated in Chapter 2. Using these previous results (page 56), we have that the imbedded chain has steady-state probabilities given by $\boldsymbol{\pi} = (\frac{3}{7}, \frac{2}{7}, \frac{2}{7})$. Combining these with the mean sojourn rates in Example 4.2 and with Eq. (4.2) yields

$$p(a) = \frac{6}{11}, \quad p(b) = \frac{2}{11}, \quad \text{and} \quad p(c) = \frac{3}{11}.$$

The salesman of the example thus spends 54.55% of his time in town a, 18.18% of his time in town b, and 27.27% of his time in town c.

The information contained in the imbedded Markov matrix and the mean sojourn rates can be combined into one matrix, which is called the *generator matrix* of the Markov process.

DEFINITION 4.4 *Let $Y = \{Y_t; t \geq 0\}$ be a Markov process with an imbedded Markov matrix* $\boldsymbol{P}$ *and a mean sojourn rate of $\lambda(i)$ for each state i in the state space. The* GENERATOR MATRIX, $\boldsymbol{G}$, *for the Markov process is given by*

$$G(i,j) = \begin{cases} -\lambda(i) & \text{for } i = j \\ \lambda(i)P(i,j) & \text{for } i \neq j \end{cases}.$$

Generator matrices are extremely important in the application of Markov processes. A generator matrix for a Markov process has two properties: (1) Each row sum is zero and (2) the off-diagonal elements are nonnegative. These properties can be seen from Definition 4.4 by remembering that the imbedded Markov matrix is nonnegative with row sums of one and has zeros on the diagonals. The physical interpretation of $\boldsymbol{G}$ is that $G(i,j)$ is the *rate* at which the process goes from state i to state j.

EXAMPLE 4.3 We again return to our initial example to illustrate the generator matrix. Using the imbedded Markov chain and sojourn times given in Example 4.2, the generator matrix for that Markov process is

$$\boldsymbol{G} = \begin{matrix} a \\ b \\ c \end{matrix} \begin{bmatrix} -\frac{1}{2} & \frac{1}{4} & \frac{1}{4} \\ \frac{3}{4} & -1 & \frac{1}{4} \\ \frac{1}{2} & \frac{1}{6} & -\frac{2}{3} \end{bmatrix}.$$

Thus, Definition 4.4 indicates that if the transition matrix for the imbedded Markov chain and the mean sojourn rates are known, the generator matrix can be computed. State diagrams for Markov processes are very similar to diagrams for Markov chains, except that the arrows representing transition probabilities represent transition rates (see Figure 4.2). ■

FIGURE 4.2 State diagram for the Markov process of Example 4.3.

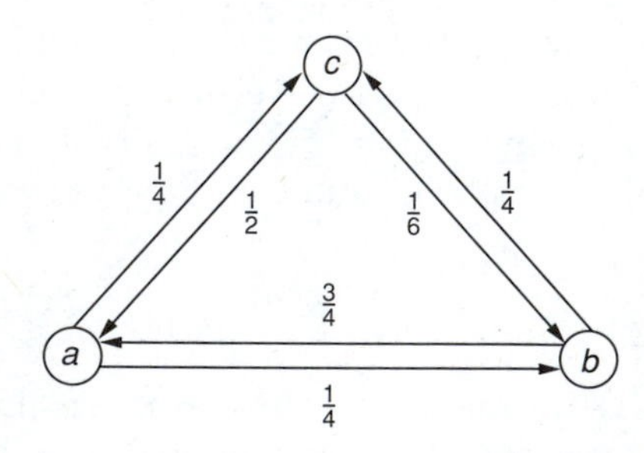

It should also be clear that not only can the transition rate matrix be obtained from the imbedded Markov matrix and mean sojourn rates, but the reverse is also true. From the generator matrix, the absolute value of the diagonal elements gives the mean sojourn rates. The transition matrix for the imbedded chain is then obtained by dividing the off-diagonals element by the absolute value of that row's diagonal element. The diagonal elements of the transition matrix for the imbedded Markov chain are zero. For example, suppose a Markov process with state space $\{a, b, c\}$ has a generator matrix given by

$$\boldsymbol{G} = \begin{matrix} a \\ b \\ c \end{matrix} \begin{bmatrix} -2 & 2 & 0 \\ 2 & -4 & 2 \\ 1 & 4 & -5 \end{bmatrix}.$$

The mean sojourn rates are thus $\lambda_a = 2$, $\lambda_b = 4$, and $\lambda_c = 5$, and the transition matrix for the imbedded Markov chain is given by

$$\boldsymbol{P} = \begin{matrix} a \\ b \\ c \end{matrix} \begin{bmatrix} 0 & 1 & 0 \\ 0.5 & 0 & 0.5 \\ 0.2 & 0.8 & 0 \end{bmatrix}.$$

▶ *Suggestion: Do Exercises 4.1 and 4.2a.*

It is customary when describing a Markov process to obtain the generator matrix directly and never specifically look at the imbedded Markov matrix. The usual technique is to obtain the off-diagonal elements first and then let the diagonal element equal the negative of the sum of the off-diagonal elements on the row. The generator is used to obtain the steady-state probabilities directly as follows.

Property 4.5 Let $Y = \{Y_t; t \geq 0\}$ be a Markov process with an irreducible, recurrent state space E and with generator matrix $\boldsymbol{G}$. Furthermore, let $\boldsymbol{p}$ be a vector of steady-state probabilities defined by Eq. (4.1). Then $\boldsymbol{p}$ is the solution to

$$\boldsymbol{pG} = \boldsymbol{0}$$

$$\sum_{j \in E} p(j) = 1.$$

Obviously, this equation is very similar to the analogous equation in Property 2.16. The difference is in the right-hand side of the first equation. The easy way to remember which equation to use is that the multiplier for the right-hand side is the same as the row sums. Since Markov matrices have row sums of one, the right-hand side is *one* times $\boldsymbol{\pi}$; since the generator matrix has row sums of zero, the right-hand side is *zero* times $\boldsymbol{p}$.

EXAMPLE 4.4 A small garage operates a tractor repair service in a busy farming community. Because the garage is small, only two tractors can be kept there at any one time and the owner is the only repairman. If one tractor is being repaired and one is waiting, then any further customers who request service are turned away. The time between the arrival of customers is an exponential random variable and an average of four customers arrive each day. Repair time varies considerably and it is also an exponential random variable. Based on previous records, the owner repairs an average of five tractors per day if kept busy. The arrival and repair of tractors can be described as a Markov process with state space $\{0, 1, 2\}$ where a state represents the number of tractors in the garage. Instead of determining the Markov matrix for the imbedded chain, the generator matrix can be calculated directly. As in Markov chain problems, drawing a state diagram is helpful. State diagrams for Markov processes show transition rates instead of transition probabilities. Figure 4.3 shows the diagram with the transition rates. The rate of going from state 0 to state 1 and from state 1 to state 2 is four per day since that is the arrival rate of customers to the system. The rate of going from state 2 to state 1 and from state 1 to state 0 is five per day because that is the repair rate. Transition rates are only to adjacent states because both repairs and arrivals only occur one at a time. From the diagram, the generator matrix is obtained as

$$\boldsymbol{G} = \begin{matrix} 0 \\ 1 \\ 2 \end{matrix} \begin{bmatrix} -4 & 4 & 0 \\ 5 & -9 & 4 \\ 0 & 5 & -5 \end{bmatrix}.$$

The steady-state probabilities are thus obtained by solving

$$\begin{aligned} -4p_0 + 5p_1 \qquad &= 0, \\ 4p_0 - 9p_1 + 5p_2 &= 0, \\ 4p_1 - 5p_2 &= 0, \\ p_0 + p_1 + p_2 &= 1, \end{aligned}$$

which yields (after deleting one of the first three equations) $p_0 = \frac{25}{61}$, $p_1 = \frac{20}{61}$, and $p_2 = \frac{16}{61}$. As with Markov chains, there will always be one redundant equation. The steady-state equations can be interpreted to indicate that the owner is idle 41% of the time and 26% of the time the shop is full. ■

An assumption was made in Example 4.4 regarding exponential random variables. The length of time that the process is in state 1 is a function of

FIGURE 4.3 State diagram for the Markov process of Example 4.4.

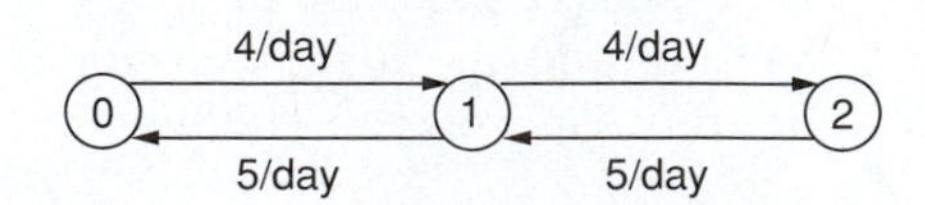

two exponentially distributed random variables. In effect, the sojourn time in state 1 depends on a "race" between the next arrival and the completion of the repair. The assumption is that the minimum of two exponential random variables is also exponentially distributed. Actually, that is not too difficult to prove, so we leave it as an exercise.

EXAMPLE 4.5 **Poisson Process.** We are interested in counting the number of vehicles that go through a tollbooth. The arriving vehicles form a Poisson process (page 18) with a mean rate of 120 cars per hour. Since a Poisson process has exponential interarrival times, a Poisson process is also a Markov process with a state space equal to the nonnegative integers. A possible realization of the process is given in Figure 4.4, and the generator matrix is infinite dimensioned given by

$$\boldsymbol{G} = \begin{matrix} 0 \\ 1 \\ 2 \\ \vdots \end{matrix} \begin{bmatrix} -120 & 120 & 0 & 0 & \cdots \\ 0 & -120 & 120 & 0 & \cdots \\ 0 & 0 & -120 & 120 & \cdots \\ \vdots & & & \ddots & \end{bmatrix}.$$

Assume that we are actually interested in the cumulative amount of tolls collected. The toll is a function of the type of vehicle going through the tollbooth. Fifty percent of the vehicles pay 25 cents, one-third of the vehicles pay 75 cents, and one-sixth of the vehicles pay \$1. If we let the state space be the cumulative number of quarters collected, the process of interest is called a compound Poisson process. A typical realization is given in Figure 4.5 and the generator matrix is given by

$$\boldsymbol{G} = \begin{matrix} 0 \\ 1 \\ 2 \\ \vdots \end{matrix} \begin{bmatrix} -120 & 60 & 0 & 40 & 20 & 0 & 0 & \cdots \\ 0 & -120 & 60 & 0 & 40 & 20 & 0 & \cdots \\ 0 & 0 & -120 & 60 & 0 & 40 & 20 & \cdots \\ \vdots & & & \ddots & & & & \end{bmatrix}.$$

FIGURE 4.4 Typical realization for a Poisson process.

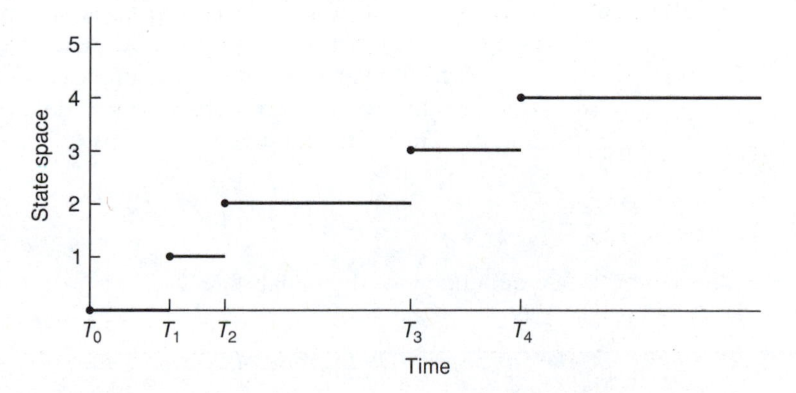

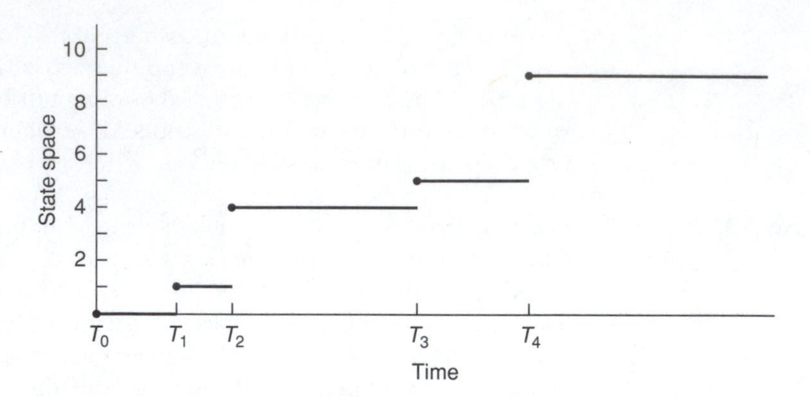

FIGURE 4.5 Typical realization for a compound Poisson process.

Notice that the only difference between the Poisson process and the compound Poisson process is that the Poisson process always increases by jump sizes of exactly one unit, whereas the compound Poisson process increases by an amount equal to a random variable. Furthermore, the sequence of random variables governing the size of the successive jumps is an independent, identically distributed sequence. ■

▶ *Suggestion: Do Exercises 4.2b, 4.3, 4.6, 4.7, 4.9, 4.13, and 4.14.*

4.3 REVENUES AND COSTS

Markov processes often represent revenue-producing situations, so the steady-state probabilities are used to determine a long-run average return. Suppose that whenever the Markov process is in state i, a profit is produced at a rate of $f(i)$. Thus, $\boldsymbol{f}$ is a vector of profit rates. The long-run profit per unit time is then simply the (vector) product of the steady-state probabilities times $\boldsymbol{f}$.

Property 4.6 Let $Y = \{Y_t; t \geq 0\}$ be a Markov process with an irreducible, recurrent state space E and a profit rate vector denoted by $\boldsymbol{f}$ [i.e., $f(i)$ is the rate that profit is accumulated whenever the process is in state i]. Furthermore, let $\boldsymbol{p}$ denote the steady-state probabilities as defined by Eq. (4.1). Then, the long-run profit per unit time is given by

$$\lim_{t\to\infty} \frac{1}{t} E\left[\int_0^t f(Y_s)\, ds\right] = \sum_{j\in E} p(j) f(j)$$

independent of the initial state.

▶ *Suggestion: Do Exercises 4.2c and 4.4a.*

The left-hand side of the preceding equation is simply the mathematical way of writing the long-run average accumulated profit. It is sometimes important to include revenue produced at transition times so that the total profit is the sum of the accumulation of profit resulting from the sojourn times and the jumps. In other words, we would not only have the profit rate vector $\boldsymbol{f}$ but also a matrix that indicates a profit obtained at each jump.

Property 4.7 Let $Y = \{Y_t; t \geq 0\}$ be a Markov process with an irreducible, recurrent state space E, a profit rate vector denoted by $\boldsymbol{f}$, and a matrix of jump profits denoted by $\boldsymbol{h}$ [i.e., $h(i,j)$ is the profit obtained whenever the process jumps from state i to state j]. Furthermore, let $\boldsymbol{p}$ denote the steady-state probabilities as defined by Eq. (4.1). Then, the long-run profit per unit time is given by

$$\lim_{t\to\infty} \frac{1}{t} E\left[\int_0^t f(Y_s)\,ds + \sum_{s\leq t} h(Y_{s-}, Y_s)\right] =$$

$$\sum_{i\in E} p(i)\left[f(i) + \sum_{k\in E} G(i,k)h(i,k)\right]$$

where $h(i,i) = 0$ and $\boldsymbol{G}$ is the generator matrix of the process.

The summation on the left-hand side of the preceding equation may appear different from what you are used to seeing. The notation Y_{s-} indicates the left-hand limit of the process; that is, $Y_{s-} = \lim_{t\to s^-} Y_t$ where t approaches s from the left, i.e., $t < s$. Since Markov processes are always right-continuous, the only times for which $Y_{s-} \neq Y_s$ is when s is a jump time. (When we say that the Markov process is right-continuous, we mean that $Y_s = \lim_{t\to s^+} Y_t$ where the limit is such that t approaches s from the right, i.e., $t > s$.) Because the diagonal elements of $\boldsymbol{h}$ are zero, the only times at which the summation includes nonzero terms are at jump times.

EXAMPLE 4.6 Returning to Example 4.1 for illustration, assume that the revenue possible from each town varies. Whenever the salesman is working town a, his profit comes in at a rate of \$80 per day. In town b, his profit is \$100 per day. In town c, his profit is \$125 per day. There is also a cost associated with changing towns. This cost is estimated at 25 cents per mile and it is 50 miles from a to b, 65 miles from a to c, and 80 miles from b to c. The functions (using 5-day weeks) are

$$\boldsymbol{f} = (400, 500, 625)^T,$$

$$\boldsymbol{h} = \begin{bmatrix} 0 & -12.50 & -16.25 \\ -12.50 & 0 & -20.00 \\ -16.25 & -20.00 & 0 \end{bmatrix}.$$

Therefore, using

$$G = \begin{bmatrix} -\frac{1}{2} & \frac{1}{4} & \frac{1}{4} \\ \frac{3}{4} & -1 & \frac{1}{4} \\ \frac{1}{2} & \frac{1}{6} & -\frac{2}{3} \end{bmatrix}$$

and

$$p(a) = \frac{6}{11}, \quad p(b) = \frac{2}{11}, \quad p(c) = \frac{3}{11},$$

the long-run weekly average profit is

$$\frac{6}{11} \times (400 - 7.19) + \frac{2}{11} \times (500 - 14.38) + \frac{3}{11} \times (625 - 11.46) = 469.88. \blacksquare$$

▶ *Suggestion: Do Exercises 4.5 and 4.8.*

It is sometimes important to consider the time-value of the profit (or cost). In the discrete case, a discount factor of α was used. We often think in terms of a rate of return or an interest rate. If i were the rate of return, then the discount factor used for Markov chains is related to i according to the formula $\alpha = 1/(1+i)$; thus, the present value of one dollar obtained one period from the present is equal to α. For a continuous-time problem, we let β denote the discount rate, which in this case would be the same as the (nominal) rate of return except that we assume that compounding occurs continuously. Thus, the present value of one dollar obtained one period from the present equals $e^{-\beta}$. When the time-value of money is important, it is difficult to include the jump time profits but the function $\boldsymbol{f}$ is easily utilized according to the following property.

Property 4.8 Let $Y = \{Y_t; t \geq 0\}$ be a Markov process with a generator matrix $\boldsymbol{G}$, a profit rate vector $\boldsymbol{f}$, and a discount factor of β. Then, the present value of the total discounted profit (over an infinite planning horizon) is given by

$$E\left[\int_0^\infty e^{-\beta s} f(Y_s)\, ds \mid Y_0 = i\right] = ((\beta \boldsymbol{I} - \boldsymbol{G})^{-1} \boldsymbol{f})(i).$$

EXAMPLE 4.7 Assume that the salesman of Example 4.1 wants to determine the present value of the total revenue using a discount rate of 5%, where the revenue rate function is given by $\boldsymbol{f} = (400, 500, 625)^T$. The present value of all future revenue is thus given by

$$\begin{bmatrix} \frac{11}{20} & -\frac{1}{4} & -\frac{1}{4} \\ -\frac{3}{4} & \frac{21}{20} & -\frac{1}{4} \\ -\frac{1}{2} & -\frac{1}{6} & \frac{43}{60} \end{bmatrix}^{-1} \begin{bmatrix} 400 \\ 500 \\ 625 \end{bmatrix} = \begin{bmatrix} 11.31 & 3.51 & 5.17 \\ 10.54 & 4.28 & 5.17 \\ 10.34 & 3.45 & 6.21 \end{bmatrix} \begin{bmatrix} 400 \\ 500 \\ 625 \end{bmatrix} = \begin{bmatrix} 9515 \\ 9592 \\ 9741 \end{bmatrix}.$$

Note that the total discounted revenue depends on the initial state. Thus, if at time 0 the salesman is in city a, then the present value of his total revenue is \$9515; whereas, if he started in city c, the present value of his total revenue would be \$9741. ■

▶ *Suggestion: Do Exercise 4.4b,c.*

4.4 TIME-DEPENDENT PROBABILITIES

A thorough treatment of the time-dependent probabilities for Markov processes is beyond the scope of this book; however, we do present a brief summary in this section for those students who may want to continue studying random processes and would like to know something of what the future might hold. For other students, this section can be skipped with no loss of continuity.

As before, the Markov process will be represented by $\{Y_t; t \geq 0\}$ with finite state space E and generator matrix $\boldsymbol{G}$. The basic relationship that probabilities for a Markov process must satisfy are analogous to the relationship that the transition probabilities of a Markov chain satisfy. As you recall from Section 2.2, the following holds for Markov chains:

$$
\begin{aligned}
P\{X_{n+m} = j \mid X_0 = i\} & \\
&= \sum_{k \in E} \Pr\{X_n = k \mid X_0 = i\} \times \Pr\{X_{n+m} = j \mid X_n = k\} \\
&= \sum_{k \in E} \Pr\{X_n = k \mid X_0 = i\} \times \Pr\{X_m = j \mid X_0 = k\}.
\end{aligned}
$$

The second equality is a result of the stationary property for Markov chains and is called the Chapman-Kolmogorov equation for Markov chains. The Chapman-Kolmogorov equation for Markov processes can be expressed similarly as

$$
\begin{aligned}
P\{Y_{t+s} = j \mid Y_0 = i\} & \\
&= \sum_{k \in E} \Pr\{Y_t = k \mid Y_0 = i\} \times \Pr\{Y_{t+s} = j \mid Y_t = k\} \\
&= \sum_{k \in E} \Pr\{Y_t = k \mid Y_0 = i\} \times \Pr\{Y_s = j \mid Y_0 = k\},
\end{aligned}
$$

for $t, s \geq 0$ and $i, j \in E$. In other words, the Chapman-Kolmogorov property indicates that the probability of going from state i to state j in $t + s$ time units is the sum of the product of probabilities of going from state i to an arbitrary state k in t time units and going from state k to state j in s time units (see Figure 4.6).

Before giving a closed-form expression for the time-dependent probabilities, it is necessary to first discuss the exponentiation of a matrix. Recall that

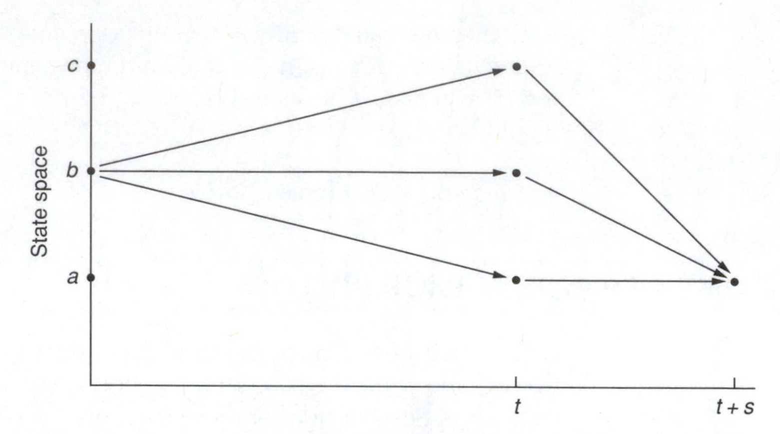

FIGURE 4.6 Graphical representation for the Chapman-Kolmogorov equations for a three-state Markov process.

for scalars, the following power series holds for all values of a:

$$e^a = \sum_{n=0}^{\infty} \frac{a^n}{n!}.$$

This same relationship holds for matrices and becomes the definition of the exponentiation of a matrix; namely,

$$e^{\mathbf{A}} = \sum_{n=0}^{\infty} \frac{\mathbf{A}^n}{n!}. \tag{4.3}$$

Note that $e^{\mathbf{A}}(i,j)$ refers to the $i-j$ element of the matrix $e^{\mathbf{A}}$; thus, in general, $e^{\mathbf{A}}(i,j) \neq e^{\mathbf{A}(i,j)}$.

Using the preceding definition and the Chapman-Kolmogorov equation, it is possible to derive the following property:

Property 4.9 Let $Y = \{Y_t; t \geq 0\}$ be a Markov process with state space E and generator matrix $\mathbf{G}$, then for $i, j \in E$ and $t \geq 0$

$$\Pr\{Y_t = j \mid Y_0 = i\} = e^{t\mathbf{G}}(i,j),$$

where $t\mathbf{G}$ is the matrix formed by multiplying each element of the generator matrix by the scalar t.

EXAMPLE 4.8 We are interested in an electronic device that is either "on standby" or "in use." The on-standby periods are exponentially distributed with a mean of 20 seconds, and the in-use periods are exponentially distributed with a mean of 12 seconds. Because the exponential assumption is satisfied, a Markov process $\{Y_t; t \geq 0\}$ with state space $E = \{0, 1\}$ can be used to model the electronic device as it alternates between being on standby and in use. State 0 denotes on standby, and state 1 denotes in use. The generator matrix is given by

$$\boldsymbol{G} = \begin{matrix} 0 \\ 1 \end{matrix}\begin{bmatrix} -3 & 3 \\ 5 & -5 \end{bmatrix},$$

where the time unit is minutes. Using Property 4.5, it is easy to show that the long-run probability of being in use is $\frac{3}{8}$. However, we are interested in the time-dependent probabilities. If the generator can be written in diagonal form, then it is not difficult to apply Property 4.9. In other words, if it is possible to find a matrix $\boldsymbol{Q}$ such that

$$\boldsymbol{G} = \boldsymbol{Q}\boldsymbol{D}\boldsymbol{Q}^{-1},$$

where $\boldsymbol{D}$ is a diagonal matrix, then the following holds

$$e^{t\boldsymbol{G}} = \boldsymbol{Q}e^{t\boldsymbol{D}}\boldsymbol{Q}^{-1}.$$

Furthermore, $e^{t\boldsymbol{D}}$ is a diagonal matrix whenever $\boldsymbol{D}$ is diagonal, in which case $e^{t\boldsymbol{D}}(i, i) = e^{tD(i,i)}$. For the generator matrix given, we have

$$\begin{bmatrix} -3 & 3 \\ 5 & -5 \end{bmatrix} = \begin{bmatrix} 1 & 3 \\ 1 & -5 \end{bmatrix} \cdot \begin{bmatrix} 0 & 0 \\ 0 & -8 \end{bmatrix} \cdot \begin{bmatrix} 0.625 & 0.375 \\ 0.125 & -0.125 \end{bmatrix};$$

therefore,

$$e^{\boldsymbol{G}t} = \begin{bmatrix} 1 & 3 \\ 1 & -5 \end{bmatrix} \cdot \begin{bmatrix} 1 & 0 \\ 0 & e^{-8t} \end{bmatrix} \cdot \begin{bmatrix} 0.625 & 0.375 \\ 0.125 & -0.125 \end{bmatrix}$$
$$= \begin{bmatrix} 0.625 + 0.375e^{-8t} & 0.375 - 0.375e^{-8t} \\ 0.625 - 0.625e^{-8t} & 0.375 + 0.625e^{-8t} \end{bmatrix}.$$

The probability that the electronic device is in use at time t given that it started in use at time 0 is thus given by

$$\Pr\{Y_t = 1 \mid Y_0 = 1\} = 0.375 + 0.625e^{-8t}.$$

(See Figure 4.7 for a graphical representation of the time-dependent probabilities.)

We would like to know the expected amount of time that the electronic device is in use during its first 1 minute of operation given that it was in use at time 0. Mathematically, the quantity to be calculated is given by

$$E\left[\int_0^1 Y_s\, ds \mid Y_0 = 1\right] = \int_0^1 \Pr\{Y_s = 1 \mid Y_0 = 1\}\, ds.$$

Because Y_s is 1 if the device is in use at time s and zero otherwise, the left-hand side of the above expression is the expected length of time that

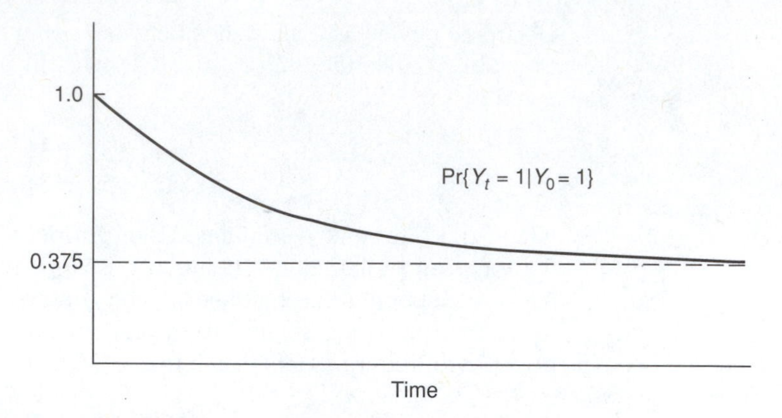

FIGURE 4.7 Time-dependent probabilities for Example 4.8.

the process is in use during its first minute of operation. The equality holds because the expected value of a Bernoulli random variable equals the probability that the random variable equals 1. (We ignore the technical details involved in interchanging the expectation operator and the integral.) Combining the preceding two equations, we have

$$E\left[\int_0^1 Y_s\, ds \mid Y_0 = 1\right] = \int_0^1 \Pr\{Y_s = 1 \mid Y_0 = 1\}\, ds$$

$$= \int_0^1 0.375 + 0.625e^{-8s}\, ds$$

$$= 0.375 + \frac{0.625}{8}(1 - e^{-8}) = 0.453 \text{ minutes}.$$

It is often useful to determine the total cost (or profit) of an operation modeled by a Markov process. In order to express the expected cost over a time interval, let $C_{(0,t)}$ be a random variable denoting the total cost incurred by the Markov process over the interval $(0, t)$. Then, for $i \in E$,

$$E[C_{(0,t)} \mid Y_0 = i] = \sum_{k \in E} f(k) \int_0^t \Pr\{Y_s = k \mid Y_0 = i\}\, ds,$$

where $f(k)$ is the cost rate incurred by the process while in state k.

For example, suppose that the electronic device costs the company 5¢ every minute it is on standby and it costs 35¢ every minute the device is in use. We may then be interested in estimating the total cost of operation for the first 2 minutes of utilizing the device given that the device was on standby when the process started; thus

$$
\begin{aligned}
E[C_{(0,2)} \mid Y_0 = 0] &= 0.05 \int_0^2 \Pr\{Y_s = 0 \mid Y_0 = 0\}\, ds \\
&\quad + 0.35 \int_0^2 \Pr\{Y_s = 1 \mid Y_0 = 0\}\, ds \\
&= 0.05 \times \left[0.625 \times 2 + \frac{0.375}{8}(1 - e^{-8\times 2})\right] \\
&\quad + 0.35 \times \left[0.375 \times 2 - \frac{0.375}{8}(1 - e^{-8\times 2})\right] \\
&= 0.31.
\end{aligned}
$$

▶ *Suggestion: Do Exercises 4.10–4.12.*

4.5 EXERCISES

4.1 The following matrix is a generator for a Markov process. Complete its entries.

$$
\boldsymbol{G} = \begin{bmatrix} - & 3 & 0 & 7 \\ 5 & -12 & 4 & - \\ 1 & - & -6 & 5 \\ 3 & 2 & 9 & - \end{bmatrix}
$$

4.2 Let Y be a Markov process with state space $\{a, b, c, d\}$ and an imbedded Markov chain having a Markov matrix given by

$$
\boldsymbol{P} = \begin{bmatrix} 0.0 & 0.1 & 0.2 & 0.7 \\ 0.0 & 0.0 & 0.4 & 0.6 \\ 0.8 & 0.1 & 0.0 & 0.1 \\ 1.0 & 0.0 & 0.0 & 0.0 \end{bmatrix}.
$$

The mean sojourn times in states a, b, c, and d are 2, 5, 0.5, and 1, respectively.

(a) Give the generator matrix for this Markov process.

(b) What is $\lim_{t\to\infty} \Pr\{Y_t = a\}$?

(c) Let $\boldsymbol{r} = (10, 25, 30, 50)^T$ be a reward vector and determine $\lim_{t\to\infty} \frac{1}{t} E[\int_0^t r(Y_t)\, dt]$.

4.3 Let T and S be exponentially distributed random variables with means $1/a$ and $1/b$, respectively. Define the random variable $U = \min\{T, S\}$. Justify the relationship $\Pr\{U > u\} = \Pr\{T > u\} \times \Pr\{S > u\}$ and derive the distribution function for U.

4.4 A revenue-producing system can be in one of four states: high income, medium income, low income, costs. The movement of the system among the states is according to a Markov process Y with state space $E = \{h, m, l, c\}$ and with a generator matrix given by

$$G = \begin{matrix} h \\ m \\ l \\ c \end{matrix} \begin{bmatrix} -0.2 & 0.1 & 0.1 & 0.0 \\ 0.0 & -0.4 & 0.3 & 0.1 \\ 0.0 & 0.0 & -0.5 & 0.5 \\ 1.5 & 0.0 & 0.0 & -1.5 \end{bmatrix}.$$

While the system is in state h, m, l, or c it produces a profit at a rate of \$500, \$250, \$100, or –\$600 per time unit. The company would like to reduce the time spent in the fourth state and has determined that by doubling the cost (i.e., from \$600 to \$1200) incurred while in that state the mean time spent in the state can be cut in half. Is the additional expense worthwhile?

(a) Use the long-run average profit for the criterion.

(b) Use a total discounted-cost criterion assuming a discount rate of 10%.

(c) Use a total discounted-cost criterion assuming the company uses a 25% annual rate of return, and the time step for this problem is assumed to be in weeks.

4.5 Let Y be a Markov process with state space $\{a, b, c, d\}$ and generator matrix given by

$$G = \begin{bmatrix} -5 & 4 & 0 & 1 \\ 6 & -10 & 4 & 0 \\ 0 & 0 & -1 & 1 \\ 2 & 0 & 0 & -2 \end{bmatrix}.$$

Costs are incurred at a rate of \$100, \$300, \$500, and \$1000 per time unit while the process is in states a, b, c, and d, respectively. Furthermore, each time a jump is made from state d to state a an additional cost of \$5000 is incurred. (All other jumps do not incur additional costs.) For a maintenance cost of \$400 per time unit, all cost rates can be cut in half and the "jump" cost can be eliminated. Based on long-run averages, is the maintenance cost worthwhile?

4.6 A small gas station has one pump and room for a total of three cars (one at the pump and two waiting). The time between car arrivals to the station is an exponential random variable with the average arrival rate of ten cars per hour. The time each car spends in front of the pump is an exponential random variable with a mean of 5 minutes (i.e., a mean rate of 12 per hour). If there are three cars in the station and another car arrives, the newly arrived car keeps going and never enters the station.

(a) Model the gas station as a Markov process Y, where Y_t denotes the number of cars in the station at time t. Give its generator matrix.

(b) What is the long-run probability that the station is empty?

(c) What is the long-run expected number of cars in the station?

4.7 A small convenience store has a room for only five people inside. Cars arrive at the store randomly, with the interarrival times between cars being an exponential random variable with a mean of ten cars arriving each hour. The number of people within each car is a random variable, N, where $\Pr\{N = 1\} = 0.1$, $\Pr\{N = 2\} = 0.7$, and $\Pr\{N = 3\} = 0.2$. People from the cars come into the store and stay in the store an exponential length of time. The mean length of stay in the store is 10 minutes and each person acts independent of all other people, leaving the store singly and waiting in their cars for the others. If a car arrives and the store is too full for everyone in the car to enter the store, the car will leave and nobody from that car will enter the store. Model the store as a Markov process Y, where Y_t denotes the number of individuals in the store at time t. Give its generator matrix.

4.8 A certain piece of electronic equipment has two components. The time until failure for component A is described by an exponential distribution function with a mean time of 100 hours. Component B has a mean life until failure of 200 hours and is also described by an exponential distribution. When one component fails, the equipment is turned off and maintenance is performed. The time to fix the component is exponentially distributed with a mean time of 5 hours if it was A that failed and 4 hours if it was B that failed. Let Y be a Markov process with state space $E = \{w, a, b\}$, where state w denotes that the equipment is working, a denotes that component A has failed, and b denotes that component B has failed.

(a) Give the generator for Y.

(b) What is the long-run probability that the equipment is working?

(c) An outside contractor does the repair work on the components when a failure occurs and charges \$100 per hour for time plus travel expenses, which is an additional \$500 for each visit. The company has determined that they can hire and train their own repairperson. If they have their own employee for the repair work, it will cost the company \$40 per hour while the machine is running as well as when it is down. Ignoring the initial training cost and the possibility that an employee who is hired for repair work can do other things while the machine is running, is it economically worthwhile to hire and train their own person?

4.9 An electronic component works as follows: Electric impulses arrive at the component with exponentially distributed interarrival times such that the mean arrival rate of impulses is 90 per hour. An impulse is "stored" until the third impulse arrives, then the component "fires" and enters a "recovery" phase. If an impulse arrives while the component is in the recovery phase, it is ignored. The length of time for which the component remains in the recovery phase is an exponential random variable with a mean time of 1 minute. After the recovery phase is over, the cycle is repeated; that is, the third arriving impulse will instantaneously fire the component.

(a) Give the generator matrix for a Markov process model of the dynamics of this electronic component.

(b) What is the long-run probability that the component is in the recovery phase?

(c) How many times would you expect the component to fire each hour?

4.10 Consider the electronic device that is either in the on-standby or in-use state as described in Example 4.8. Find the following quantities.

(a) The expected cost incurred during the time interval between the third and fourth minutes, given that at time zero the device was in use.

(b) The expected cost incurred during the time interval between the third and fourth minutes, given that at time zero the device was in use with probability 0.8 and on standby with probability 0.2.

4.11 Let $\boldsymbol{\mu}$ be a vector of initial probabilities for a Markov process Y, and let $\boldsymbol{f}$ denote a cost rate vector associated with Y. Write a general expression for the expected cost incurred by the Markov process during the time interval $(t, t+s)$.

4.12 Let $\boldsymbol{f}$ denote a profit rate vector associated with a Markov process Y, and let β denote a discount rate. Write a general expression for the profit returned by the process during the interval $(t, t+s)$ given that the process was in state i at time 0.

4.13 Simulate the Markov process defined by the following generator matrix:

$$\mathbf{G} = \begin{matrix} a \\ b \\ c \end{matrix}\begin{bmatrix} -1.0 & 0.3 & 0.7 \\ 0.1 & -0.2 & 0.1 \\ 0.3 & 0.2 & -0.5 \end{bmatrix}.$$

(Example 3.6 may provide a guide for simulating Markov processes.) From the simulation, estimate the probability that the process will be in state b at time 10 given that the process started in state a.

4.14 Write a computer program to determine the long-run probability of being in state a for the Markov process of Problem 4.13. (Remember that the long-run probability for a particular state can be estimated by the fraction of time that the process spends in that state. The estimate can also be improved slightly if the initial conditions are ignored. In other words,

$$p(j) \approx \left(\int_{t_0}^{t} I(Y_s, j)\, ds \right) \Big/ (t - t_0),$$

where t_0 is small and t is large. Part of your problem will be to determine quantitative values for "small" and "large.")

5 Queueing Processes

Many phenomena for which mathematical descriptions are desired involve waiting lines either of people or material. A queue is a waiting line, and queueing processes are those stochastic processes arising from waiting line phenomena. For example, the modeling of the arrival process of grain trucks to an elevator, the utilization of data processing services at a computer center, and the flow of jobs at a job shop facility all involve waiting lines. Although queues are ubiquitous, they are usually ignored when deterministic models are developed to describe systems. Furthermore, the random fluctuations inherent in queueing processes often cause systems to act in a counterintuitive fashion. Therefore, the study of queues is extremely important for the development of system models and an understanding of system behavior.

In this chapter we present modeling techniques employed for queueing systems governed by the exponential process. The final section of the chapter deals with some approximation techniques that are useful for implementing these models within complex systems when the exponential assumptions are not satisfied. The next chapter (Chapter 6) presents additional simulation techniques necessary when queueing systems are to be simulated. The final chapter (Chapter 10) in the textbook presents some advanced analytical techniques useful for modeling nonexponential queueing systems.

5.1 BASIC DEFINITIONS AND NOTATION

A queueing process involves the arrival of customers to a service facility and the servicing of those customers. All customers that have arrived but are not yet being served are said to be in the *queue*. The queueing *system* includes all customers in the queue and all customers in service (see Figure 5.1).

Several useful conventions have evolved over the last 20 to 40 years that help in specifying the assumptions used in a particular analysis. D. G. Kendall[1] is usually given credit for initiating the basic notation of today, and

[1] D. G. Kendall, "Stochastic Processes Occurring in the Theory of Queues and Their Analysis by the Method of Imbedded Markov Chains," *Ann. Math. Statist.* **24** (1953), pp. 338–354.

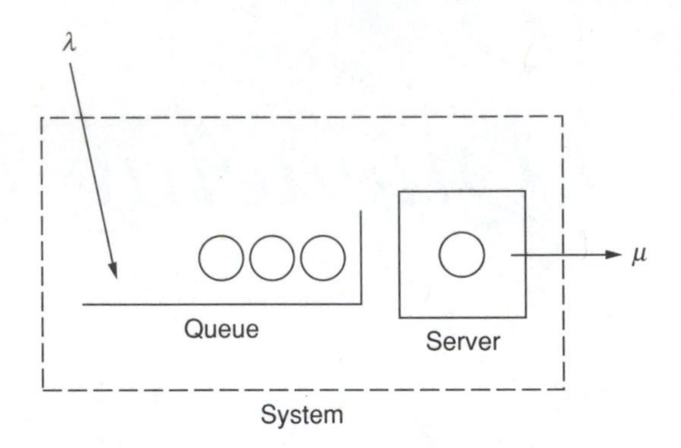

FIGURE 5.1 Representation of a queueing system containing four customers, three in the queue and one in the server.

it was standardized in 1971 (*Queueing Standardization Conference Report*, May 11, 1971). Kendall's notation is a shorthand that quickly indicates the assumptions used in a particular queueing model. For example, a formula developed for a G/D/1/∞/FIFO queue would be a formula that could be used for any general arrival process, only a deterministic service time, one server, an unlimited system capacity, and a discipline that works on a first-in/first-out (FIFO) basis. The general notation has the following form:

$$\left(\begin{matrix}\text{arrival}\\\text{process}\end{matrix}\Big/\begin{matrix}\text{service}\\\text{process}\end{matrix}\Big/\begin{matrix}\text{number}\\\text{of servers}\end{matrix}\Big/\begin{matrix}\text{maximum}\\\text{possible in system}\end{matrix}\Big/\begin{matrix}\text{queue}\\\text{discipline}\end{matrix}\right)$$

Table 5.1 gives the common abbreviations used with this notation. Whenever an infinite capacity and FIFO discipline are used, the last two descriptors can be left off; thus, an M/M/1 queue would refer to exponential interarrival and service times, one server, unlimited capacity, and a FIFO discipline.

Our purpose in this chapter is to provide an introduction to queueing processes and introduce the types of problems commonly encountered while studying queues. To maintain the introductory level of this material, arrival processes to the queueing systems will be assumed to be Poisson [see Chapter 1, Eq. (1.12)] and service times will be exponential. Thus, this chapter is devoted to investigating M/M/c/K systems for various values of c and K.

5.2 SINGLE-SERVER SYSTEMS

The simplest queueing system to analyze is the M/M/1 (or, equivalently, M/M/1/∞/FIFO) structure. The M/M/1 system assumes customers arrive according to a Poisson process with mean rate λ and are served by a single server whose time for service is random with an exponential distribution of

TABLE 5.1 Queueing symbols.

CHARACTERISTICS	SYMBOLS	EXPLANATION
Interarrival time distribution	M	Exponential (Markov)
	D	Deterministic
	E_k	Erlang type k
	G	General
Service time distribution	M	Exponential (Markov)
	D	Deterministic
	E_k	Erlang type k
	G	General
Number of servers	$1, 2, \cdots, \infty$	Parallel servers
System capacity	$1, 2, \cdots, \infty$	Maximum allowable in system
Queue discipline	FIFO	First in/first out
	LIFO	Last in/first out
	SIRO	Service in random order
	PRI	Priority
	GD	General discipline

Table taken in part from *Fundamentals of Queueing Theory* by D. Gross and C. M. Harris, copyright © 1974 by John Wiley & Sons, Inc. Reprinted by permission of John Wiley & Sons, Inc.

mean $1/\mu$. If the server is idle and a customer arrives, then that customer enters the server immediately. If the server is busy and a customer arrives, then the arriving customer enters the queue, which has infinite capacity. When service for a customer is completed, the customer leaves and the customer who has been in the queue the longest instantaneously enters the service facility and service begins again. Thus, the flow of customers through the system is a Markov process with state space $\{0, 1, \cdots\}$. The Markov process is denoted by $\{N_t; t \geq 0\}$ where N_t denotes the number of customers in the system at time t. The steady-state probabilities are

$$p_n = \lim_{t\to\infty} \Pr\{N_t = n\}.$$

We let N be a random variable with probability mass function $\{p_0, p_1, \cdots\}$. The random variable N thus represents the number of customers in the system at steady state, and p_n represents the long-run probability that there are n customers in the system. (You might also note that another way to view p_n is as the long-run fraction of time that the system contains n customers.) Sometimes we will be interested in the number of customers that are in the queue and thus waiting for service; therefore, let the random variable N_q denote the steady-state number in the queue. In other words, if the system is idle, $N_q = N$; when the system is busy, $N_q = N - 1$.

Our immediate goal is to derive an expression for $p_n, n = 0, 1, \cdots$, in terms of the mean arrival and service rates. This derivation usually involves two steps: (1) Obtain a system of equations defining the probabilities and (2) solve the system of equations. After some experience, you should find step 1 relatively easy; it is usually step 2 that is difficult. In other words, the system of equations defining p_n is not hard to obtain, but it is sometimes hard to solve.

An intuitive approach for obtaining the system of equations is to draw a state diagram and then use a rate balance approach, which is a system of equations formed by setting "rate in" equal to "rate out" for each node or state of the queueing system. Figure 5.2 shows the state diagram for the M/M/1 system. Referring to this figure, we see that the rate into the box around node 0 is μp_1; the rate out of the box around node 0 is λp_0; thus, "rate in" = "rate out" yields

$$\mu p_1 = \lambda p_0 .$$

The rate into the box around node 1 is $\lambda p_0 + \mu p_2$; the rate out of the box around node 1 is $(\mu + \lambda)p_1$; thus

$$\lambda p_0 + \mu p_2 = (\mu + \lambda)p_1 .$$

Continuing in a similar fashion and rearranging, we obtain the system

$$\begin{aligned} p_1 &= \frac{\lambda}{\mu} p_0 \quad \text{and} \\ p_{n+1} &= \frac{\lambda + \mu}{\mu} p_n - \frac{\lambda}{\mu} p_{n-1} \quad \text{for } n = 1, 2, \cdots . \end{aligned} \tag{5.1}$$

A more rigorous approach for obtaining the system of equations of Eq. (5.1) is to first give the generator matrix (from Chapter 4) for the M/M/1 system. Since the interarrival and service times are exponential, the queueing system is a Markov process and thus Property 4.5 can be used. The rate at which the process goes from state n to state $n+1$ is λ, and the rate of going from state n to state $n-1$ is μ; therefore, the generator is the infinite dimensioned matrix given as

$$\mathbf{G} = \begin{bmatrix} -\lambda & \lambda & & & \\ \mu & -(\mu + \lambda) & \lambda & & \\ & \mu & -(\mu + \lambda) & \lambda & \\ & & \ddots & \ddots & \ddots \end{bmatrix} . \tag{5.2}$$

The system of equations formed by $\boldsymbol{pG} = \mathbf{0}$ then yields Eqs. (5.1) again.

FIGURE 5.2 State diagram for an M/M/1 queueing system illustrating the rate balance approach.

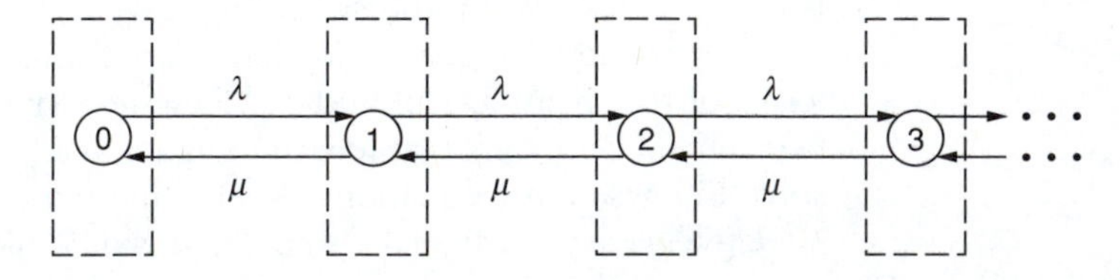

The system of equations of Eq. (5.1) can be solved by successively forward substituting solutions and expressing all variables in terms of p_0. Because we already have $p_1 = (\lambda/\mu)p_0$, we look at p_2 and then p_3:

$$\begin{aligned} p_2 &= \frac{\lambda+\mu}{\mu}p_1 - \frac{\lambda}{\mu}p_0 \\ &= \frac{\lambda+\mu}{\mu}\left(\frac{\lambda}{\mu}p_0\right) - \frac{\lambda}{\mu}\frac{\mu}{\mu}p_0 \\ &= \frac{\lambda^2}{\mu^2}p_0, \\ p_3 &= \frac{\lambda+\mu}{\mu}p_2 - \frac{\lambda}{\mu}p_1 \\ &= \frac{\lambda+\mu}{\mu}\left(\frac{\lambda^2}{\mu^2}p_0\right) - \frac{\lambda}{\mu}\frac{\mu}{\mu}\left(\frac{\lambda}{\mu}p_0\right) \\ &= \frac{\lambda^3}{\mu^3}p_0. \end{aligned} \tag{5.3}$$

At this point, a pattern begins to emerge and we can assert that

$$p_n = \frac{\lambda^n}{\mu^n}p_0 \quad \text{for } n \geq 0. \tag{5.4}$$

The assertion is proven by mathematical induction; that is, use of the induction hypothesis together with the general equation in Eq. (5.1) yields

$$\begin{aligned} p_{n+1} &= \frac{\lambda+\mu}{\mu}\left(\frac{\lambda^n}{\mu^n}p_0\right) - \frac{\lambda}{\mu}\frac{\mu}{\mu}\left(\frac{\lambda^{n-1}}{\mu^{n-1}}p_0\right) \\ &= \frac{\lambda^{n+1}}{\mu^{n+1}}p_0, \end{aligned}$$

and thus Eq. (5.4) is shown to hold. The ratio λ/μ is called the *traffic intensity* for the queueing system and is denoted by ρ for the M/M/1 system. (More generally, ρ is usually defined as the arrival rate divided by the maximum system service rate.)

We now have p_n for all n in terms of p_0 so the long-run probabilities become known as soon as p_0 can be obtained. If you review the material on Markov processes, you should see that we have taken advantage of Property 4.5, except we have not yet used the second equation given in the property. Thus, an expression for p_0 can be determined by using the norming equation, namely:

$$\begin{aligned} 1 &= \sum_{n=0}^{\infty} p_n = p_0 \sum_{n=0}^{\infty} \frac{\lambda^n}{\mu^n} = p_0 \sum_{n=0}^{\infty} \rho^n \\ &= \frac{p_0}{1-\rho}. \end{aligned} \tag{5.5}$$

The equality in this expression made use of the geometric progression,[2] so it is valid only for $\rho < 1$. If $\rho \geq 1$, the average number of customers and time spent in the system increase without bound and the system becomes unstable. In some respects this is a surprising result. Based on deterministic intuition, a person might be tempted to design a system such that the service rate is equal to the arrival rate, thus creating a "balanced" system. This is false logic for random interarrival or service times, since in that case the system will never reach steady state.

The preceding value for p_0 can be combined with Eq. (5.4) to obtain for the M/M/1 system

$$p_n = (1 - \rho)\rho^n \quad \text{for } n = 0, 1, \cdots, \tag{5.6}$$

where $\rho = \lambda/\mu$ and $\rho < 1$.

The steps followed in deriving Eq. (5.6) are the pattern for many other Markov queueing systems. Once the derivation of the M/M/1 system is known, all other queueing system derivations in this text will be easy. It is, therefore, good to review these steps so that they become familiar:

1. Form the Markov generator matrix, $\boldsymbol{G}$ [Eq. (5.2)].
2. Obtain a system of equations by solving $\boldsymbol{pG} = \boldsymbol{0}$ [Eq. (5.1)].
3. Solve the system of equations in terms of p_0 by successive forward substitution and induction if possible [Eq. (5.4)].
4. Use the norming equation to find p_0 [Eq (5.5)].

Once the procedure becomes familiar, the only difficult step will be the third step. It is not always possible to find a closed-form solution to the system of equations, and often techniques other than successive forward substitution must be used. However, these techniques are beyond the scope of this text and will not be presented.

EXAMPLE 5.1 An operator of a small grain elevator has a single unloading dock. Arrivals of trucks during the busy season form a Poisson process with a mean arrival rate of four per hour. Because of varying loads (and the desire of the drivers to talk) the length of time each truck spends in front of the unloading dock is approximated by an exponential random variable with a mean time of 14 minutes. Assuming that the parking spaces are unlimited, the M/M/1 queueing system describes the waiting lines that form. Accordingly, we have

$$\lambda = 4/\text{hr}, \quad \mu = (60/14)/\text{hr}, \quad \rho = 0.9333.$$

The probability of the unloading dock being idle is

$$p_0 = 1 - \rho = 0.0667.$$

The probability that there are exactly three trucks waiting is

$$\Pr\{N_q = 3\} = \Pr\{N = 4\} = p_4 = 0.9333^4 \times 0.0667 = 0.05.$$

[2]The geometric progression is $\sum_{n=0}^{\infty} r^n = 1/(1 - r)$ for $|r| < 1$.

Finally, the probability that four or more trucks are in the system is

$$\Pr\{N \geq 4\} = \sum_{n=4}^{\infty} p_n = (1-\rho)\sum_{n=4}^{\infty} \rho^n = \rho^4 = 0.759.$$

[In this expression, the second equality is obtained by using Eq. (5.6) to substitute out p_n. The third equality comes by observing that ρ^4 is a multiplicative factor in each term of the series, so it can be "moved" outside the summation sign, making the resulting summation a geometric progression.] ■

Several measures of effectiveness are useful as descriptors of queueing systems. The most common measures are the expected number of customers in the system, denoted by L, and the expected number in the queue, denoted by L_q. These expected values are obtained by utilizing Eq. (5.6) and the derivative of the geometric progression,[3] yielding an expression for the expected number in an M/M/1 system as

$$\begin{aligned} L = E[N] &= \sum_{n=0}^{\infty} n p_n \\ &= \sum_{n=1}^{\infty} n p_n = \sum_{n=1}^{\infty} n\rho^n(1-\rho) \\ &= (1-\rho)\rho \sum_{n=1}^{\infty} n\rho^{n-1} = \frac{\rho}{1-\rho}. \end{aligned} \tag{5.7}$$

The expected number of customers waiting within an M/M/1 queueing system is obtained similarly:

$$\begin{aligned} L_q &= 0 \times (p_0 + p_1) + \sum_{n=1}^{\infty} n p_{n+1} \\ &= \sum_{n=1}^{\infty} n\rho^{n+1}(1-\rho) \\ &= (1-\rho)\rho^2 \sum_{n=1}^{\infty} n\rho^{n-1} = \frac{\rho^2}{1-\rho}. \end{aligned} \tag{5.8}$$

When describing a random variable, it is always dangerous simply to use its mean value as the descriptor. For this reason, we also give the variance of the number in the system and queue:

$$\text{var}(N) = \frac{\rho}{(1-\rho)^2}, \tag{5.9}$$

[3]Taking the derivative of both sides of the geometric progression yields $\sum_{n=1}^{\infty} nr^{n-1} = 1/(1-r)^2$ for $|r| < 1$.

$$\mathrm{var}(N_q) = \frac{\rho^2(1+\rho-\rho^2)}{(1-\rho)^2}. \tag{5.10}$$

Waiting times are another important measure of a queueing system. Fortunately for our computational effort, there is an easy relationship between the mean number waiting and average length of time a customer waits. J. D. C. Little[4] showed that *almost all steady-state queueing systems* satisfy the following:

$$\begin{aligned} L &= \lambda_e W, \\ L_q &= \lambda_e W_q, \end{aligned} \tag{5.11}$$

where W and W_q are the expected waiting time a customer spends in the system and in the queue, respectively, and λ_e is the *effective* mean arrival rate into the system. Some queueing systems do not allow all arriving customers the privilege of entering the system; therefore λ_e is the mean arrival rate of actual entries into the system. For the M/M/1 system, the effective arrival rate is the same as the arrival rate (i.e., $\lambda_e = \lambda$); thus

$$\begin{aligned} W &= E[T] = \frac{1}{\mu-\lambda}, \\ W_q &= E[T_q] = \frac{\rho}{\mu-\lambda}, \end{aligned} \tag{5.12}$$

where T is the random variable denoting the time a customer (in steady state) spends in the system and T_q is the random variable for the time spent in the queue.

When the arrival process is Poisson, there is a version of Little's formula that holds for variances, at least for single-server systems:

$$\begin{aligned} \mathrm{var}(N) - E[N] &= \lambda_e^2 \mathrm{var}(T), \\ \mathrm{var}(N_q) - E[N_q] &= \lambda_e^2 \mathrm{var}(T_q). \end{aligned} \tag{5.13}$$

Again, we emphasize that the power of Little's formulas [Eqs. (5.11) and (5.13)] arises because of their generality. Equation (5.11) holds for any steady-state queueing system, and Eq. (5.13) holds for M/G/1/∞ systems. Applying Eq. (5.13) to the M/M/1 system we obtain

$$\begin{aligned} \mathrm{var}(T) &= \frac{1}{(\mu-\lambda)^2} = \frac{1}{\mu^2(1-\rho)^2}, \\ \mathrm{var}(T_q) &= \frac{2\rho-\rho^2}{(\mu-\lambda)^2} = \frac{\rho(2-\rho)}{\mu^2(1-\rho)^2}. \end{aligned} \tag{5.14}$$

In Example 5.1, the mean number of trucks in the system is 14 [Eq. (5.7)] with a standard deviation of 14.5 [Eq. (5.9)]; the mean number of trucks in the queue is approximately 13.1 [Eq. (5.8)] with standard deviation of 14.4

[4] J. D. C. Little, "A Proof for the Queuing Formula $L = \lambda W$," *Operations Research*, **9** (1961), pp. 383–387.

[Eq. (5.10)]; the mean time each truck spends in the system is 3.5 hours [Eq. (5.12)] with a standard deviation of 3.5 hours [Eq. (5.14)]; and the mean time each truck waits in line until its turn at the dock is 3 hours and 16 minutes [Eq. (5.12)] with a standard deviation of 3 hours and 29 minutes [Eq. (5.14)].

EXAMPLE 5.2 A large car dealer has a policy of providing cars for its customers that have car problems. When a customer brings the car in for repair, that customer has use of a dealer's car. The dealer estimates that the dealer cost for providing the service is \$10 per day for as long as the customer's car is in the shop. (Thus, if the customer's car was in the shop for 1.5 days, the dealer's cost would be \$15.) Arrivals to the shop of customers with car problems form a Poisson process with a mean rate of one every other day. There is one mechanic dedicated to those customers' cars. The time that the mechanic spends on a car can be described by an exponential random variable with a mean of 1.6 days. We would like to know the expected cost per day of this policy to the car dealer. Assuming infinite capacity, we have the assumptions of the M/M/1 queueing system satisfied, with $\lambda = 0.5/\text{day}$ and $\mu = 0.625/\text{day}$, yielding a $\rho = 0.8$. (Note the mean rate is the *reciprocal* of the mean time.) Using the M/M/1 equations, we have $L = 4$ and $W = 8$ days. Thus, whenever a customer comes in with car problems, it will cost the dealer \$80. Since a customer comes in every other day (on the average) the total cost to the dealer for this policy is \$40 per day. In other words, cost is equal to $\$10 \times W \times \lambda$. But by Little's formula, this is equivalent to $\$10 \times L$. The cost structure illustrated with this example is a very common occurrence for queueing systems. In other words, if c is the cost per item per time unit that the item spends in the system, the expected system cost per time unit is cL.

▶ *Suggestion: Do Exercises 5.1–5.4.*

Finite Capacity for Single-Server Systems. The assumption of infinite capacity is often not suitable. When a finite system capacity is necessary, the state probabilities and measures of effectiveness presented in Eqs. (5.6) through (5.14), with the exception of Eq. (5.11), are inappropriate to use; thus, new probabilities and measures of effectiveness must be developed for the M/M/1/K system (see Figure 5.3).

As will be seen, the equations for the state probabilities are identical except for the norming equation. If K is the maximum number of customers possible in the system, the generator matrix is of dimension $(K+1) \times (K+1)$ and has the form

$$\boldsymbol{G} = \begin{bmatrix} -\lambda & \lambda & & & \\ \mu & -(\lambda+\mu) & \lambda & & \\ & \ddots & \ddots & \ddots & \\ & & \mu & -(\lambda+\mu) & \lambda \\ & & & \mu & -\mu \end{bmatrix}.$$

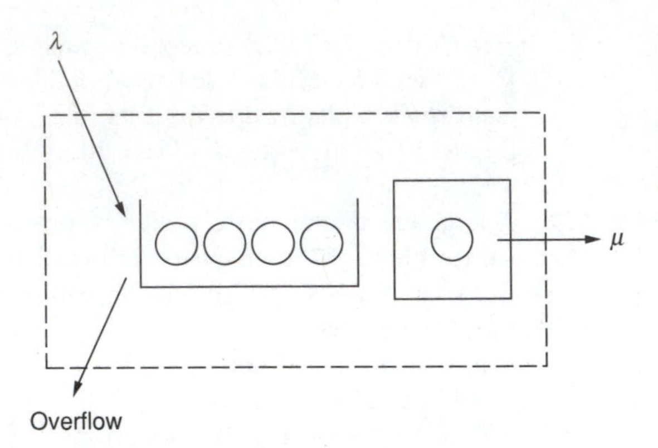

FIGURE 5.3 Representation of a full M/M/1/5 queueing system.

The system $\boldsymbol{pG} = \mathbf{0}$ yields

$$\begin{aligned} \mu p_1 &= \lambda p_0, \\ \mu p_{n+1} &= (\lambda + \mu)p_n - \lambda p_{n-1} \quad \text{for } n = 1, \cdots, K-1, \\ \mu p_K &= \lambda p_{K-1}. \end{aligned} \tag{5.15}$$

Using successive substitution, we again have

$$p_n = \rho^n p_0 \quad \text{for } n = 0, 1, \cdots, K,$$

where $\rho = \lambda/\mu$. The last equation in Eq. (5.15) is ignored because there is always a redundant equation in a finite irreducible Markov system. The norming equation is now used to obtain p_0 as follows:

$$\begin{aligned} 1 &= \sum_{n=0}^{K} p_n = p_0 \sum_{n=0}^{K} \rho^n \\ &= \begin{cases} p_0 \frac{1-\rho^{K+1}}{1-\rho} & \text{for } \rho \neq 1 \\ p_0(K+1) & \text{for } \rho = 1 \end{cases}. \end{aligned}$$

Because the preceding sum is finite, we used the finite geometric progression[5] so that ρ may be larger than one. Therefore, for an M/M/1/K system,

$$p_n = \begin{cases} \rho^n(1-\rho)/(1-\rho^{K+1}) & \text{for } \rho \neq 1 \\ 1/(K+1) & \text{for } \rho = 1 \end{cases}, \tag{5.16}$$

for $n = 0, 1, \cdots, K$.

[5]The finite geometric progression is $\sum_{n=0}^{k-1} r^n = (1 - r^k)/(1 - r)$ if $r \neq 1$.

The means for the numbers in the system and in the queue are

$$L = \sum_{n=0}^{K} np_n = \sum_{n=1}^{K} np_n = p_0\rho \sum_{n=1}^{K} n\rho^{n-1}$$

$$= \begin{cases} \rho[1 + K\rho^{K+1} - (K+1)\rho^K]/[(1-\rho)(1-\rho^{K+1})] & \text{for } \rho \neq 1 \\ K/2 & \text{for } \rho = 1 \end{cases}, \quad (5.17)$$

$$L_q = \sum_{n=1}^{K-1} np_{n+1} = p_0\rho^2 \sum_{n=1}^{K-1} n\rho^{n-1}$$

$$= \begin{cases} L - [\rho(1-\rho^K)/(1-\rho^{K+1})] & \text{for } \rho \neq 1 \\ K(K-1)/[2(K+1)] & \text{for } \rho = 1 \end{cases}. \quad (5.18)$$

The variances for these quantities for the M/M/1/K system are

$$\text{var}(N) = \begin{cases} [\rho/(1-\rho^{K+1})(1-\rho)^2] \\ \quad \times [1 + \rho - (K+1)^2\rho^K \\ \quad + (2K^2 + 2K - 1)\rho^{K+1} - K^2\rho^{K+2}] - L^2 & \text{for } \rho \neq 1 \\ K(K+2)/12 & \text{for } \rho = 1 \end{cases}, \quad (5.19)$$

$$\text{var}(N_q) = \text{var}(N) - p_0(L + L_q).$$

The probability that an arriving customer enters the system is the probability that the system is not full. Therefore, to utilize Little's formula, we set $\lambda_e = \lambda(1 - p_K)$ for the effective arrival rate to obtain the waiting time equations as follows:

$$W = \frac{L}{\lambda(1 - p_K)},$$
$$W_q = W - \frac{1}{\mu}. \quad (5.20)$$

The temptation is to use the formulas given in Eq. (5.13) to obtain expressions for the variances of the waiting times. However, the "Little-like" relationships for variances are based on the assumption that the system sees Poisson arrivals. The finite capacity limitation prohibits Poisson arrivals *into* the system, so Eq. (5.13) cannot be used.

EXAMPLE 5.3 A corporation must maintain a large fleet of tractors. They have one mechanic, who works on the tractors as they break down on a first-come/first-serve basis. The arrival of tractors to the shop needing repair work is approximated by a Poisson distribution with a mean rate of three per week. The length of time needed for repair varies according to an exponential distribution with a mean repair time of one-half week per tractor. The current corporate policy is to utilize an outside repair shop whenever more than two tractors are in the company shop, so that, at most, one tractor is allowed to wait. Each week that a tractor spends in the shop costs the company \$100. To utilize the outside shop costs \$500 per tractor. (The \$500 includes lost time.) We wish to review corporate policy and determine the optimum cut-off point for the outside shop; that is, we will determine the maximum number allowed in the company shop before sending tractors to the outside repair facility. The total operating costs per week are

$$\text{Cost} = 100L + 500\lambda p_K .$$

For the current policy, which is an M/M/1/2 system, we have $L = 1.26$ and $p_2 = 0.474$; thus, the cost is

$$\text{Cost}_{K=2} = 100 \times 1.26 + 500 \times 3 \times 0.474 = \$837/\text{week}.$$

If the company allows three in the system, then $L = 1.98$ and $p_3 = 0.415$, which yields

$$\text{Cost}_{K=3} = 100 \times 1.98 + 500 \times 3 \times 0.415 = \$820/\text{week}.$$

If a maximum of four are allowed in the shop, then

$$\text{Cost}_{K=4} = 100 \times 2.76 + 500 \times 3 \times 0.384 = \$852/\text{week}.$$

Therefore, the recommendation is to send a tractor to an outside shop only when more than three are in the system. ■

► *Suggestion: Do Exercises 5.6 and 5.7.*

5.3 MULTIPLE-SERVER QUEUES

A birth–death process is a special type of Markov process that is applicable to many types of Markov queueing systems. The birth–death process is a process in which changes of state are only to adjacent states (Figure 5.4). The generator matrix for a general birth–death process is given by

$$\boldsymbol{G} = \begin{bmatrix} -\lambda_0 & \lambda_0 & & & \\ \mu_1 & -(\mu_1 + \lambda_1) & \lambda_1 & & \\ & \mu_2 & -(\mu_2 + \lambda_2) & \lambda_2 & \\ & & \ddots & \ddots & \ddots \end{bmatrix},$$

where λ_n and μ_n are the birth rate (arrival rate) and death rate (service rate), respectively, when the process is in state n.

FIGURE 5.4 State diagram for a birth–death process.

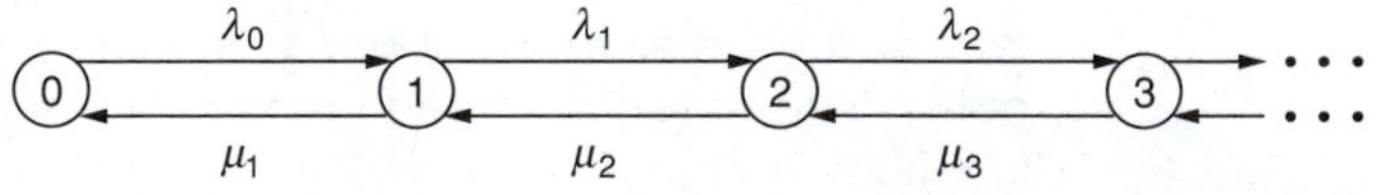

λ_0 λ_1 λ_2
0 1 2 3 ...
μ_1 μ_2 μ_3

As before, the long-run probabilities are obtained by first forming the system of equations defined by $\boldsymbol{pG} = \mathbf{0}$. (Or, equivalently, the system of equations may be obtained using the rate-balance approach applied to the state diagram of Figure 5.4.) The resulting system is

$$\begin{aligned} \lambda_0 p_0 &= \mu_1 p_1 \\ (\lambda_1 + \mu_1) p_1 &= \lambda_0 p_0 + \mu_2 p_2 \\ (\lambda_2 + \mu_2) p_2 &= \lambda_1 p_1 + \mu_3 p_3 \\ &\vdots \end{aligned}$$

This system is solved in terms of p_0 by successive substitution, and we have

$$\begin{aligned} p_1 &= p_0 \frac{\lambda_0}{\mu_1} \\ p_2 &= p_0 \frac{\lambda_0 \lambda_1}{\mu_1 \mu_2} \\ &\vdots \\ p_n &= p_0 \frac{\lambda_0 \times \cdots \times \lambda_{n-1}}{\mu_1 \times \cdots \times \mu_n} \\ &\vdots \end{aligned} \tag{5.21}$$

where

$$p_0 = \frac{1}{1 + \sum_{n=1}^{\infty} \prod_{k=0}^{n-1} (\lambda_k / \mu_{k+1})} .$$

The birth–death process can be used for many types of queueing systems. Equation (5.6), for the M/M/1 system, arises from Eq. (5.21) by letting the birth rates be the constant λ and the death rates be the constant μ. The M/M/1/K system equations can be obtained by letting $\lambda_n = 0$ for all $n > K$.

The M/M/c queueing system is a birth–death process (see Figure 5.5) with the following values for the birth rates and death rates:

$$\begin{aligned} \lambda_n &= \lambda \quad \text{for } n = 0, 1, \cdots, \\ \mu_n &= \begin{cases} n\mu & \text{for } n = 1, \cdots, c-1 \\ c\mu & \text{for } n = c, c+1, \cdots. \end{cases} \end{aligned} \tag{5.22}$$

The reason that $\mu_n = n\mu$ for $n = 1, \cdots, c-1$ is that when there are fewer than c customers in the system, each customer in the system is being served; thus, the service rate would be equal to the number of customers, since unoccupied servers remain idle (i.e., free servers do not help busy servers). If there are more than c customers in the system, then exactly c servers are busy, and thus the service rate must be $c\mu$. Substituting Eq. (5.22) into Eqs. (5.21) for the M/M/c system, yields

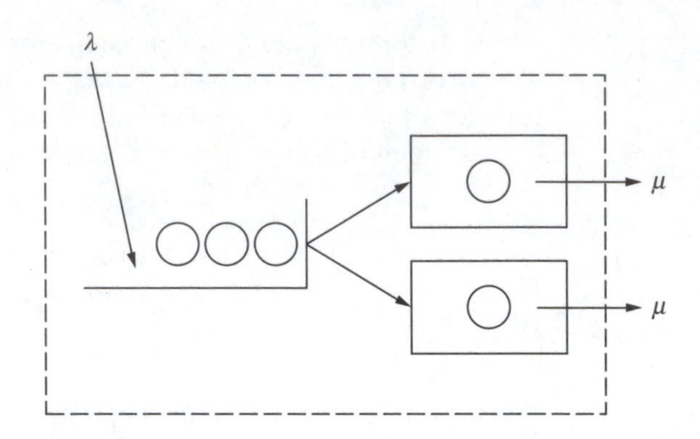

FIGURE 5.5 Representation of an M/M/2 queueing system. (Transfer from queue to server is instantaneous.)

$$p_n = \begin{cases} p_0 r^n / n! & \text{for } n = 0, 1, \cdots, c-1 \\ p_0 r^n / (c^{n-c} c!) & \text{for } n = c, c+1, \cdots \end{cases}, \tag{5.23}$$

$$p_0 = \frac{1}{\left[cr^c / c!(c-r) \right] + \sum_{n=0}^{c-1} [r^n / n!]},$$

where $r = \lambda / \mu$ and $\rho = r/c < 1$.

The measures of effectiveness for the M/M/c queueing system involve slightly more manipulations, but it can be shown that

$$L_q = \sum_{n=c}^{\infty} (n-c) p_n = \frac{p_0 r^c \rho}{c!(1-\rho)^2}.$$

Little's formula is then applied to obtain

$$W_q = \frac{L_q}{\lambda},$$

and by summing mean values

$$W = W_q + \frac{1}{\mu},$$

and finally applying Little's formula again

$$L = L_q + r.$$

We let $r = \lambda/\mu$ because most textbooks and technical papers usually reserve ρ to be the arrival rate divided by the maximum service rate, namely, $\rho = \lambda / c\mu$. It can then be shown that ρ gives the server utilization; that is, ρ is the fraction of time during which an arbitrarily chosen server is busy.

For completeness, we also give the formula needed to obtain the variances for the M/M/c system as

$$E[N_q(N_q - 1)] = \frac{2 p_0 r^c \rho^2}{c!(1-\rho)^3}, \tag{5.24}$$

$$E[T_q^2] = \frac{2p_0 r^c}{\mu^2 c^2 c!(1-\rho)^3}, \tag{5.25}$$

$$\text{var}(T) = \text{var}(T_q) + \frac{1}{\mu^2}, \tag{5.26}$$

$$E[N(N-1)] = \frac{E[N_q(N_q-1)]}{\rho^c} + p_0 \sum_{n=1}^{c-1} n(n-1)r^n \left(\frac{1}{n!} - \frac{c^{c-n}}{c!}\right). \tag{5.27}$$

EXAMPLE 5.4 The corporation from the previous example has implemented the policy of never allowing more than three tractors in its repair shop. For \$600 per week, it can hire a second mechanic. Is it worthwhile to do so if the expected cost is used as the criterion? To answer this question, the old cost for the M/M/1/3 system (refer to page 130) is compared to the proposed cost for an M/M/2/3 system. The birth–death equations [Eq. (5.21)] are used with

$$\lambda_n = \begin{cases} \lambda & \text{for } n = 0, 1, 2 \\ 0 & \text{for } n = 3, 4, \cdots \end{cases},$$

$$\mu_n = \begin{cases} \mu & \text{for } n = 1 \\ 2\mu & \text{for } n = 2 \text{ and } 3 \end{cases},$$

where $\lambda = 3/\text{week}$ and $\mu = 2/\text{week}$. This gives

$$p_1 = 1.5p_0,$$
$$p_2 = 1.125p_0,$$
$$p_3 = 0.84375p_0,$$
$$p_0 = \frac{1}{1 + 1.5 + 1.125 + 0.84375} = 0.224.$$

The expected number in the system is

$$L = 0.224 \times (1 \times 1.5 + 2 \times 1.125 + 3 \times 0.84375)$$
$$= 1.407.$$

The cost of the proposed system is

$$\text{Cost}_{c=2} = 100 \times 1.407 + 500 \times 3 \times 0.189 + 600$$
$$= \$824.20/\text{week}.$$

Therefore, it is not worthwhile to hire a second mechanic, since this cost is greater than the \$820 calculated in the previous example. ■

This section closes with a common example that illustrates the versatility of the birth–death equation. Arrivals to a system usually come from an infinite (or very large) population, so that an individual arrival does not affect the overall arrival rate. However, in some circumstances the arrivals come from a finite population; thus, the arrival rate cannot be assumed constant.

EXAMPLE 5.5 A small corporation has three old machines that continually break down. Each machine breaks down on the average of once a week. The corporation

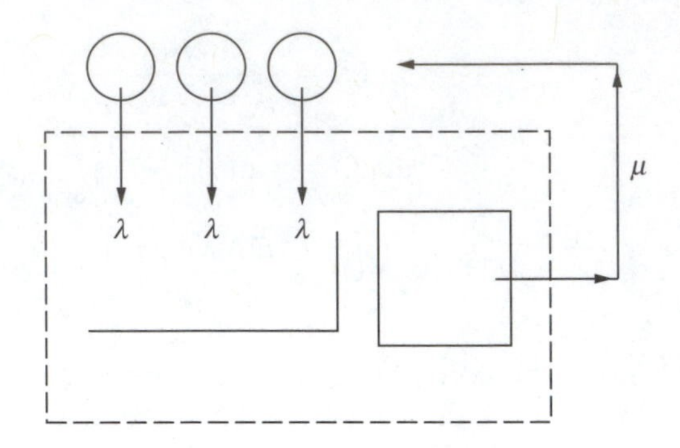

FIGURE 5.6 Representation of the machine-repair queueing system of Example 5.5.

has one mechanic, who takes, on the average, one-half of a week to repair a machine. (See Figure 5.6 for a schematic of this "machine-repair" queueing system.) Assuming that breakdowns and repairs are exponential random variables, the birth–death equations can be used as follows:

$$\lambda_n = \begin{cases} (3-n)\lambda & \text{for } n = 0, 1, 2 \\ 0 & \text{for } n = 3, \cdots \end{cases},$$

$$\mu_n = \mu \quad \text{for } n = 1, 2, 3$$

with $\lambda = 1$ per week and $\mu = 2$ per week. This yields

$$\begin{aligned} p_1 &= 1.5p_0, \\ p_2 &= 1.5p_0, \\ p_3 &= 0.75p_0, \\ p_0 &= \frac{1}{1 + 1.5 + 1.5 + 0.75} = 0.21, \quad \text{and} \\ L &= 0.21 \times (1 \times 1.5 + 2 \times 1.5 + 3 \times 0.75) = 1.42. \end{aligned}$$

Let us further assume that for every hour that a machine is tied up in the repair shop, the corporation loses \$25. The cost of this system per hour due to the unavailability of the machines is calculated as

$$\begin{aligned} \text{Cost} &= 25 \times L = 25 \times 1.42 \\ &= \$35.50/\text{hr}. \end{aligned}$$

■

▶ *Suggestion: Do Exercises 5.5, 5.8–5.14, 5.21, and 5.22.*

5.4 JACKSON NETWORKS

Many systems for which queueing models are appropriate involve multiple queues where the output from one queue becomes the input to another. The modeling of such systems can be very difficult unless the network of queues has some special structure. In particular, if the system has Poisson arrivals, and each queue within the network has unlimited capacity and exponential servers, the system is called a Jackson network, and some of its measures of effectiveness can be computed using the methodology developed in the previous sections. In particular, the following formally defines those networks that can be easily analyzed.

DEFINITION 5.1 *A network of queues is called a* JACKSON NETWORK *if the following conditions are satisfied:*

1. *All outside arrivals to each queueing system in the network must arrive according to a Poisson process.*
2. *All service times must be exponentially distributed.*
3. *All queues must have unlimited capacity.*
4. *When a job leaves one queueing system, the probability that it will go to another queueing system is independent of that job's past history and is independent of the location of any other job.*

The effect of these conditions is that the M/M/c formulas, developed in the previous sections, can be used for the analysis of networks. As can be seen, the first three conditions are from the standard M/M/c assumptions. The fourth condition is not quite as straightforward, but it can be viewed as a Markov assumption; that is, its effect is that the movement of jobs through the network can be described as a Markov chain.

■ **EXAMPLE 5.6** Jobs arrive at a job shop according to a Poisson process with a mean rate of eight jobs per hour. Upon arrival, the jobs first go through a drill press, then 75% of them go to the paint shop, and then they go to a final packaging step. Those jobs that do not go to the paint shop instead go directly to the final packaging step. The system has one drill press, two painters, and one packer. The mean time spent on the drill press is 6 minutes, the mean time spent during the painting operation is 15 minutes, and the mean time for packaging is 5 minutes; all times are exponentially distributed. (See Figure 5.7 for a schematic representation of the network.) Thus, this is a Jackson network, and the movement of jobs within the network can be described by the sub-Markov matrix (called the *switching probability matrix*)

$$\boldsymbol{P} = \begin{bmatrix} 0 & 0.75 & 0.25 \\ 0 & 0 & 1 \\ 0 & 0 & 0 \end{bmatrix},$$

where the drill press is node 1, the paint shop is node 2, and packaging is node 3. ■

The basic idea for the analysis of Jackson networks is that an arrival rate is determined for each node, and then each queueing system (node) is

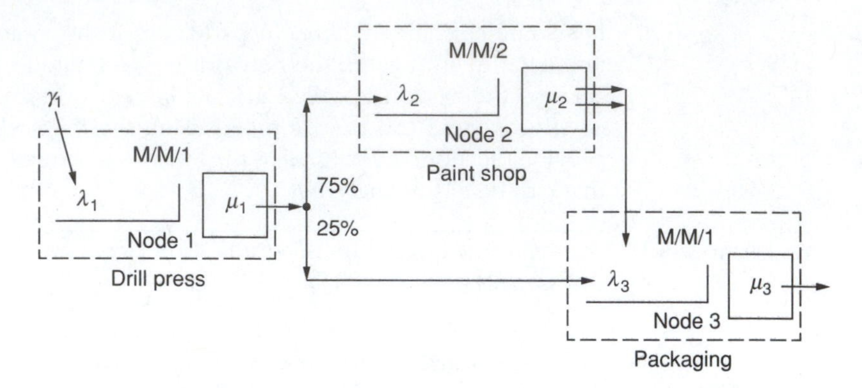

FIGURE 5.7 Representation of the queueing network of Example 5.6.

treated as if it were an independent system. To illustrate, we return to the job shop example (Example 5.6) and let λ_i denote the arrival rate into node i; thus $\lambda_1 = 8$, $\lambda_2 = 0.75 \times 8 = 6$, and $\lambda_3 = 0.25 \times 8 + 6 = 8$. The other data of the example indicate that $\mu_1 = 10$, $\mu_2 = 4$, and $\mu_3 = 12$; thus, the traffic intensities are $\rho_1 = 0.8$, $\rho_2 = 0.75$ (note that node 2 has two servers), and $\rho_3 = 0.667$. With these parameters, we apply the previously developed M/M/1 [Eqs. (5.6) and (5.7)] and M/M/2 (see Exercise 5.4) formulas. The average number of jobs at each node is now easily calculated as $L_1 = 4$, $L_2 = 3.43$, and $L_3 = 2$, and the probabilities that there are no jobs in the various nodes are given as $p_0^1 = 0.2$, $p_0^2 = 0.143$, and $p_0^3 = 0.333$.

The only potentially difficult part of the analysis is determining the input rate to each queueing node. As in the above example, we let λ_i denote the total input to node i. First, observe that for a steady-state system, the input must equal the output, so the total output from node i also equals λ_i. Second, observe that the input to node i must equal the input from the outside, plus any output from other nodes that are routed to node i. Thus, we have the general relation

$$\lambda_i = \gamma_i + \sum_k \lambda_k P(k, i),$$

where γ_i is the arrival rate to node i from outside the network, and $P(k, i)$ is the probability that output from node k is routed to node i. There is one such equation for each node in the network, and the resulting system of equations can be solved to determine the net input rates. In general, the following property holds.

Property 5.2 Let the switching probability matrix $\boldsymbol{P}$ describe the flow of jobs within a Jackson network, and let γ_i denote the mean arrival rate of jobs going directly into node i from outside the network with $\boldsymbol{\gamma}$ being the vector of these rates. Then

$$\boldsymbol{\lambda} = \boldsymbol{\gamma}(\boldsymbol{I} - \boldsymbol{P})^{-1}$$

where the components of the vector $\boldsymbol{\lambda}$ give the arrival rate into the various nodes; that is, λ_i is the net rate into node i.

After the net rate into each node is known, the network can be decomposed and each node treated as if it were a set of independent queueing systems with Poisson input. (If there is any feedback within the queueing network, the input streams to the nodes will not only be dependent on each other, but there will also be input processes that are not Poisson. However, these aspects can be ignored when computing the quantities described in the remainder of this section.)

Property 5.3 Consider a Jackson network containing m nodes. Let N_i denote a random variable indicating the number of jobs at node i (the number in the queue plus the number in the system). Then,

$$P\{N_1 = n_1, \cdots, N_m = n_m\} = P\{N_1 = n_1\} \times \cdots \times P\{N_m = n_m\},$$

and the probabilities $P\{N_i = n_i\}$ for $n_i = 0, 1, \cdots$ can be calculated using Eq. (5.23).

EXAMPLE 5.7 Now let us change the previous example slightly. Assume that after the painting operation, 10% of the jobs are discovered to be defective and must be returned to the drill press operation, and after the packaging operation, 5% of the jobs are discovered to be defective and must also be returned to the beginning of the drill press operation (see Figure 5.8). Now the switching probability matrix describing the movement of jobs within the network is

$$\boldsymbol{P} = \begin{bmatrix} 0 & 0.75 & 0.25 \\ 0.10 & 0 & 0.90 \\ 0.05 & 0 & 0 \end{bmatrix}.$$

FIGURE 5.8 Representation of the queueing network of Example 5.7.

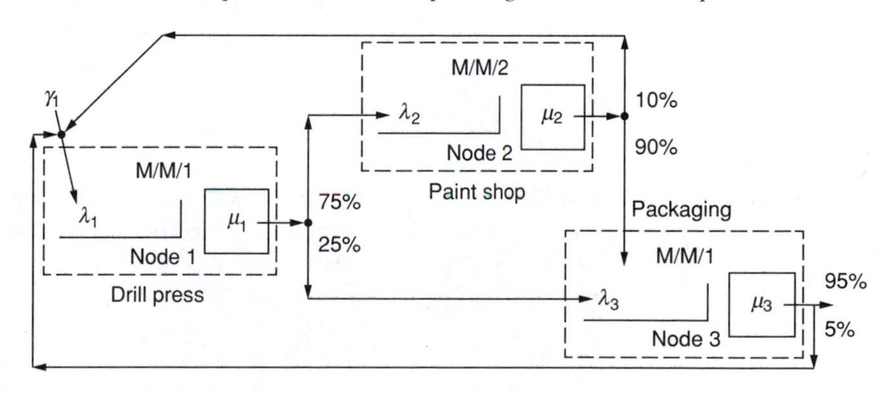

The equations defining the net arrival rates are

$$\begin{aligned}\lambda_1 &= 8 + 0.1\lambda_2 + 0.05\lambda_3,\\ \lambda_2 &= 0.75\lambda_1,\\ \lambda_3 &= 0.25\lambda_1 + 0.9\lambda_2.\end{aligned}$$

The vector $\boldsymbol{\gamma}$ is $(8, 0, 0)$ and solving the above system yields $\lambda_1 = 9.104$, $\lambda_2 = 6.828$, and $\lambda_3 = 8.421$, which implies traffic intensities of $\rho_1 = 0.9104$, $\rho_2 = 0.8535$, and $\rho_3 = 0.702$. Again we use the M/M/1 and M/M/2 formulas and obtain the average number in the network to be 18.8. Thus, the inspection and resulting feedback doubles the mean number of jobs in the system. (Note that the number of jobs in the system is the sum of the number of jobs at the individual nodes.) The probability that there are no jobs in the system is 0.002. (Note that the probability that there are no jobs in the system equals the product of the probabilities that there are no jobs at each of the individual nodes.) ■

A major reason for developing a network model for a queueing system is to estimate the length of time needed for a job to travel through the queueing network. The basic mathematical tool for analyzing the flow time through a network involves a random sum of random variables. In particular, suppose that $X_1, X_2, \cdots$, are independent, identically distributed random variables with mean and variance given by μ and σ^2, respectively. Furthermore, let N be a random variable having a nonnegative integer value independent from the X random variables. A random sum, S, is defined by $S = X_1 + \cdots + X_N$, where if $N = 0$ then $S = 0$. It is possible to show that

$$\begin{aligned}E[S] &= \mu E[N],\\ \mathrm{var}(S) &= \sigma^2 E[N] + \mu^2\,\mathrm{var}(N).\end{aligned}$$

The two important components needed for determining the flow time through a network are (1) the time spent at each node and (2) the number of times each node is visited. The mean time a job spends at node i per visit is denoted by W_i, where the value of W_i (mean waiting time in a queueing system) can be determined according to the formulas on page 132. The mean number of visits to each node can be calculated by Property 2.18 after the initial probability vector is obtained, which denotes the probability mass function for the initial node of an arriving job.

Property 5.4 Consider a Jackson network containing m nodes with T_{net} being a random variable denoting the time that a job spends in the network. Let $\boldsymbol{P}$ be the switching probability matrix and $\boldsymbol{\gamma}$ be the vector giving the external arrival rates. Define the potential matrix $\boldsymbol{R}$ to be

$$\boldsymbol{R} = (\boldsymbol{I} - \boldsymbol{P})^{-1}$$

and the initial probability vector $\boldsymbol{\nu}$ by

$$\nu(i) = \gamma(i) \Big/ \sum_{j=1}^{m} \gamma(j).$$

Then, the mean flow time through the network is given by

$$E[T_{\text{net}}] = \sum_{i=1}^{m} W_i \boldsymbol{\nu R}(i),$$

where W_i is the mean system waiting time as defined on page 132 and $\boldsymbol{\nu R}(i)$ is the i'th component of the vector resulting from the product $\boldsymbol{\nu R}$.

It is also possible to give the variance for the flow time through a network, but it is more involved in its calculation. The main reason for the additional complication is that the number of visits to one node is not independent from the number of visits to another node. In particular, we have the following covariance relationship:

$$\text{cov}(N_j, N_k) = \begin{cases} \boldsymbol{\nu R}(j)[2R(j,j) - \boldsymbol{\nu R}(j) - 1] & \text{for } j = k \\ \boldsymbol{\nu R}(j)R(j,k) + \boldsymbol{\nu R}(k)R(k,j) - \boldsymbol{\nu R}(j)\boldsymbol{\nu R}(k) & \text{for } j \neq k \end{cases},$$

where N_i is a random variable denoting the number of visits to node i. With this property giving the relationship among the various nodes, we derive the following property.

Property 5.5 Consider a network as in Property 5.4. The variance of the flow time through the network is given by

$$\text{var}(T_{\text{net}}) = \sum_{i=1}^{m} \text{var}(T_i)\boldsymbol{\nu R}(i) + \sum_{j=1}^{m}\sum_{k=1}^{m} W_j W_k \,\text{cov}(N_j, N_k),$$

where $\text{var}(T_i)$ is the variance of the waiting time in a queueing system as given by Eq. (5.26).

EXAMPLE 5.8 We return to Example 5.7 to illustrate these properties. The first step is to obtain W_1, W_2, and W_3. Using the M/M/1 formula for W with $\lambda = 9.104/\text{hr}$ and $\mu = 10/\text{hr}$, we obtain $W_1 = 1.12$ hr. Using the M/M/2 formula for W with $\lambda = 6.828/\text{hr}$ and $\mu = 4/\text{hr}$ yields $W_2 = 0.92$ hr. Finally, returning to the M/M/1 formula with $\lambda = 8.421/\text{hr}$ and $\mu = 12/\text{hr}$ yields $W_3 = 0.28$ hr. After obtaining the node waiting times, the next step is to obtain the number of visits to each node; specifically, we obtain the potential matrix as

$$\boldsymbol{R} = \begin{bmatrix} 1.00 & -0.75 & -0.25 \\ -0.10 & 1.0 & -0.90 \\ -0.05 & 0 & 1.0 \end{bmatrix}^{-1} = \begin{bmatrix} 1.138 & 0.853 & 1.053 \\ 0.165 & 1.124 & 1.053 \\ 0.057 & 0.043 & 1.053 \end{bmatrix}.$$

The initial probability vector is $\boldsymbol{\nu} = (1, 0, 0)$ indicating that all jobs enter the network through node 1; therefore,

$$E[T_{\text{net}}] = 1.12 \times 1.138 + 0.92 \times 0.853 + 0.28 \times 1.053 \approx 2 \text{ hr } 21 \text{ min.} \quad \blacksquare$$

▶ *Suggestion: Do Exercises 5.15–5.18.*

5.5 APPROXIMATIONS

The models developed in the previous sections depended strongly on the use of the exponential distribution. Unfortunately, the exponential assumption is not appropriate for many practical systems for which queueing models are desired. However, without the exponential assumption, exact results are much more difficult. In response to the need for numerical results for nonexponential queueing systems, a significant amount of research has been conducted that deals with approximations in queueing theory. In this section we report on some of these approximations. We recommend that the interested student read the survey paper by Ward Whitt[6] for a thorough review.

5.5.1 The Allen-Cunneen Approximation

Assume that we are interested in modeling a G/G/c queueing system. As before, the mean arrival rate is denoted by λ, and the mean service rate is μ. An approximation developed by Allen and Cunneen[7] uses the squared coefficient of variation; therefore, we define c_a^2 to be the variance of the interarrival times divided by the square of the mean of the interarrival times, and c_s^2 to be the variance of the service times divided by the square of the mean of the service times.

To obtain an approximation for the mean waiting time, W, spent in a G/G/c system, we first determine the mean time spent in the queue and then add the mean service time to the queue time; namely, we use the fact that $W = W_q + (1/\mu)$. The value for W_q can be obtained according to the following property.

Property 5.6 Let λ and μ be the mean arrival rate and mean service rate, respectively, for a G/G/c queueing system. Let $W_{q,\text{M/M/}c}$ denote the mean waiting time for an M/M/c queue with the same mean arrival and service rates. Then the time spent in the queue for the G/G/c system is approximated by

$$W_q \approx \left(\frac{c_a^2 + c_s^2}{2} \right) W_{q,\text{M/M/}c}. \qquad \textbf{(5.28)}$$

[6]W. Whitt, "Approximations for the GI/G/m Queue," *Production and Operations Management* **2** (1993), pp. 114–161.

[7]A. O. Allen, *Probability, Statistics and Queueing Theory, with Computer Science Applications,* 2nd ed. (Boston, Mass.: Academic Press, 1990), p. 341.

Notice that the equation holds as an equality if the interarrival and service times are indeed exponential, since the exponential has a squared coefficient of variation equal to one. Equation (5.28) is also exact for an M/G/1 queueing system, and it is known to be an excellent approximation for the M/G/c system. In general, the approximation works best for $c_a \geq 1$ with c_a and c_s being close to the same value.

EXAMPLE 5.9 Suppose we wish to model the unloading dock at a manufacturing facility. At the dock, only one crew member does the unloading. The questions of interest are the average waiting time for arriving trucks and the average number of trucks at the dock at any point in time. Unfortunately, we do not know the underlying probability laws governing the arrivals of trucks or the service times; therefore, data are collected over a representative time period to obtain the necessary statistical estimates. The data yield a mean of 31.3 minutes and a standard deviation of 35.7 minutes for the interarrival times. For the service times, the results of the data give 20.4 minutes for the mean and 10.0 minutes for the standard deviation. Thus, $\lambda = 1.917/\text{hr}$, $c_a^2 = 1.30$, $\mu = 2.941/\text{hr}$, and $c_s^2 = 0.24$ with a traffic intensity of $\rho = 0.652$. (It would be appropriate to collect more data and perform a statistical "goodness-of-fit" test to help determine whether or not these data may come from an exponential distribution. Assuming a reasonable number of data points, it becomes intuitively clear that at least the service times are not exponentially distributed since an exponential distribution has the property that its mean is equal to its standard deviation.)

To obtain W, we first determine W_q. From Eq. (5.12), we get that $W_{q,\text{M/M/1}} = 0.652/(2.941 - 1.917)$ hr $= 40$ min. Now using Eq. (5.28), the queue time for the trucks is

$$W_q \approx \frac{1.30 + 0.24}{2} 40 = 30.8 \text{ min},$$

yielding a waiting time in the system of

$$W \approx 30.8 + \frac{60}{2.941} = 51.2 \text{ min}.$$

(Notice that since the mean service rate is 2.941 per hour, the reciprocal is the mean service time, which must be multiplied by 60 to convert it to minutes.) It now follows from Little's formula [Eq. (5.11)] that the mean number in the system is $L \approx 1.636$. (Notice, before using Little's formula, that care must be taken to ensure that the units are consistent; that is, the quantity W must be expressed in terms of hours, because λ is in terms of hours.) ■

▶ *Suggestion: Do Exercise 5.19.*

5.5.2 Network Approximations

Many systems that involve networks are not Jackson networks but can be modeled through approximations. A very common scenario is that instead of each job being routed through the network according to a Markov chain, there are classes of jobs for which each job within a class has a fixed route. For example, consider a job shop with two machines. Further suppose that jobs arrive at a rate of 10 per hour, where arriving jobs always start on machine 1, proceed to machine 2, then return to machine 1 for a final processing step (see the left-hand schematic in Figure 5.9). Such a route is deterministic, and thus does not satisfy the Markov assumption (condition 4) of the Jackson network definition. However, a Jackson network can be used to approximate a network with deterministic routing by fixing the switching probabilities so that the mean flow rates in the Jackson network are identical to the flow rates in the network with deterministic routes. In this job shop example, we could assume that all jobs start at machine 1, then there is a 50% probability that a job leaving machine 1 exits the system or proceeds to machine 2; thus we have the switching probability matrix

$$\boldsymbol{P} = \begin{bmatrix} 0 & 0.5 \\ 1 & 0 \end{bmatrix},$$

governing the flow of jobs within the two-node network (see the right-hand schematic in Figure 5.9). Notice that in both networks, an input rate of $\gamma = 10/\text{hr}$ results in a net flow of 20 per hour into node 1 and 10 per hour into node 2. The difficulty with the Jackson network approximation is that the variances of the two networks will not be the same.

A Jackson network model will always produce a high variability in flow times; thus, if flow times are deterministic as in the preceding example, a Jackson network model would be inappropriate. However, consider a network system with several distinct classes. Even if each class has deterministic flow times, the actual flow times through the network would be random because of the variability in classes. Thus, when a system contains several different classes, it may be reasonable to model the aggregate as a Jackson network.

The purpose of this subsection is to show how to model a network in which there are several different classes of jobs, but all jobs within one class have a fixed route through the network. The only assumption from Definition 5.1 that we enforce ahead of time is the unlimited capacity assumption at each node. The assumption of Poisson input is used as an approximation,

FIGURE 5.9 Comparison between a network with deterministic routing and a network with stochastic routing.

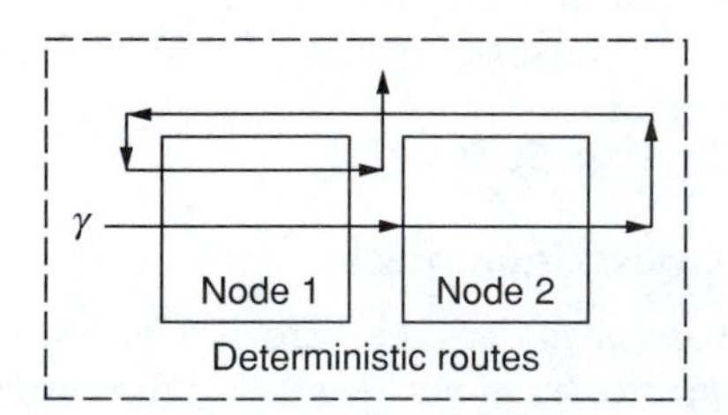

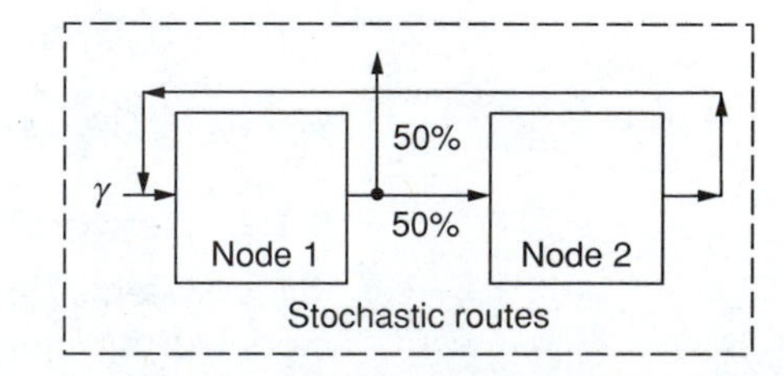

although the arrival rates of the individual classes need not be Poisson. Other authors have shown that when many independent arrival processes are added together, the resultant arrival process tends to be Poisson even if the individual processes are not. The two main issues we need to deal with are how to obtain each node's arrival rate and how to obtain the mean service times.

Calculating the Arrival Rate to a Node. We assume that an m-node queueing network has $\bar{k}$ classes of jobs. Each class of jobs has associated with it an arrival rate, γ_k, and a "routing" vector, $(n_{k,1}, n_{k,2}, \cdots, n_{k,\bar{n}_k})$, where $\bar{n}_k$ is the total number of steps within that route. For example, assume that the routing vector for the first class is (1,2,1). Thus, we have $\bar{n}_1 = 3$, and first class jobs start at node 1, then go to node 2, and then back to node 1 for final processing.

EXAMPLE 5.10 Consider the network illustrated in Figure 5.10. It is a queueing network with three classes of jobs, each class having its own deterministic route. Class 1 jobs have the routing vector (1,2,3,2) with $\bar{n}_1 = 4$; class 2 jobs have the routing vector (2,3) with $\bar{n}_2 = 2$; and, class 3 jobs have the routing vector (1,2,1,3) with $\bar{n}_3 = 4$. With deterministic routes it is also easy to determine the total flows through each node: $\lambda_1 = \gamma_1 + 2\gamma_3$, $\lambda_2 = 2\gamma_1 + \gamma_2 + \gamma_3$, and $\lambda_3 = \gamma_1 + \gamma_2 + \gamma_3$. The following property makes these calculations specific for the general case. ■

Property 5.7 Let γ_k be the arrival rate of class k jobs to the network, and let those jobs have a route defined by $(n_{k,1}, n_{k,2}, \cdots, n_{k,\bar{n}_k})$. Let $\eta_{k,i}$ be the number of times that route k passes through node i; that is,

$$\eta_{k,i} = \sum_{l=1}^{\bar{n}_k} I(i, n_{k,l}),$$

where $\boldsymbol{I}$ is the identity matrix so that $I(i, n) = 1$ if and only if $i = n$. The net arrival rate to node i in the network is given by

$$\lambda_i = \sum_{k=1}^{\bar{k}} \gamma_k \eta_{k,i},$$

or in matrix notation $\boldsymbol{\lambda} = \boldsymbol{\gamma\eta}$.

Calculating the Service Rate at a Node. After the net arrival rate has been determined, each node in the network can be considered as a separate queueing system. Although the arrival process is not Poisson, it can be assumed Poisson because it is the superposition of several different arrival streams. Each class of jobs typically has its own distribution for service times. Therefore, we need to determine the probability that a given job within the node is from a particular class. Let $q_{k,i}$ denote the probability that a randomly

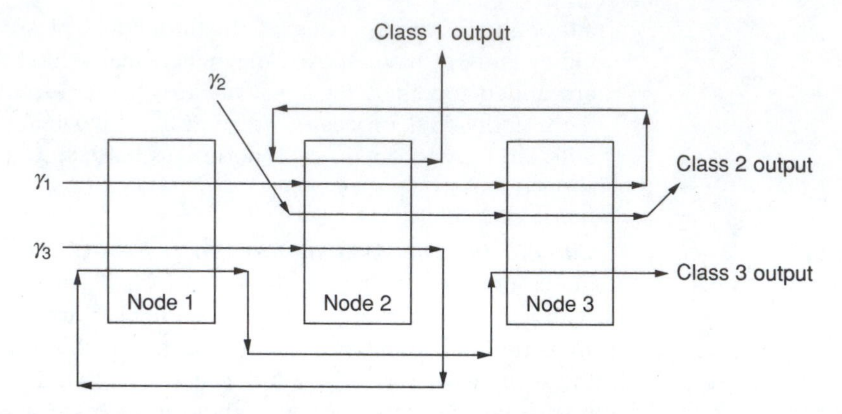

FIGURE 5.10 Routings within a queueing network containing three classes.

chosen job at node i is from class k; then

$$q_{k,i} = \frac{\gamma_k \eta_{k,i}}{\lambda_i}. \tag{5.29}$$

EXAMPLE 5.11 We return to the network of Example 5.10 and Figure 5.10. The matrix that contains the values for $\eta_{k,i}$ is given by

$$\begin{array}{c} \\ \\ \boldsymbol{\eta} = \end{array} \begin{array}{c} \text{job} \\ 1 \\ 2 \\ 3 \end{array} \begin{array}{c} \text{node} \\ \begin{array}{ccc} 1 & 2 & 3 \end{array} \\ \begin{bmatrix} 1 & 2 & 1 \\ 0 & 1 & 1 \\ 2 & 1 & 1 \end{bmatrix} \end{array}.$$

(Note that, in general, the matrix of the η's need not be square.) Assume that $\gamma_1 = 10$, $\gamma_2 = 20$, and $\gamma_3 = 50$. The initial probability vector for the network is given as $\boldsymbol{\nu} = (0.125, 0.250, 0.625)$, and the input rates to each node are $\lambda_1 = 110$, $\lambda_2 = 90$, and $\lambda_3 = 80$. The matrix of probabilities containing the terms $q_{k,i}$ is given by

$$\boldsymbol{q} = \begin{bmatrix} 0.091 & 0.222 & 0.125 \\ 0 & 0.222 & 0.250 \\ 0.909 & 0.556 & 0.625 \end{bmatrix}.$$

The service time distribution is a mixture of the service times from the separate classes so that the mean and variance can be determined as follows.

Property 5.8 Assume that the mean and variance of the service times at node i for jobs of class k are given by $1/\mu_{k,i}$ and $\sigma^2_{k,i}$, respectively. Let S_i be a random variable denoting the service time at node i of a randomly selected job. Then

$$E[S_i] = \sum_{k=1}^{\bar{k}} q_{k,i} \frac{1}{\mu_{k,i}},$$

$$E[S_i^2] = \sum_{k=1}^{\bar{k}} q_{k,i}[\sigma_{k,i}^2 + (1/\mu_{k,i})^2].$$

Flow Time Through a Network. We can now combine Properties 5.7 and 5.8 with the Allen-Cunneen approximation [Eq. (5.28)] to obtain an estimate for the mean flow time of a job through a non-Jackson network. The two major assumptions are that there are no queue capacity limitations and there are several classes of jobs arriving at the system.

Property 5.9 Consider a network containing m nodes and $\bar{k}$ distinct classes of jobs. Let $T_{\text{net},k}$ be a random variable denoting the time that a k-class job spends in the network. Define $W_{q,i,\text{M/M/}c}$ to be the mean waiting time in an M/M/c queue using an arrival rate of λ_i (from Property 5.7), a service rate of $1/E[S_i]$ (from Property 5.8), and a value of c equal to the number of servers at node i. Then, the mean flow time through the network for a job of class k is given by

$$E[T_{\text{net},k}] \approx \sum_{i=1}^{m} \left(\frac{1}{\mu_{k,i}} + \frac{E[S_i^2]}{2E[S_i]^2} W_{q,i,\text{M/M/}c} \right) \eta_{k,i}.$$

EXAMPLE 5.12 We illustrate the use of Property 5.9 with a production center that consists of four departments responsible for the following operations: turret lathe (T), milling machine (M), drill press (D), and inspection (I). There are five different products with their routes given in the following table and illustrated in Figure 5.11:

Product	Route	Production Rate
A	T – M – D – I	65 units per 8-hour shift
B	M – T – M – D – T – I	45 units per 8-hour shift
C	T – M – I	30 units per 8-hour shift
D	M – D – T – D – I	75 units per 8-hour shift
E	T – M – D – T – I	120 units per 8-hour shift

Each department has only one machine except that there are two inspectors. The mean and standard deviation of the processing times (in minutes) within the various departments are given in the following table:

	T		M		D		I	
Product	Mean	S.D.	Mean	S.D.	Mean	S.D.	Mean	S.D.
A	1.3	0.2	1.1	0.3	0.9	0.2	2.1	1.4
B	0.5	0.1	0.6	0.4	1.1	0.3	3.1	2.5
C	1.2	0.5	1.5	0.4	0.0	0.0	1.9	1.1
D	1.1	0.8	1.2	0.4	1.5	0.8	2.8	1.9
E	0.8	0.3	1.0	1.0	1.0	0.5	2.7	2.0

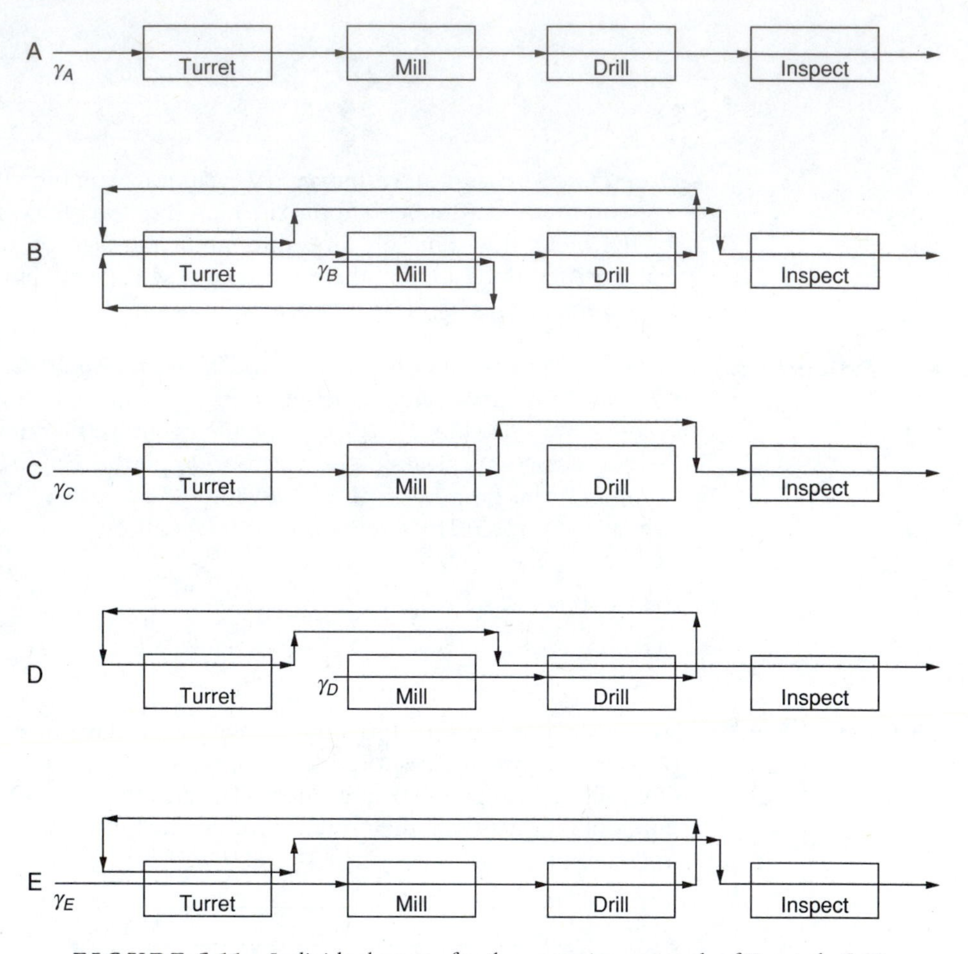

FIGURE 5.11 Individual routes for the queueing network of Example 5.12.

We present most of the calculations needed to obtain the mean flow times through the network by way of several tables. The mean production rates for the different classes are converted to a mean hourly arrival rate (right-hand column in Table 5.2). The number of visits to each node is given in Table 5.2 so that the net arrival rate to each node can be calculated (last row of the table). Then Table 5.3 gives the probabilities needed to define the service time moments.

With the mixture probabilities for the service times (the $q_{k,i}$), the main measures of effectiveness at each node can be calculated. The main quantities associated with the nodes are given in Table 5.4. (Note that the final row in the table is the result using the Allen-Cunneen approximation applied to the second-to-last row of the table.)

The final step in the calculations combines the waiting time in the queue with the time in the server, yielding the time spent at each node. These times are given in Table 5.5. These values are then multiplied by the values in

TABLE 5.2 Values for $\eta_{k,i}$ (in main body of table) and λ_i (in bottom row).

	Nodes				
Classes	*T*	*M*	*D*	*I*	γ_k (/hr)
A	1	1	1	1	8.125
B	2	2	1	1	5.625
C	1	1	0	1	3.750
D	1	1	2	1	9.375
E	2	1	1	1	15.000
λ_i (/hr)	62.5	47.5	47.5	41.875	

TABLE 5.3 Values for $q_{k,i}$ and the moments of the service times.

	Nodes			
Classes	*T*	*M*	*D*	*I*
A	0.13	0.171	0.171	0.194
B	0.18	0.237	0.118	0.134
C	0.06	0.079	0	0.090
D	0.15	0.197	0.395	0.224
E	0.48	0.316	0.316	0.358
$E[S_i]$	0.880	1.001	1.192	2.588
$E[S_i^2]$	1.001	1.483	1.835	10.401

TABLE 5.4 Waiting time in the queue at each node.

	Nodes			
	T	*M*	*D*	*I*
Arrival rate/hr	62.5	47.5	47.5	41.875
Service rate/hr	68.2	59.9	50.3	23.2
Number servers	1	1	1	2
$W_{q,i,M/M/c}$ (min)	9.65	3.84	20.24	11.353
$W_{q,i}$ (min)	6.24	2.84	13.07	8.82

TABLE 5.5 Values for the node waiting times and network flow times.

	Nodes				
Classes	*T*	*M*	*D*	*I*	$T_{\text{net},k}$
A	7.54	3.94	13.97	10.92	36.37
B	6.74	3.44	14.17	11.92	46.45
C	7.44	4.34	0	10.72	22.50
D	7.34	4.04	14.57	11.62	52.14
E	7.04	3.84	14.07	11.52	43.51

Table 5.2 to obtain the mean flow time through the network for each class of jobs (right-hand column of Table 5.5). ■

▶ *Suggestion: Do Exercise 5.20.*

5.6 EXERCISES

5.1 Cars arrive at a tollbooth 24 hours per day according to a Poisson process with a mean rate of 15 per hour.

(a) What is the expected number of cars that will arrive at the booth between 1:00 P.M. and 1:30 P.M.?

(b) What is the expected length of time between two consecutively arriving cars?

(c) It is now 1:12 P.M. and a car has just arrived. What is the expected number of cars that will arrive between now and 1:30 P.M.?

(d) It is now 1:12 P.M. and a car has just arrived. What is the probability that two more cars will arrive between now and 1:30 P.M.?

(e) It is now 1:12 P.M. and the last car to arrive came at 1:05 P.M.. What is the probability that no additional cars will arrive before 1:30 P.M.?

(f) It is now 1:12 P.M. and the last car to arrive came at 1:05 P.M. What is the expected length of time between the last car to arrive and the next car to arrive?

5.2 A large hotel has placed a single fax machine in an office for customer services. The arrival of customers needing to use the fax follows a Poisson process with a mean rate of eight per hour. The time each person spends using the fax is highly variable and is approximated by an exponential distribution with a mean time of 5 minutes.

(a) What is the probability that the fax office will be empty?

(b) What is the probability that nobody will be waiting to use the fax?

(c) What is the average time that a customer must wait in line to use the fax?

(d) What is the probability that an arriving customer will see two people waiting in line?

5.3 A drill press in a job shop has parts arriving to be drilled according to a Poisson process with mean rate of 15 per hour. The average length of time it takes to complete each part is a random variable with an exponential distribution function whose mean is 3 minutes.

(a) What is the probability that the drill press is busy?

(b) What is the average number of parts waiting to be drilled?

(c) What is the probability that at least one part is waiting to be drilled?

(d) What is the average length of time that a part spends in the drill press room?

(e) It costs the company 8 cents for each minute that each part spends in the drilling room. For an additional expenditure of \$10 per hour, the company can decrease the average length of time for the drilling operation to 2 minutes. Is the additional expenditure worthwhile?

5.4 Derive results for the M/M/2 system using the methodology developed in Section 5.2 (i.e., ignore the general birth–death derivations of Section 5.3). Denote the mean arrival rate by λ and the mean service rate for each server by μ.

(a) Give the generator matrix for the system.

(b) Solve the system of equations given by $\boldsymbol{pG} = \boldsymbol{0}$ by using successive substitution to obtain $p_n = 2\rho^n p_0$ for $n = 1, 2, \cdots$, and

$$p_0 = \frac{1-\rho}{1+\rho}, \quad \text{where } \rho = \frac{\lambda}{2\mu}.$$

(c) Show that $L = 2\rho/(1-\rho^2)$ and $L_q = 2\rho^3/(1-\rho^2)$.

5.5 Derive results for the M/M/3 system using the birth–death equations in Section 5.3. Denote the mean arrival rate by λ, the mean service rate for each server by μ, and the traffic intensity by $\rho = \lambda/3\mu$.

(a) Show that $p_1 = 3\rho p_0$, $p_n = 4.5\rho^n p_0$ for $n = 2, 3, \cdots$, and $p_0 = (1-\rho)/(1 + 2\rho + 1.5\rho^2)$.

(b) Show that $L_q = 9\rho^4/[(1-\rho)(2 + 4\rho + 3\rho^2)]$.

5.6 A small gasoline service station next to an interstate highway is open 24 hours per day and has one pump and room for two other cars. Furthermore, we assume that the assumptions for an M/M/1/3 queueing system are appropriate. The mean arrival rate of cars is eight per hour and the mean service time at the pump is 6 minutes. The expected profit received from each car is \$5. For an extra \$60 per day, the owner of the station can increase the capacity for waiting cars by one (thus, becoming an M/M/1/4 system). Is the extra \$60 worthwhile?

5.7 A repair center within a manufacturing plant is open 24 hours a day and there is always one person present. The arrival of items needing to be fixed at the repair center is according to a Poisson process with a mean rate of six per day. The length of time it takes for the items to be repaired is highly variable and follows an exponential distribution with a mean time of 5 hours. The current management policy is to allow a maximum of three jobs in the repair center. If three jobs are in the center and a fourth job arrives, then the job is sent to an outside contractor who will return the job 24 hours later. For each day that an item is in the repair center, it costs the company \$30. When an item is sent to the outside contractor, it costs the company \$30 for the lost time, plus \$75 for the repair.

(a) It has been suggested that management change the policy to allow four jobs in the center; thus jobs would be sent to the outside contractor only when four are present. Is this a better policy?

(b) What would be the optimum cut-off policy? In other words, at what level would it be best to send the overflow jobs to the outside contractor?

(c) The cost of staffing and maintaining the repair center 24 hours per day is $400 per day. Is that a wise economic policy or would it be better to shut down the repair center and use only the outside contractor?

(d) We assume that parts (a), (b), and (c) were answered using a minimum long-run expected cost criterion. Discuss the appropriateness of other considerations besides the long-run expected cost.

5.8 A small computer store has two clerks to help customers (but infinite capacity to hold customers). Customers arrive at the store according to a Poisson process with a mean rate of five per hour. Fifty percent of the arrivals want to buy hardware and 50% want to buy software. The current policy of the store is that one clerk is designated to handle only software customers, and one clerk is designated to handle only hardware customers; thus, the store actually acts as two independent M/M/1 systems. Whether the customer wants hardware or software, the time spent with one of the store's clerks is exponentially distributed with a mean of 20 minutes. The owner of the store is considering changing the operating policy of the store and having the clerks help with both software and hardware; thus, there would never be a clerk idle when two or more customers are in the store. The disadvantage is that the clerks would be less efficient because they would have to deal with some things with which they are unfamiliar. It is estimated that the change would increase the mean service time to 21 minutes.

(a) If the goal is to minimize the expected waiting time of a customer, which policy is best?

(b) If the goal is to minimize the expected number of customers in the store, which policy is best?

5.9 In a certain manufacturing plant, the final operation is a painting operation. The painting center is always staffed by two workers operating in parallel, although because of the physical setup they cannot help each other. Thus, the painting center acts as an M/M/2 system where arrivals occur according to a Poisson process with a mean arrival rate of 100 per day. Each worker takes an average of 27 minutes to paint each item. Concern has been raised recently about excess work in process, so management is considering two alternatives to reduce the average inventory in the painting center. The first alternative is to expand the painting center and hire a third worker. (The assumption is that the third worker, after a training period, will also average 27 minutes per part.) The second alternative is to install a robot that can paint automatically. However, because of the variability of the parts to be painted, the painting time would still be exponentially distributed, but with the robot the mean time would be 10 minutes per part.

(a) Which alternative reduces the inventory the most?

(b) The cost of inventory (including the part that is being worked on) is estimated to be $0.50 per part per hour. The cost per worker (salary and overhead) is estimated to be \$40,000 per year, and the cost of installing and maintaining a robot is estimated to be \$100,000 per year. Which alternative, if any, is justifiable using a long-term expected cost criterion?

5.10 In the gasoline service station of Exercise 4.6, consider the alternative of adding an extra pump for \$90 per day. In other words, is it worthwhile to convert the M/M/1/3 system to an M/M/2/3 system?

5.11 A parking facility for a shopping center is large enough so that we can consider its capacity infinite. Cars arrive at the parking facility according to a Poisson process with a mean arrival rate of λ. Each car stays in the facility an exponentially distributed length of time, independent from all other cars, with a mean time of $1/\mu$. Thus, the parking facility can be viewed as an M/M/∞ queueing system.

(a) What is the probability that the facility contains n cars?

(b) What is the long-run expected number of cars in the facility?

(c) What is the long-run variance of the number of cars in the facility?

(d) What is the long-run expected queue length for the M/M/∞ system?

(e) What is the mean expected time spent in the system?

5.12 A company offers a correspondence course for students not passing high school algebra. People sign up to take the course according to a Poisson process with a mean of two per week. Students taking the course progress at their own rate, independent of how many other students are also taking the correspondence course. The actual length of time that a student remains in the course is an exponential random variable with a mean of 15 weeks. What is the long-run expected number of students in the course at an arbitrary point in time?

5.13 A company has assigned one worker to be responsible for the repair of a group of five machines. The machines break down independent of each other according to an exponential random variable. The mean length of working time for a machine is 4 days. The time it takes the worker to repair a machine is exponentially distributed with a mean of 2 days.

(a) What is the probability that none of the machines is working?

(b) What is the expected number of machines working?

(c) When a machine fails, what is the expected length of time until it will be working again?

5.14 A power plant operating 24 hours each day has four turbine-generators it uses to generate power. All turbines are identical and are capable of generating 3 megawatts of power. The company needs 6 megawatts of power, so that when all turbines are in working condition, it keeps one turbine on "warm-standby," one turbine on "cold-standby," and two turbines operating. If one turbine is down, then two are operating and one is on warm-standby. If two turbines are down, both working turbines are operating. If only one turbine is working, then the company must purchase 3 megawatts of power from another source. If all turbines are down, the company must purchase 6 megawatts.

If a turbine is in the operating mode, its time until failure is 3 weeks. If a turbine is in warm-standby, its time until failure is 9 weeks. If a turbine is in cold-standby, it cannot fail. (We assume all switchovers from warm standby to working or cold standby to warm standby are instantaneous.) The company has two workers that can serve to repair a failed turbine, and it takes a worker one-half week, on average, to repair a failed turbine. Assuming all times are exponentially distributed, determine the expected megawatt hours that must be purchased each year.

5.15 The simplest type of queueing network is an M/M/1 queueing system with Bernoulli feedback. Model the queueing system illustrated in Figure 5.12 and answer the following questions.

(a) What is the probability that the system is empty?

(b) What is the expected number of customers in the system?

(c) What is the probability that a customer will pass through the server exactly twice?

(d) What is the expected number of times that a customer will pass through the server?

(e) What is the expected amount of time that a customer will spend in the system?

5.16 Consider the queueing network illustrated in Figure 5.13. The first node of the network is an M/M/1 system and the second node of the network is an M/M/2 system. Answer the following questions.

(a) What is the probability that node 1 is empty?

(b) What is the probability that the entire network is empty?

(c) What is the probability that there is exactly one job in node 1's queue?

(d) What is the probability that there is exactly one job within the network?

(e) What is the expected number of jobs within the network?

(f) What is the expected flow time through the system for an arbitrarily selected job? (Answer this question using Property 5.4, then using Little's formula.)

(g) What is the expected flow time through the system for a job that starts at node 1? Can Little's formula be used for this?

(h) A cost of \$100/hr is associated with each job within the system. A proposal has been made that a third node could be added to the system.

FIGURE 5.12 M/M/1 queueing system with Bernoulli feedback.

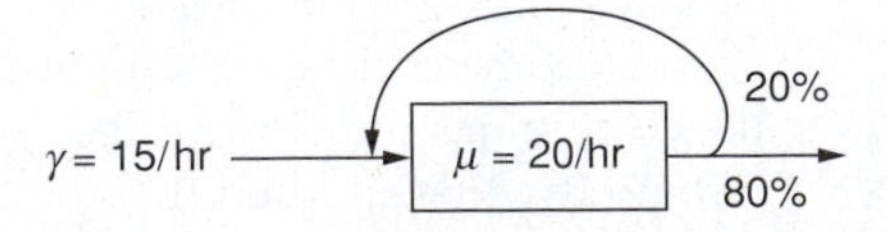

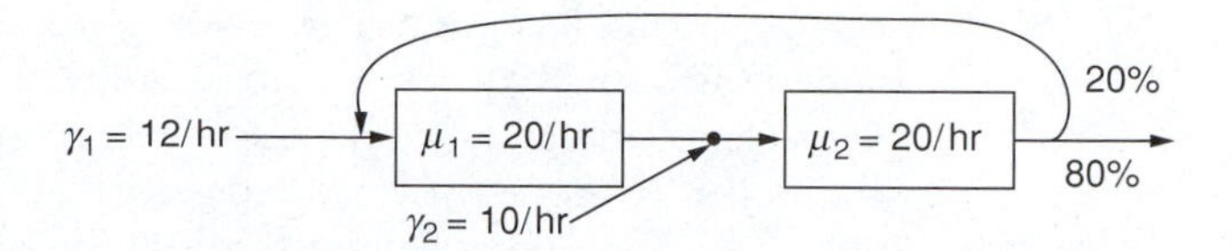

FIGURE 5.13 Two-node queueing system with one feedback loop.

The third node would be an M/M/c system, and with the addition of this node, there would be *no* feedback. Each server added to the third node operates at a rate of 20/hr and costs \$200/hr. Is the third node worthwhile? If so, how many servers should the third node use?

5.17 Customers arrive at a bank according to a Poisson process with a mean rate of 24 customers per hour. Customers go to the receptionist 20% of the time to describe the type of transaction they want to make, 60% of the customers go directly to a teller for common bank transactions, and the rest go directly to see a bank counselor for service. After talking to the receptionist, 75% of the customers are sent to a teller, 20% to a counselor, and 5% leave the bank. After completing service at a teller, 90% of the customers leave the bank and the rest go to see a bank counselor. A counselor can handle all the requests for their customers, so that when a customer is finished with a counselor, the customer leaves the bank.

One receptionist takes an exponential amount of time with each customer averaging 5 minutes per customer. There are three tellers, each with an exponential service time with mean 6 minutes. Two bank counselors serve customers with an exponential service time with an average 12 minutes each. Assume that the Jackson network assumptions are appropriate and respond to the following.

(a) Draw a block diagram to represent the network of the bank service.

(b) Write the switching probability matrix for the Jackson network.

(c) For a customer who desires a teller, what is the mean amount of time spent in line waiting for the teller?

(d) What is the average number of customers in the bank?

(e) Compute the mean flow time through the network; that is, compute the average time that a customer spends inside the bank.

(f) Compute the variance of the flow time through the network.

(g) Bank counselors often have to make telephone calls to potential customers. The bank president wants to know, on the average, the amount of time per hour that at least one counselor is available to make telephone calls.

(h) Answer parts (a) through (d) again with one change in the customer flow dynamics. Assume that after spending time with a counselor, 15% of the customers go to a teller, and 85% leave the bank. (Thus, for example, a small fraction of customers go to a teller, then talk to a counselor, and then return to a teller.)

5.18 A switchboard receptionist receives a number of requests in a day to ask trained personnel questions about different areas of computer software. Clients explain their problem to the receptionist who routes the calls to whomever is the most knowledgeable consultant in that area. There are four consultants representing three different areas of expertise. Two of the consultants (A) can handle questions regarding problems on how to operate the software, one consultant (B) can answer questions related to interaction conflicts with other software packages, and the other consultant (C) can help with software-hardware questions. The receptionist tries to route the calls to the correct person, but the question sometimes cannot be answered and the consultant may suggest another consultant. When a switch is made from one consultant to another, there is an equal probability that the consultant will suggest either of the other two consultants.

Calls arrive at the switchboard according to a Poisson distribution with a mean of 10 calls per hour. It takes the receptionist an exponential amount of time with mean of 1.5 minutes to process a call. The receptionist routes 50% of the calls to consultants A, 25% to consultant B, 15% to consultant C, and 10% of the calls cannot be helped by any consultant. It takes each consultant an exponential amount of time to process a call. The average service time for each consultant A is 15 minutes, 14 minutes for consultant B, and 20 minutes for consultant C. Of the customers routed to each consultant 15% are sent to another consultant, 5% are sent back to the receptionist, and the rest get their answer and go out of the system. Assume that the telephone lines can hold an unlimited number of callers, and that the receptionist treats each call as if it were a new call.

(a) Compute the expected length of time that a customer has to wait before his/her call is answered by the receptionist.

(b) Compute the average number of phone calls in the system.

(c) Compute the expected time a caller spends in the network.

5.19 A store manager with training in queueing theory wants to take quick action on the first day at work. One of the biggest complaints that has been heard is the length of the waiting time and the length of the line. The manager asked one of the employees to record arrival times of customers to the cashier arriving roughly between 8:00 A.M. and noon. The following arrival times were collected: 8:05, 8:07, 8:17, 8:18, 8:19, 8:25, 8:27, 8:32, 8:35, 8:40, 8:45, 8:47, 8:48, 8:48, 9:00, 9:02, 9:14, 9:15, 9:17, 9:23, 9:27, 9:29, 9:35, 9:37, 9:45, 9:55, 10:01, 10:12, 10:15, 10:30, 10:32, 10:39, 10:47, 10:50, 11:05, 11:07, 11:25, 11:27, 11:31, 11:33, 11:43, 11:49, 12:05.

Another employee measured the service times of the cashier. The service times had a mean of 3.5 minutes and a standard deviation of 5.0 minutes. Only one cashier services the customers.

(a) Using the Allen-Cunneen approximations, estimate the expected length of time that a customer has to wait before service and the expected number of customers in front of the cashier.

(b) Suppose that the manager's knowledge of queueing theory was marginal and that an M/M/1 queueing model was wrongly used assuming that the arrival rate was Poisson with the mean estimated from the data and that the service times were exponential with mean 3.5 minutes. Determine the approximate difference between the estimates obtained in part (a) and the estimates that would be obtained using the (incorrect) Markovian assumptions.

(c) The store manager has two alternatives to reduce the waiting time at the cashier. One alternative is to buy a bar reader machine that would reduce the standard deviation of the service time to 1.7 minutes and pay a monthly service fee of $50. The other alternative is to hire a second cashier to work with a currently available cashier machine. The new cashier would work at the same rate as the other and will cost the store $500 per month. Assuming that it costs $0.25 per minute per customer waiting to pay, the manager wants to know what is the best strategy. (Assume that this difficulty is for 4 hours per day, 5 days per week.)

5.20 Consider the problem given in Example 5.12. The manager of the production center has told a consultant that product D could also be made by eliminating the first drilling operation. However, the manager estimates that if this is the case about 3% of the production will be lost to defective production. The cost of scrapping a unit of product D is $50, but the work-in-process (WIP) costs are $1/hr, $1.15/hr, $1.25/hr, $5/hr, and $1/hr for products A, B, C, D and E, respectively. (For example, if we were to follow one item of part C through the manufacturing system and discover that it took 5 hours, then that part would incur a WIP carrying cost of $6.25.)

(a) In the old system (using both drilling operations for product D), what is the size of the WIP by part type?

(b) How does the proposed change to the process of product D affect the length of the production time for the other products?

(c) Is it profitable to modify the process for product D?

5.21 Simulate an M/M/1/3 queueing system with a mean arrival rate and service rate of four per hour. How many customers are turned away during the first 2 hours of operation? (Assume the system starts empty.)

5.22 Simulate a G/G/1/3 queueing system for 2 hours. Let the interarrival times and service times have a continuous uniform distribution between 0 and 30 minutes. How many customers are turned away during the first 2 hours of operation? (Assume the system starts empty.) Compare the answer to this exercise with that of Exercise 5.21.

6

Event-Driven Simulation and Output Analyses

The versatility and popularity of discrete simulation makes simulation modeling a very powerful and, sometimes, dangerous tool. Discrete simulation is powerful because it can be used to model a wide diversity of situations, and it may be dangerous because its relative simplicity can lead to misuse and misinterpretation of results. However, with good training in the use of simulation and careful statistical analysis, discrete simulation can be used for great benefit.

In this chapter, we build on the basics that were introduced in Chapter 3. A key feature that was postponed in Chapter 3 is the requirement that we deal with the timing of various concurrent activities. We introduce the major concepts through the use of simple examples that can be generated by hand.

6.1 EVENT-DRIVEN SIMULATIONS

When the timing of activities is a component of a system being simulated, a method is needed to maintain and increment time in a dynamic fashion. For example, suppose we wish to simulate a queueing system with five servers. The five-server system, when all servers are full, would have six different activities (five service completions and one arrival) to maintain. To handle the many activities in a complex system correctly and efficiently, a simulation clock is maintained and updated as events occur. In other words, if we are simulating what happens within a manufacturing plant during a 10-hour day, the simulation study might take 5 seconds of real time to execute, but the simulation clock will show an elapsed time of 10 hours.

The movement of the simulation clock is extremely critical, and there are two common methods for updating time: (1) The simulation clock could be updated by a fixed amount at each increment or (2) the clock could be updated whenever something happened to change the status of the system. The advantage of the first method is that it is conceptually simple; the disadvantage is that if the increment is too small then there are inefficiencies in "wasted" calculations, and if the increment is too large then there are mathematical difficulties in representing everything that happens during the increment. Therefore, most simulation procedures are based on the second method of time management, called *event-driven simulations*. In other words, the simulation clock is updated whenever an event occurs that might change the system. As we illustrate this concept with the following examples, observe that the simulation-by-hand is simply building tables of information. If our purpose were to develop a technique for simulation-by-hand, the following procedure would contain some clearly redundant steps; however, our purpose is to illustrate some of the basic steps that are necessary for simulating complex systems. Therefore, we suggest that you follow the basic steps outlined next even though shortcuts can be easily identified.

EXAMPLE 6.1 Consider a simplified manufacturing process (illustrated in Figure 6.1) where widgets are processed through a production center. One widget arrives at the facility every 12 minutes; however, because of poor quality control, 25% of these widgets are defective and must be scrapped. The remaining 75% of the widgets are sent to a workstation where they are processed one at a time through a single machine. The chemical composition of the widgets varies; thus, the processing time required on the machine varies. Twenty-five percent of the widgets require 17 minutes of machining time, 50% of the widgets require 16 minutes, 12.5% of the widgets require 15 minutes, and 12.5% require 14 minutes. While a widget is on the machine, other widgets that arrive at the workstation are placed in a special holding area until

FIGURE 6.1 Schematic of the manufacturing process of Example 6.1.

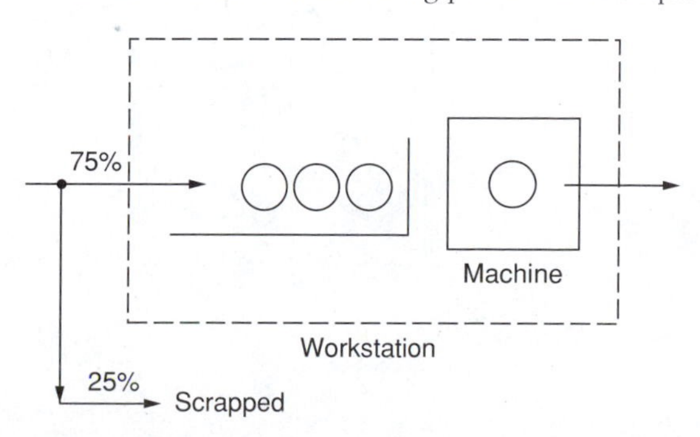

the widget being processed is completed and removed. The next widget in line is then placed on the machine and the process is repeated. Widgets in the holding area are said to be in a queue (waiting line) waiting for the machine to become available. To characterize this process in traditional queueing terminology, we would say that we are dealing with a single-server queueing system (the machining process operates on widgets one at a time) having a deterministic arrival process, random service times, and utilizing a first-in/first-out selection scheme (FIFO queueing discipline).

We wish to simulate this production center in order to answer the question, "What is the long-run average number of widgets in the facility?" For this system, two random variables must be simulated (see Table 6.1). The status of arriving widgets (defective or nondefective) will be decided by tossing two fair coins: two tails indicate scrap (25% of the outcomes) and any other combination indicates a good widget. (In Chapter 3, we discussed reproducing randomness with computer-generated random variates; however, for purposes of the examples in this chapter, we return to simulation-by-hand.) The second random variable, namely the machining time, will be decided by tossing three fair coins: three tails (with a chance of 1 in 8 or 12.5%) will indicate 14 minutes, three heads (again, 1 chance in 8) will indicate 15 minutes, the combinations THT, TTH, HTT, and THH (with a chance of 4 in 8 or 50%) will indicate 16 minutes, and HHT and HTH (with a chance of 2 in 8 or 25%) will indicate 17 minutes of machining time.

Our simulation is actually a description of the processing of widgets within the manufacturing system. The term "entity" is used in simulations to denote the generic objects that "move" through a system and are operated on by the system. Hence, a widget is an example of an entity. If a simulation were designed for the movement of people within a bank, the people would be called entities. If a simulation were designed to model packets of bits moving within a communication network, the packets would be called entities. Thus, a simulation deals with entities and events (events are things that happen that change the state of the system). In the widget

TABLE 6.1. Procedure for determining the quality of arriving widgets and the machining time for those widgets that are actually processed.

Random Value	Simulated Outcome
T-T	Scrap
T-H	Good
H-T	Good
H-H	Good

Random Value	Simulated Value
T-T-T	14 minutes
H-H-H	15 minutes
T-H-T	16 minutes
T-T-H	16 minutes
H-T-T	16 minutes
T-H-H	16 minutes
H-H-T	17 minutes
H-T-H	17 minutes

example, there are two types of events: arrivals and service completions. (Departures from the system are also events, but in our example, departures need not be identified separately because they occur at the same time as service completions.) Events are important because it is events that move the simulation clock. As the simulation progresses, at least two tables must be maintained: a table describing the state of the manufacturing process at each point in time and a table keeping track of all scheduled future events. It is the table of future events (called the "Future Events" list) that will drive the simulation.

A simulation of a large system involves many different types of lists. In this chapter, our goal is to illustrate only a few of the basic types of lists. For example, an understanding of the basic mechanics of the Future Events list is essential for a well-designed simulation model. In the next example, some additional lists and concepts will be discussed.

A simulation begins by initializing the Future Events list. For this problem, we assume that an entity (i.e., widget) arrives at the system at time 0, so the Future Events list is initialized as shown in Table 6.2. The first column in the Future Events list indicates the time that the entity is to be removed from the list and added to the system, thus moving the simulation clock to the indicated time. (In this case, we move the clock from no time to time equal to zero.) The second column gives the entity number (entities are numbered sequentially as they are created), and the third column indicates the type of event that is the cause for the update of the clock. In the widget example, there are two types of events: arrivals (*a*) and service completions (*s*).

With the initialization of the Future Events list, we are ready to begin the simulation. Conceptually, we remove the entity from the Future Events list and update the simulation clock accordingly. The table describing the state of the system is then updated as shown in Table 6.3. Column 1 designates the clock time as determined from the Future Events list. If the event causing the advance of the simulation clock is an arrival, the first "order of business" is to determine the next arrival time. As soon as this next arrival time is determined, the entity that is scheduled to arrive at that time is placed in the Future Events list. In this example, there are no calculations for determining the next arrival time, but the general principle is important: As soon as an arrival occurs, the next arrival time is immediately determined and a new entity is placed on the Future Events list. After determining the next arrival, we return to describing everything that happens at the current time. Two coins are tossed to determine if the arriving widget will be scrapped or

TABLE 6.2 Future Events list: initialization.

Future Events List		
Time	*Entity no.*	*Type*
0	1	*a*

processed. Since the coin toss results in two tails, the widget is bad and is not accepted for processing (columns 4 through 6). Since the entity does not stay in the system, the description of the system at time 0 is as depicted in Table 6.3; thus, we are ready to update the clock. We look to the Future Events list (Table 6.4) in order to increment the clock. (A description of columns 8 and 9 is delayed until more data are generated so that the explanation will be easier.)

Notice that the first item on the Future Events list (as of time 0) is crossed off, indicating that the entity was removed (to start the simulation). We now remove the next entity from the list in order to update (i.e., advance the simulation clock) the table (Table 6.5) giving the state of the system. The entity is added to Table 6.5 representing an arriving widget. As always, as soon as an arrival occurs, the next arrival time is determined (row 2, column 3) and the new entity is placed on the Future Events list (entity 3 in Table 6.6). Next, we toss two coins to determine the acceptability of the arriving entity. Since a tail-head occurred and the widget is considered good, it enters the production facility (row 2, column 6) and the machining process begins. It is now necessary to determine how long the widget will be on the machine. The result of tossing three coins (row 2, column 7) is a 16-minute machining process. Since the current clock is 12, the 16 minutes are added to the current time and the widget is scheduled to finish the machining process and leave the facility at time $12 + 16 = 28$. Since this is a future event (with respect to a clock time of 12), it is placed on the Future Events list (Table 6.6). All of the activities that occur at time 12 have been completed so it is time to return to the Future Events list (Table 6.6) to increment the simulation clock. Note that during time 12, two events were placed on the Future Events list: the arrival to occur at time 24 and the service completion that will occur at time 28.

The next (future) event within the Future Events list, as the list stands after time 12, is the arrival of entity 3, which occurs at time 24; thus, entity 3 is removed from the Future Events list and placed in the table giving the current state of the system (Table 6.7). Before processing entity 3, the next arrival event is scheduled yielding entity 4, which is immediately placed on the Future Events list. Returning to entity 3, we see that it is considered acceptable for machining after the two-coin toss; however, this time the machine is busy, so the widget (entity 3) must wait until the processing of entity 2 is finished. Note that the number in the system (row 3, column 6) is updated to 2, but a service time is not determined until the widget is actually placed on the machine. In other words, although entity 3 is waiting in a queue for the server to become available, we do not yet know when the entity will leave the system, so it is not yet placed on the Future Events list.

The last column in the table is used to keep track of the (time-weighted) average number of widgets in the system. The time-averaged number of widgets is determined by summing the number of widgets within the facility multiplied by the length of time each widget spends in the facility and then dividing by the total length of time. The last column of the final table (Table 6.11, shown later) accumulates this sum, which will be divided at the

TABLE 6.3 State of the system after time 0.

(1) Clock Time	(2) Type of Event	(3) Time Next Arrival	(4) Two-coin Toss	(5) Good or Bad Item	(6) Number in System	(7) Three-coin Toss	(8) Time Service Complete	(9) Sum of Time × Number
0	*a*	12	TT	bad	0	—	—	0

TABLE 6.4 Future Events list: after time 0.

Future Events List		
Time	*Entity no.*	*Type*
~~0~~	~~1~~	~~*a*~~
12	2	*a*

TABLE 6.5 State of the system after time 12.

(1) Clock Time	(2) Type of Event	(3) Time Next Arrival	(4) Two-coin Toss	(5) Good or Bad Item	(6) Number in System	(7) Three-coin Toss	(8) Time Service Complete	(9) Sum of Time × Number
0	*a*	12	TT	bad	0	—	—	0
12	*a*	24	TH	good	1	THT	28	0

TABLE 6.6 Future Events list: after time 12.

Future Events List		
Time	*Entity no.*	*Type*
~~0~~	~~1~~	~~*a*~~
~~12~~	~~2~~	~~*a*~~
24	3	*a*
28	2	*s*

TABLE 6.7 State of the system after time 24.

(1) Clock Time	(2) Type of Event	(3) Time Next Arrival	(4) Two-coin Toss	(5) Good or Bad Item	(6) Number in System	(7) Three-coin Toss	(8) Time Service Complete	(9) Sum of Time × Number
0	*a*	12	TT	bad	0	—	—	0
12	*a*	24	TH	good	1	THT	28	0
24	*a*	36	HH	good	2	—	—	12

end of the simulation by the total elapsed time to obtain the average number. At time 24, one item has been in the system for 12 minutes, so the last column contains a 12. To help understand this averaging process, consider Figure 6.2, which represents a graph of the number in the system versus time for the first 106 minutes. An estimate for the average number of entities in the system is calculated by taking the area under that graph and dividing it by the total time of the graph. Therefore, if we wish to determine the average number in the system, the main quantity that must be maintained is the area under the graph. Each time the clock advances, a new rectangle of area is added to the total area, and the accumulation of the area of these rectangles is maintained in the last column. Note that the accumulation of area is accomplished by looking *backwards*. Thus, the 12 in row 3, column 9, is the area under the curve in Figure 6.2 from time 0 through time 24 (or equivalently, the area of rectangle A in the figure).

To continue the simulation, we again look for the time that the next change in the system will occur (Table 6.8). Thus, the clock is advanced to time 28. The event causing the advance of the clock is a service completion, and thus the widget being processed leaves the system and the widget in the queue (entity 3) is placed on the machine. Three coins are tossed to determine the processing time, and since they come up head-head-head (row 4, column 7, Table 6.9), we assign 15 minutes of machining time to this widget. The 15 minutes of machining time are added to the current clock time of 28 minutes, and we record in column 8 (and in the Future Events list, Table 6.10) that the widget will leave the facility at time 43. To obtain the number in the last column (row 4, column 9, Table 6.9), we need to multiply the time increment by the number in the system (4×2) and add it to the previous total. Or equivalently, we calculate the area of the rectangle B in Figure 6.2 and add it to the previous area, yielding a current total area of 20.

As a summary, Table 6.11 shows the updating of the state of the system for the first 106 minutes. As illustrated, the process is to describe completely the state of the system at a particular time. During that description, one or more future events will become known, and those will be placed on the Future Events list. After completely describing the state of the system, the next event on the Future Events list is removed and the simulation clock is updated accordingly.

The final sum in the last column equals 121, so the estimate for the average number of widgets in the facility is $121/106 = 1.14$. To understand this average, assume that the system is observed at 100 random points in time. Each time the system is observed, the number of entities in the system is recorded. After the 100 observations are finished, we take the average of the resulting 100 numbers. This average should be close to 1.14. Actually, the 1.14 will have very little significance because the simulation must be run for a long time before the estimate is good. Such an estimate is highly variable unless it involves a large number of entities. This again emphasizes the fact that a simulation is a statistical experiment, and the final numbers are only estimates of the desired information, not actual values. ■

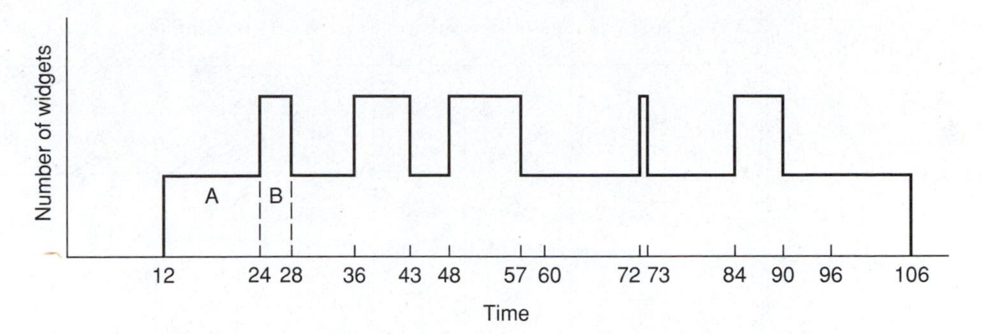

FIGURE 6.2 Plot of the data from Table 6.11. The average number of entities in the system equals the area under the curve divided by the length (time) of the curve.

TABLE 6.8 Future Events list: after time 24.

Future Events List		
Time	*Entity no.*	*Type*
~~0~~	~~1~~	~~*a*~~
~~12~~	~~2~~	~~*a*~~
~~24~~	~~3~~	~~*a*~~
28	2	*s*
36	4	*a*

TABLE 6.9 State of the system after time 28.

(1) Clock Time	(2) Type of Event	(3) Time Next Arrival	(4) Two-coin Toss	(5) Good or Bad Item	(6) Number in System	(7) Three-coin Toss	(8) Time Service Complete	(9) Sum of Time × Number
0	*a*	12	TT	bad	0	—	—	0
12	*a*	24	TH	good	1	THT	28	0
24	*a*	36	HH	good	2	—	—	12
28	*s*	—	—	—	1	HHH	43	20

TABLE 6.10 Future Events list: after time 28.

Future Events List		
Time	*Entity no.*	*Type*
~~0~~	~~1~~	~~*a*~~
~~12~~	~~2~~	~~*a*~~
~~24~~	~~3~~	~~*a*~~
~~28~~	~~2~~	~~*s*~~
36	4	*a*
43	3	*s*

TABLE 6.11 Simulated production process for 106 minutes.

(1) CLOCK TIME	(2) TYPE OF EVENT	(3) TIME NEXT ARRIVAL	(4) TWO-COIN TOSS	(5) GOOD OR BAD ITEM	(6) NUMBER IN SYSTEM	(7) THREE-COIN TOSS	(8) TIME SERVICE COMPLETE	(9) SUM OF TIME × NUMBER
0	*a*	12	TT	bad	0	—	—	0
12	*a*	24	TH	good	1	THT	28	0
24	*a*	36	HH	good	2	—	—	12
28	*s*	—	—	—	1	HHH	43	20
36	*a*	48	HT	good	2	—	—	28
43	*s*	—	—	—	1	TTT	57	42
48	*a*	60	HH	good	2	—	—	47
57	*s*	—	—	—	1	THH	73	65
60	*a*	72	TT	bad	1	—	—	68
72	*a*	84	HT	good	2	—	—	80
73	*s*	—	—	—	1	HTH	90	82
84	*a*	96	TH	good	2	—	—	93
90	*s*	—	—	—	1	HTT	106	105
96	*a*	108	TT	bad	1	—	—	111
106	*s*	—	—	—	0	—	—	121

a denotes an arrival event; *s* denotes a service completion event.

EXAMPLE 6.2 Consider a bank, illustrated in Figure 6.3, having two drive-in windows with one queue. (That is, waiting cars form one line, and the car at the head of the line will proceed to whichever window first becomes available.) Two types of customers come to the bank: customers with personal accounts and customers with commercial accounts. The time between personal-account customers coming to the bank for drive-in window service is 3 minutes with probability 0.5 or 5 minutes with probability 0.5. The length of time each car spends in front of the drive-in window varies and is 7 or 9 minutes with a 0.50 probability for each possibility. The time between arrivals of commercial-account customers is 8 minutes with probability 0.5 and 11 minutes with probability 0.5. The length of time that a commercial customer spends in front of the drive-in window is either 10 or 12 minutes with a

FIGURE 6.3 Schematic of the bank with two drive-in windows and two types of customers for Example 6.2.

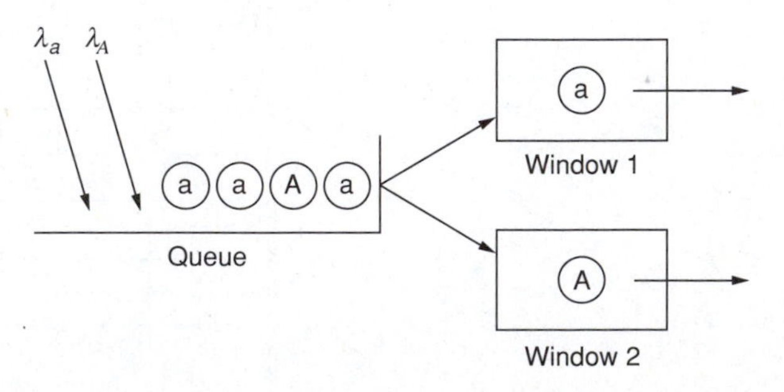

0.50 probability for each possibility. We would like to simulate this process with the purpose of determining the average length of time that a customer spends in the system. We shall again use a coin toss to reproduce the relevant random events, letting a tail represent the smaller number and a head the larger number.

There are two complications in simulating this system when we compare it to the previous example. First, there are two arrival processes: one arrival stream of entities representing personal-account customers, and the other arrival stream of entities representing commercial-account customers. Second, in order to collect data on the time each entity spends in the system, we need to mark *on the entity* the time at which that entity enters the system. When the entity leaves the system, we will subtract the entry time stored on the entity from the current clock time, thus obtaining the total time the entity spent in the system.

As before, we start by initializing the Future Events list; that is, each arrival stream generates its first arriving entity and places the entity on the list (Table 6.12). In this example, lowercase letters represent personal-account customers, and uppercase letters represent commercial customers.

The simulation begins by taking the first entity off the Future Events list and recording the state of the system at the indicated time. In case of a tie (two or more events that are scheduled to come off the Future Events list at the same time), it does not matter which entity is chosen.

In the table describing the state of the system (Table 6.13), column 2 contains not only the type of event causing the increment to the simulation clock, but also the entity number as a subscript to help keep track of the entities. After recording that the entity has entered the system (column 5), the (simulation clock) time at which the entity entered the system is recorded in an *attribute*.

The concept of an attribute is very important for simulations. An attribute is a local variable unique to each entity. For example, in some circumstances, we may want entities to represent people and record a random weight and height for each entity. In such a case, we would define two attributes in which the values of the weight and height random variables would be stored. In our drive-in window banking example, we want to record the time of entry for each entity. To do this, we define an attribute and as soon as the entity enters the system, we record the value of the simulation clock in that attribute. Now, in addition to the two previously defined lists, we add a third one called an Entities list to hold the value of the attribute for each entity.

TABLE 6.12 Future Events list: initialization.

Future Events List		
Time	*Entity no.*	*Type*
0	1	*a*
0	2	*A*

TABLE 6.13 State of the system after first entity processed.

(1) Clock time	(2) Type of event	(3) Coin toss	(4) Time next arrival	(5) Number in system	(6) Coin toss	(7) Time service complete	(8) Sum of number departs	(9) Sum of times in system
0	a_1	T	3	1	H	9	0	0

The Entities list (Table 6.14) and Future Events list (Table 6.15) are shown as they would appear after processing the first entity.

We continue the simulation by processing entities one-at-a-time in a similar manner as in the previous section. The following tables are written after simulating the arriving and processing of customers for 16 minutes. As you look at Tables 6.16, 6.17, and 6.18, there are several important observations to make. In the processing of the first two entities (rows 1 and 2, Table 6.16), random service times are generated because there are two drive-in windows, so both arriving entities immediately proceed to the window. However, there is no service time generated in the third row, because that entity enters a queue; thus, its service will not be known until after a service completion occurs. At time 9, entity 1 departs, and the queued entity enters the server (i.e., gets in front of the drive-in window). Thus, a service time is generated in row five. The next service completion occurs at time 12. The departing entity (i.e., entity 2) represents a commercial-account customer, but the random service time generated while the clock is at time 12 is for a personal-account customer, because it is entity 5 that is next in line. Thus, it is important to keep track of which entity came first to a queue. If there are several different types of entities within a simulation, it can become complicated to keep track of which type of entity will next perform a particular action. This is an advantage of special simulation languages (as compared to a language like Pascal or C); the languages designed specifically for simulation take care of much of the bookkeeping automatically.

At time 9, when entity 1 departs, we look at its attribute (in the Entities list table) to determine when it arrived so that the total waiting time can be determined. This also occurs at time 12 and time 16. For example, consider time 16. It is entity 3 that is departing, so when it departs we observe from Table 6.17 that entity 3 arrived at time 3, and since the current time is 16,

TABLE 6.14 Entities list: after first entity processed.

Entities List	
Entity Number	*Value of Attribute*
1	0

TABLE 6.15 Future Events list: after first entity processed.

Future Events List		
Time	*Entity no.*	*Type*
~~0~~	~~1~~	~~*a*~~
0	2	*A*
3	3	*a*
9	1	*s*

TABLE 6.16 State of the system after time 16.

(1) Clock Time	(2) Type of Event	(3) Coin Toss	(4) Time Next Arrival	(5) Number in System	(6) Coin Toss	(7) Time Service Complete	(8) Sum of Number Departs	(9) Sum of Times in System
0	a_1	T	3	1	H	9	0	0
0	A_2	H	11	2	H	12	0	0
3	a_3	H	8	3	—	—	0	0
8	a_5	T	11	4	—	—	0	0
9	s_1	—	—	3	T	16	1	9
11	A_4	H	22	4	—	—	1	9
11	a_6	H	16	5	—	—	1	9
12	S_2	—	—	4	H	21	2	21
16	s_3	—	—	3	T	26	3	34
16	a_8	H	21	4	—	—	3	34

the total time in the system is 13. The value of 13 is added (column 9) to the previous sum of 21 yielding a new total of 34. If the simulation were to end after time 16, our estimate for the average time in the system would be the ratio of column 9 to column 8, or $34/3 = 11.333$ minutes.

Notice that the Future Events list in Table 6.18 is arranged in the order in which the events were generated. To determine which event to choose next, we search all the future events and select the one with the minimum time. In a computer, the computer automatically deletes those entities that we have crossed out, and whenever an entity is placed in the list, the list is sorted so that the first entity always has the minimum time. ■

▶ *Suggestion: Do Exercises 6.1–6.6.*

TABLE 6.17 Entities list: after time 16.

Entities List	
Entity Number	*Value of Attribute*
1	0
2	0
3	3
4	11
5	8
6	11
8	16

TABLE 6.18 Future Events list: after time 16.

Future Events List		
Time	*Entity No.*	*Type*
~~0~~	~~1~~	~~*a*~~
~~0~~	~~2~~	~~*A*~~
~~3~~	~~3~~	~~*a*~~
~~9~~	~~1~~	~~*s*~~
~~11~~	~~4~~	~~*A*~~
~~8~~	~~5~~	~~*a*~~
~~11~~	~~6~~	~~*a*~~
~~12~~	~~2~~	~~*S*~~
~~16~~	~~3~~	~~*s*~~
22	7	*A*
~~16~~	~~8~~	~~*a*~~
21	5	*s*
26	4	*S*
21	7	*a*

6.2 STATISTICAL ANALYSIS OF OUTPUT

Simulation modeling is a very powerful tool for analysis that can be learned by most technically inclined individuals and, thus, its use has become widespread. The typical approach is to learn the basic simulation concepts and a simulation language, but pay very little attention to the proper analysis of the output. The result is that many simulation studies are performed improperly. Simulation programs do not serve much purpose without a proper interpretation of the output. In fact, an improper understanding of the results of a simulation can be detrimental in that the analysis could lead to an incorrect decision. Therefore, now that the basic concepts of simulation have been introduced, it is necessary for us to discuss how to analyze the output.

Output analysis of simulation is often given only a cursory review because it is a more difficult process than simple programming. Furthermore, there remain unanswered questions in the area of output analysis. In this section, we discuss enough of the concepts to equip the analyst with the minimum set of tools. We caution that there is much more to this area than we are able to cover at the level of this introductory text. Many of the concepts presented here are based on the text by Law and Kelton,[1] which we recommend for further reading.

Simulation output can be divided into two categories: *terminating simulations* and *steady-state simulations*. A terminating simulation is one in which there is a fixed time reference of interest. An example of a terminating simulation is Exercise 6.6, which deals with a store that only operates for 4 hours during the day; thus the simulation terminates after 4 hours of simulated clock time. Steady-state simulations do not have a fixed length of time; they deal with questions of long-run behavior. If we assume that the widgets from Example 6.1 are produced 24 hours a day seven days a week, then questions dealing with the long-run behavior of the process would be of interest. Output analyses of terminating and steady-state simulations are considered separately.

6.2.1 Terminating Simulations

Consider a small service station with only one gasoline pump located along a major road leading into a large metropolitan area. It is open from 6:00 A.M. to 10:00 P.M. every day, with a rush hour occurring from 4:00 P.M. to 6:00 P.M. (Access to the station is difficult for people going into the city; therefore, a morning rush hour does not cause congestion at the gasoline pump.) Because of space limitations, there is room for only four customers in the station; in other words, if one customer is at the pump, a maximum

[1]A. M. Law and W. D. Kelton, *Simulation Modeling and Analysis* (New York: McGraw-Hill Book Company, 1982).

of three vehicles can be waiting. The owner of the station is concerned that too many customers are lost because people find the station full; therefore, a simulation study is desired to determine how many customers are lost due to the limited capacity. If the number lost is significant, the owner will consider purchasing a second pump. All interarrival times are observed to be exponentially distributed random variables, with an average of 6 arrivals per hour from 6:00 A.M. to 4:00 P.M., 20 per hour from 4:00 P.M. to 6:00 P.M., and 6 per hour from 6:00 P.M. to 10:00 P.M. The time each customer spends in front of the pump is also exponentially distributed, with an average time of 4 minutes. (In other words, we are interested in an M/M/1/4 system, except that the input is a nonhomogenous Poisson process, and we are interested in transient results instead of steady-state results. However, it is possible to use the material from the previous chapter to obtain a quick-and-dirty estimate for the mean number turned away. The M/M/1/4 model with $\lambda = 20$ and $\mu = 15$ yields an overflow rate of 6.6 customers per hour. The difficulty is that steady state is never reached, but as an estimate, this would yield an overflow of 13.2 customers during the daily rush hour.)

This service station example illustrates a situation suitable for a terminating simulation study. (A steady-state analysis does not make sense, since the process is constantly being "restarted" every morning.) The first step in determining the feasibility of obtaining a second pump is to simulate the above situation to estimate the expected number of daily customers lost under the current configuration of one pump. To illustrate this, we simulated the service station and executed the program yielding the number of lost customers in a day as 25, 10, and 12, respectively, for three different runs. With variations of this magnitude, it is clear that care must be taken in estimating the mean number of customers lost. Furthermore, after an estimate is made, it is important to know the degree of confidence that we can place in the estimate. Therefore, instead of using a single value for an estimate, it is more appropriate to give a *confidence interval*. For example, a 95% confidence interval for the expected number of customers lost each day is an interval that we can say with 95% confidence contains the true mean. That is, we expect to be wrong only 5% of the time. (The choice of using 95% confidence instead of 97.2% or 88.3% confidence is, for the most part, one of historical precedence.)

To illustrate the building of confidence intervals, the simulation of the one-pump service station was executed 30 times and the daily total number of lost customers was recorded in Table 6.19. The histogram of the results is shown in Figure 6.4. (It might appear to be redundant to plot the histogram since the table gives all the information; however, we recommend that you make a practice of plotting your data in some fashion because quantitative information cannot replace a visual impression for producing an intuitive feel for the simulated process.)

Confidence Intervals for Means. The usual techniques for constructing confidence intervals assume that the data are observations of independent,

TABLE 6.19 Results from 30 simulations of the one-pump service station.

25	10	12	16	31	8
14	21	7	6	21	6
7	13	10	13	0	21
12	9	13	13	16	9
14	23	4	12	8	13

normally distributed random variables. Under such an assumption, the $1-\alpha$ confidence interval for a mean value is given by

$$\left[\overline{X} - t_{(n-1,\frac{\alpha}{2})}\frac{S}{\sqrt{n}}, \quad \overline{X} + t_{(n-1,\frac{\alpha}{2})}\frac{S}{\sqrt{n}}\right], \tag{6.1}$$

where n is the number of data points, $t_{(n-1,\frac{\alpha}{2})}$ is a critical value (see Table C.3) based on the Student-t probability distribution, and $\overline{X}$ and S are the sample mean and standard deviation using the n data points given by

$$\begin{aligned} \overline{X} &= \frac{X_1 + X_2 + \cdots + X_n}{n}, \\ S^2 &= \frac{(X_1^2 + X_2^2 + \cdots + X_n^2) - n\overline{X}^2}{n-1}. \end{aligned} \tag{6.2}$$

The $n-1$ subscript in the t value refers to the degrees of freedom, and $\alpha/2$ refers to the amount of "right-hand" error we are willing to risk. To

FIGURE 6.4 Histogram from 30 simulations of the one-pump service station using intervals of five units.

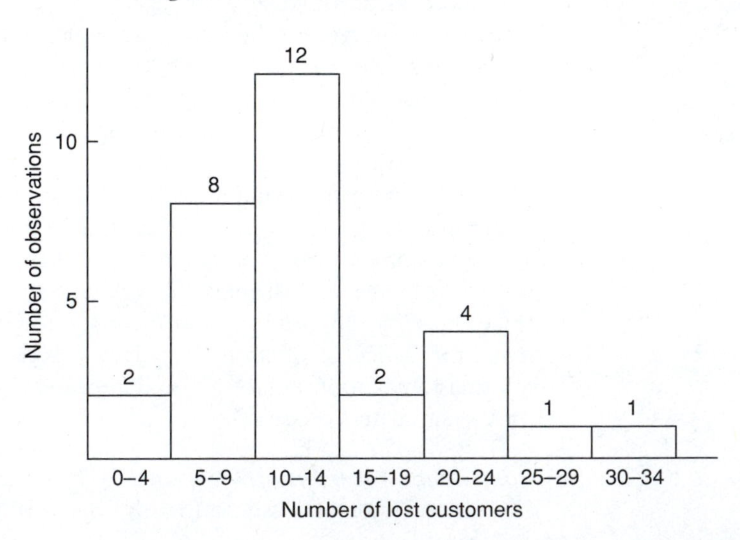

understand why the $1 - \alpha$ confidence interval uses a critical value associated with $\alpha/2$, consider a 95% confidence interval; such a confidence interval has an associated risk of 5% error. Since the error is divided evenly on both sides of the interval, there is a 2.5% chance of error to the left of the interval and a 2.5% chance of error to the right of the interval.

EXAMPLE 6.3 The sample mean and standard deviation for the results obtained from the 30 simulations (Table 6.19) are $\bar{x} = 12.9$ and $s = 6.7$, respectively. (Notice the use of the lowercase letters. The uppercase letters indicate a random variable; after an observation is made, a lowercase letter must be used indicating a particular value.) The critical t value for a 95% confidence interval with 29 degrees of freedom is 2.045. (Observe that the Student-t distribution limits to the normal distribution as the degrees of freedom increase.) Equation (6.1) thus yields

$$12.9 \pm 2.045 \times \frac{6.702}{\sqrt{30}} = (10.4, 15.4).$$

That is, we are 95% confident that the actual expected number of customers lost each day is between 10.4 and 15.4. ■

We need to consider the two assumptions under which Eq. (6.1) theoretically holds. The assumption of independent observations is adequately satisfied as long as each replicate execution of the simulation program begins with a different initial random number seed. (Chapter 3 discusses the dependence of random variate generation on the initial number seed.) The normality assumption is troublesome and can cause errors in the confidence interval. This is especially true for the service station example, since the histogram of Figure 6.4 does not give the appearance of a symmetric bell-shaped curve. However, regardless of the actual underlying distribution (assuming a finite mean and variance), we can take advantage of the *central limit theorem* and assume normally distributed random variables when the sample size is large. The central limit theorem states that sample means, when properly normed, become normally distributed as the sample size increases. A commonly used rule of thumb is to have a sample size of at least 25 to 30 individual observations.

Confidence Intervals for Variances. Two common measures estimated with a simulation are mean values and variances. Equation (6.1) gives the confidence interval used for means. The $1 - \alpha$ confidence interval for the variance is

$$\left[\frac{(n-1)S^2}{\chi^2_{(n-1,\frac{\alpha}{2})}}, \frac{(n-1)S^2}{\chi^2_{(n-1,1-\frac{\alpha}{2})}}\right], \qquad (6.3)$$

where $\chi^2_{n,\alpha}$ is the critical chi-square value (see Table C.4) using n degrees of freedom with a "right-hand" error of α. Notice that the larger χ^2 value is used in the left-hand limit of the interval, whereas the smaller value is used in the right-hand limit.

For a large sample, a confidence interval for the standard deviation is obtained by a normal approximation to the chi-square distribution. Let $z_{\alpha/2}$ represent the critical value from the standardized normal table (either Table C.2 or the last line from Table C.3) with a right-hand error of $\alpha/2$. Then the large sample $1-\alpha$ confidence interval for the standard deviation is

$$\left(\frac{s}{1+\left(z_{\alpha/2}/\sqrt{2n}\right)},\quad \frac{s}{1-\left(z_{\alpha/2}/\sqrt{2n}\right)}\right). \tag{6.4}$$

EXAMPLE 6.4 We again use the data of Table 6.19 to illustrate 95% confidence intervals for the variance and standard deviation. The sample variance was $s^2 = 44.89$ and the two critical values from Table C.4 are $\chi^2_{(29,0.975)} = 16.05$ and $\chi^2_{(29,0.025)} = 45.72$. Therefore, the 95% confidence interval for the variance is

$$\left(\frac{29\times 44.92}{45.72},\quad \frac{29\times 44.92}{16.05}\right) = (28.49, 81.16).$$

Many practitioners will use "large sample size" statistics for samples of size 25 to 30; therefore, we can use the data of Table 6.19 to illustrate confidence intervals for standard deviations. The critical z value (from Table C.2 or the last line of Table C.3) is 1.960, which yields a 95% confidence interval of

$$\left(\frac{6.702}{1+\frac{1.960}{\sqrt{60}}},\quad \frac{6.702}{1-\frac{1.960}{\sqrt{60}}}\right) = (5.35, 8.97).$$

Confidence Intervals for Proportions. It is sometimes important to obtain information about the tails of the distribution. For example, the average number of customers turned away may not be as important as knowing the probability that 20 or more customers per day will be turned away. (In other words, we might want to minimize the probability of "bad" things happening, instead of concentrating on expected values.) A major problem with tail probabilities is that the sample size must be larger to estimate tail probabilities than the sample size needed to estimate means. (A rule of thumb is to have more than 5 occurrences of the event for which the probability statement is being made. Thus, the 30 simulations contained in Table 6.19 are not enough to estimate the probability that 25 or more customers per day are turned away.) Suppose we are interested in the probability that some event will occur. Let M_n denote the number of times the event occurs from a sample of size n. The estimate of the probability is then given by $\hat{P} = M_n/n$ and the $1-\alpha$ confidence interval for the probability (assuming n is large and $M_n > 5$) is

$$\left(\hat{P} - z_{\alpha/2}\sqrt{\frac{\hat{P}(1-\hat{P})}{n}},\quad \hat{P} + z_{\alpha/2}\sqrt{\frac{\hat{P}(1-\hat{P})}{n}} \right). \tag{6.5}$$

EXAMPLE 6.5 We are interested in estimating the probability that 20 or more customers will be turned away on any given day from the service station. The data of Table 6.19 indicate that 6 out of 30 runs resulted in 20 or more customers being turned away; therefore, our estimate for the probability is $p = 6/30 = 0.2$ and thus the 95% confidence interval is

$$0.2 \pm 1.960 \times \sqrt{\frac{0.16}{30}} = (0.06, 0.34).$$

▶ *Suggestion: Do Exercises 6.7, 6.9 and 6.16a.*

6.2.2 Steady-State Simulations

A large manufacturing company has hundreds of a certain type of machine that it uses in its manufacturing process. These machines are critical to the manufacturing process, so the company maintains a repair center that is open 24 hours a day, 7 days a week. At any time, day or night, there is one technician who takes care of fixing machines. Each machine that is in the repair center costs the company \$75 per hour in lost production due to its unavailability. It has been determined that the arrival of failed machines to the repair center can be approximated by a Poisson arrival process with a mean rate of 4 per day, i.e., independent and exponentially distributed interarrival times with a mean of $1/4$ day between arrivals. (We have assumed that the company has such a large number of these machines in use that the failure of a few of the machines does not significantly affect the arrival rate resulting from the failure of any of the remaining machines.) The time it takes to repair a machine is highly variable because of the many different things that can go wrong; in fact, the repair times are also described by an exponentially distributed random variable with a mean time of 4 hours. The repair center thus operates as a single-server, infinite capacity, FIFO queueing system. The company would like to determine whether it would be economically advantageous, in the long run, to add another technician per shift so that two machines could be under repair at the same time when the need arises. Each technician costs the company \$30 per hour, including salary and overhead.

A program was written to simulate the costs incurred with this repair facility using only one technician. The high degree of variability in daily costs is illustrated in Figure 6.5, in which daily costs for three simulations are shown. Figure 6.6 shows the plot of the running averages of those costs. The variability between simulations for this repair facility is illustrated by comparing the three curves in Figure 6.6.

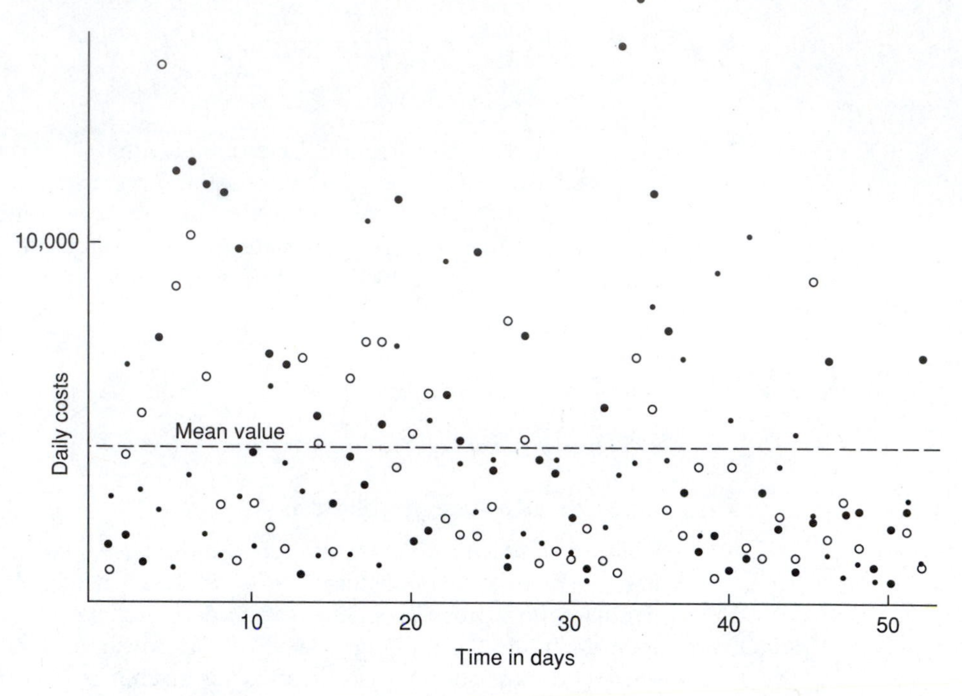

FIGURE 6.5 Daily costs versus time for three runs of the machine-repair problem simulation.

FIGURE 6.6 Running average of cost versus time for three runs of the machine-repair problem simulation.

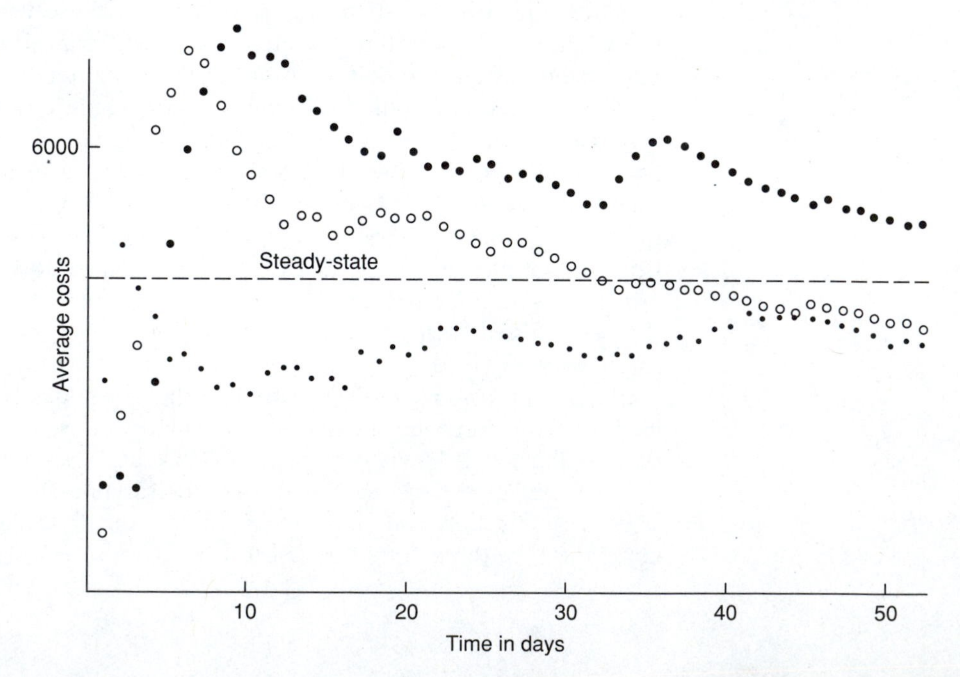

In our example, we are interested in a long-range policy, and it is the long-run average cost per day on which the decision will be based. Consequently, we need to look at the behavior of this quantity over time. (The first question that actually needs to be asked when a steady-state analysis is contemplated is "Do steady-state values exist?" It is easy to construct situations in which no steady state exists. For example, if a slow server is trying to handle fast arrivals, the backlog might continually increase and never reach steady state. A rigorous answer to the "existence" question is beyond the scope of this textbook. We shall simply assume throughout this section that the processes under consideration do reach steady-state behavior.) Notice that the graphs of the average costs in Figure 6.6 initially increase rapidly, and then they approach the straight line drawn on the graph. Extrapolating beyond the shown graph, the plot of the average costs would eventually cease to change. Its limiting value with respect to time is called its *steady-state value.* The straight dotted line in the figure is the theoretical steady-state value, which in this case is the cost per day incurred by the machines being in the repair room.

One of the primary purposes of a steady-state simulation is to estimate the long-run values of performance measures, including means, variances, and in some cases, probabilities. But, because values from simulations are statistical estimates, we also would like to determine the accuracy (i.e., confidence limits) of the estimates. While both obtaining an estimate and determining its accuracy present challenges, it is usually easier to derive the estimates than it is to determine their confidence limits. If a steady-state value exists, the time-averaged value from any single simulation will limit to the steady-state value as the run length increases to infinity. The obvious practical difficulty is to know when to stop the simulation. Since the preceding simulation was stopped short of infinity, we do not expect the simulation's final average cost to be exactly equal to the true long-run value; therefore, it is necessary to determine how much confidence we have in the estimate. To illustrate the problem, assume that the simulations illustrated in Figure 6.6 were stopped after 52 days. Taking any one of the simulations in the figure, the analyst would possibly conclude that the simulation is reasonably close to steady state because the graphs indicate that the major fluctuations have ceased; however, none of the three examples has yet reached the asymptotic value. Of course, this particular example was chosen because we could easily determine the theoretically correct value, and with this knowledge we have the ability to measure the accuracy of our estimators. Without the theoretical values (we would not normally simulate if the theoretic values were known), we must turn to statistical methods, which usually involve confidence limits.

Before discussing specific procedures for obtaining confidence intervals, let us consider some of the underlying concepts. For illustrative purposes, our context will be the determination of an estimate for mean cost. Remember that constructing a confidence interval for a parameter requires an estimator *and* the variance of the estimator.

Let μ_{cost} denote the long-run mean, and let σ^2_{cost} be the long-run variance of the daily cost. Further, let $C_1, C_2, \cdots$, be the sequence of daily costs obtained by the simulation on days $1, 2, \cdots$, respectively. Suppose n such observations have been obtained by a run of the simulation, and let

$$\overline{C}(n) = \frac{1}{n}\sum_{i=1}^{n} C_i; \tag{6.6}$$

thus, $\overline{C}(n)$ is the average of the first n data values. Figure 6.6 plots $\overline{C}(n)$ versus n for three different realizations, where n is the number of days. In theory,

$$\lim_{n\to\infty} \overline{C}(n) = \mu_{\text{cost}}, \tag{6.7}$$

which means that if we let n grow indefinitely, the estimate of Eq. (6.6) converges to the true value of μ_{cost}. The intuitive meaning of Eq. (6.7) is that we can take a single simulation, run it a long time, then use the time-averaged value for an estimate of the long-run expected value.[2] In similar fashion, the long-run sample variance,

$$S^2(n) = \frac{\left(\sum_{i=1}^{n} C_i^2\right) - n\left[\overline{C}(n)\right]^2}{n-1}, \tag{6.8}$$

converges to σ^2_{cost} as n limits to ∞. Consequently, we can control the quality of our estimates of the population mean and variance by the run lengths of the simulation.

The quality of the estimator $\overline{C}(n)$ of μ_{cost} is measured by the variance of $\overline{C}(n)$. (Actually, we need its distribution too, but as we shall see, difficulties will arise with just its variance.) You might recall that if we assume $C_1, \cdots, C_n$ are independent and identically distributed, an unbiased estimator for the variance of $\overline{C}(n)$ is $S^2(n)/n$. However, in a simulation study, the sequential random variables $C_1, \cdots, C_n$ are often highly dependent, which introduces a (possibly large) bias into $S^2(n)/n$ as an estimator for the variance of $\overline{C}(n)$. Therefore, the sample variance [Eq. (6.8)] of the sequentially generated data should not be used with Eq. (6.1) to build a confidence interval for μ_{cost}. These facts should be stressed: When n is large, $\overline{C}(n)$ is a valid estimator for μ_{cost}, $S^2(n)$ is a valid estimator for σ^2_{cost}, but $S^2(n)/n$ is *not* a valid estimator for the variance of $\overline{C}(n)$.

Replicates. The most straightforward method of obtaining confidence intervals is to use several replicates of the simulation run (using different initial random number seeds), just as was done for the terminating simulation analyses. What this does is provide a way to create truly random samples by sequential estimation. To illustrate, consider the steady-state simulation repair

[2]When a process has the property that the limiting time-averaged value and the value averaged over the state space are the same, the process is said to have the *ergodic property*.

center mentioned earlier in this section. We might decide that a "long time" is 52 days and make 30 different simulations of 52 days each to obtain a random sample, where the 30 individual data points are the 52-day averaged daily cost values. (The end points on the graph in Figure 6.6 yield three such data points.) A major difficulty with the replicate approach, and the reason that it is usually not recommended for steady-state simulation problems, is that the results are highly dependent on the decision as to what constitutes a long time. In theory, the steady-state results of a simulation are not dependent on the initial conditions; however, if a run is too short, the steady-state estimator will be biased due to the initial conditions.

For example, we ran the simulation program of the repair center 30 times. If we use (ridiculously short) runs of 2 days and Eq. (6.1) to estimate the daily cost due to machine failure, it would be $2276 per day with a 95% confidence interval of ($1822, $2730). The true theoretical steady-state value for this problem is $4320, so it is apparent that this estimate is too low. Using a run length of 52 days yields an estimate of $4335 per day with a 95% confidence interval of ($3897, $4773). From these results, it is tempting to say that 52 days is long enough to be at steady state. However, the property given by Eq. (6.7) indicates that if the simulation is at steady state, there should be very little variation between runs. In other words, the maximum error represented by a confidence interval based on several simulations at steady state should be close to zero. But the maximum error of the above interval based on 52 days is 438 or 10.1% of the mean value—not at all negligible. If we use 2002 days to simulate a long time, 30 runs would yield a 95% confidence interval of ($4236, $4462), which has a maximum error less than 3% of the mean.

Start-Up Time. A common approach taken to eliminate the bias of the simulation due to its initial conditions is to establish a "start-up" period and only calculate statistics after the start-up period is finished. For example, we might take 2 days for the start-up period and then calculate the average daily costs based on the costs from day 3 through day 52. Again, the difficulty with this approach is the accurate determination of the best values for the start-up time and the overall run length. If the start-up period is too short, the estimate will be biased by the initial period. If the start-up period is too long, the estimate will have more inherent variability, producing a larger confidence interval than would be expected from the length of the run. We should also mention that practical considerations do enter the picture. A large number of repetitions of long simulations takes a great deal of time. Time is usually at a premium, which causes a serious obstacle to the use of the replication technique. With these considerations in mind, we investigate an alternative method for obtaining confidence intervals for mean values.

Batch Means. Let us review again the context of our estimation problem. We make a long simulation run (preferably containing a start-up period) obtaining m data points, $C_1, \cdots, C_m$. We desire to estimate μ_{cost}, which is the long-run mean value; that is,

$$\mu_{\text{cost}} = E\left[C_m\right],$$

where m is assumed large. The most common estimator for μ_{cost} is based on Eq. (6.2), which indicates that the sample mean [Eq. (6.2)] is a good estimator. However, in most simulations C_i and C_{i+1} are highly dependent random variables, which is why estimating the standard deviation of the estimator for μ_{cost} is difficult.

As mentioned previously, steady-state results are independent of the initial conditions.[3] This fact implies that if two costs are sufficiently far apart in time, they are essentially independent; that is, C_i and C_{i+k} can be considered independent if k is large enough. This leads to one of the most popular forms of obtaining confidence intervals: the method of batch means. The concept is that the data will be grouped into batches, and then the mean of each batch will be used for the estimators. If the batches are large enough, the batch means can be considered independent because, intuitively, they are far apart. Since the batch means are considered independent, they can be used to estimate not only the overall mean, but also the standard deviation. Specifically, let us take the m data points and group them into n batches, each batch containing $l = m/n$ data points, and let $\overline{C}_j$ denote the mean of the j'th batch; namely,

$$\begin{aligned}
\overline{C}_1 &= \frac{C_1 + \cdots + C_l}{l}, \\
\overline{C}_2 &= \frac{C_{l+1} + \cdots + C_{l+l}}{l}, \\
&\vdots \\
\overline{C}_n &= \frac{C_{(n-1)l+1} + \cdots + C_{nl}}{l},
\end{aligned} \tag{6.9}$$

where $m = nl$. (Figure 6.7 illustrates this concept using a simulation of the repair center example. In the figure, an initial period of two observations is ignored, then five batches of seven each are formed. Notice that the figure is for illustration purposes only; usually a single batch includes hundreds or thousands of observations.) As l gets large, $\overline{C}_j$ and $\overline{C}_{j+1}$ become independent. Using the definitions in Eq. (6.2), we have the following:

$$\begin{aligned}
\overline{X}_{\text{batch}} &= \frac{\overline{C}_1 + \cdots + \overline{C}_n}{n} = \frac{C_1 + \cdots + C_m}{m}, \\
S^2_{\text{batch}} &= \frac{\overline{C}_1^2 + \cdots + \overline{C}_n^2 - n\overline{X}^2_{\text{batch}}}{n-1}.
\end{aligned} \tag{6.10}$$

[3]To be technically correct in this paragraph, we need to discuss concepts and assumptions related to the ergodic property, but this is beyond the scope of this textbook. Here we intend to give only an intuitive feel for these principles.

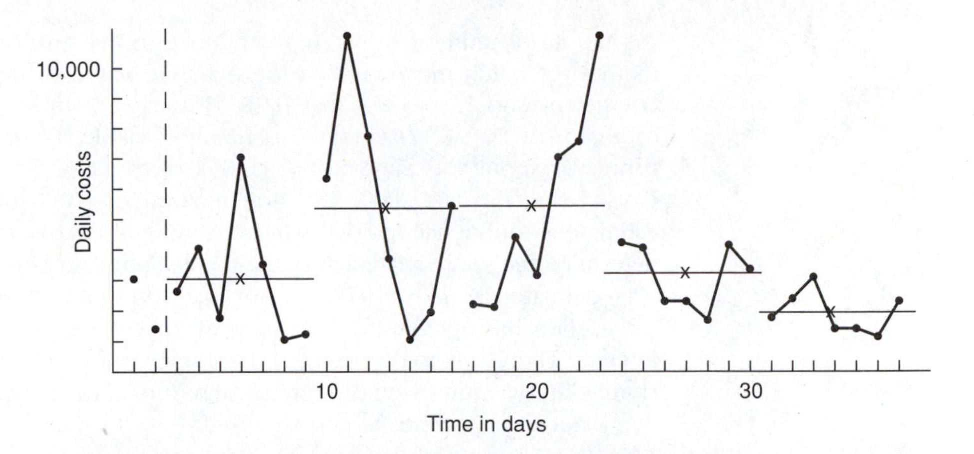

FIGURE 6.7 Batch means from a simulation of the repair center example.

The confidence interval is, thus, given by

$$\left(\overline{X}_{\text{batch}} - t_{(n-1,\alpha/2)}\frac{S_{\text{batch}}}{\sqrt{n}},\ \ \overline{X}_{\text{batch}} + t_{(n-1,\alpha/2)}\frac{S_{\text{batch}}}{\sqrt{n}}\right), \tag{6.11}$$

where $t_{(n-1,\frac{\alpha}{2})}$ has the same meaning as it does in Eq. (6.1).

The two major assumptions made in the development of the confidence interval of Eq. (6.11) are that (1) the random variables $\overline{C}_1, \cdots, \overline{C}_n$ are distributed according to a normal distribution and (2) the random variables are independent. As long as the batch size is large, the central limit theorem again permits the assumption of normality (a rule of thumb for using the central limit theorem is to have a batch sample size of at least 30). The independence assumption is more difficult to verify. One test for independence of the sequence is to estimate the correlation coefficient (Definition 1.26) for adjacent values, namely, the correlation between $\overline{C}_i$ and $\overline{C}_{i+1}$, called the lag 1 correlation coefficient. (A zero correlation between adjacent values does not guarantee independence, but a nonzero value guarantees lack of independence.)

The most common estimator for the lag 1 correlation coefficient for the sequence $\overline{C}_1, \cdots, \overline{C}_n$ is given by

$$\hat{\rho} = \frac{\sum_{i=1}^{n-1}[(\overline{C}_i - \overline{X}_{\text{batch}})(\overline{C}_{i+1} - \overline{X}_{\text{batch}})]}{\sum_{i=1}^{n}[(\overline{C}_i - \overline{X}_{\text{batch}})^2]}, \tag{6.12}$$

where $\overline{X}_{\text{batch}}$ is defined in Eq. (6.10). A slightly easier formula [again using Eq. (6.10)] for computing this estimator is

$$\hat{\rho} = \frac{\sum_{i=1}^{n-1} \overline{C}_i\overline{C}_{i+1} - n\overline{X}_{\text{batch}}^2 + \overline{X}_{\text{batch}}(\overline{C}_1 + \overline{C}_n - \overline{X}_{\text{batch}})}{(n-1)S_{\text{batch}}^2}. \tag{6.13}$$

A rule of thumb for having confidence in the results obtained from the method of batch means is that the absolute value of the lag 1 correlation coefficient should be less that 0.05. However, a major problem with the estimator of Eq. (6.12) is that it is highly variable. For example, the repair center example was simulated for 2002 days using a start-up period of 2 days. Using the last 2000 days, the method of batch means was used to estimate a confidence interval where the size of each batch was 50 days. To determine the appropriateness of the method, the lag 1 correlation coefficient was estimated as $\hat{\rho} = -0.04$. If the -0.04 number is an accurate estimate for ρ, then the one simulation run is all that is needed for the confidence interval. Therefore, to determine the accuracy of the $\hat{\rho} = -0.04$ estimate, the simulation was run five more times, each time being a run of 2002 days. The five values for $\hat{\rho}$ were 0.21, 0.25, -0.04, -0.13, and 0.07. The conclusion from this example is that the estimator for ρ is too variable to be reliable when only 40 batches are used; thus, more batches are needed to obtain a reliable estimator.

The difficulty with choosing a batch size is that if the batch size is large enough to obtain a small lag 1 correlation coefficient, then the number of batches will be too small to produce a reliable estimate; however, if the number of batches is large enough to produce a reliable estimate, then the batch size will be too small to maintain independence between batch means.

When the modeler has the freedom to choose the run length based on values dynamically determined within the simulation, he or she may use the textbook by Law and Kelton (mentioned at the beginning of this chapter) that contains a suggested approach for confidence intervals. The authors' procedure is to extend the run length until the absolute value of the lag 1 correlation coefficient determined from 400 batches is less than 0.4. In other words, for a given run length, the batch size necessary to produce 400 batches is used, and the lag 1 correlation coefficient is calculated based on those 400 batches. If the absolute value of the coefficient is greater than 0.4, one can infer that the batches are too small, and the run length is increased. Once the absolute value of the coefficient is less than 0.4, the run length is fixed and the batch size is increased by a factor of 10. The resulting 40 batches are then used to obtain the confidence interval. The justification for this approach is that when 400 batches produce a coefficient of less than 0.4 and then the batches are made 10 times larger, the resulting 40 batches should produce a coefficient of less than 0.05.

More commonly, the run length of the simulation cannot be dynamically determined, but must be fixed at the start of the simulation. A good quick-and-dirty estimate of confidence intervals may be made using a short start-up period, dividing the remaining run length into five to ten batches, and then using those batches to estimate the confidence interval.[4] The idea of using

[4]A. M. Law and W. D. Kelton, "Confidence Intervals for Steady-State Simulations: I. A Survey of Fixed Sample Size Procedures," *Operations Research* **32** (1984), pp. 1221–1239, suggests five batches. B. Schmeiser, "Batch Size Effects in the Analysis of Simulation Output," *Operations Research* **30** (1982), pp. 556–568, suggests at least ten batches.

a small number of batches is flawed since it is difficult to justify using the central limit theorem; in practice, however, it does seem to perform adequately.

EXAMPLE 6.6 Table 6.20 (and Figure 6.7) contains simulation output to be used in a numerical example. (We use a run length that is too small for accuracy, but it is better for illustrating the computations. Do not infer from this example that estimates should be made based on such short runs.) This example uses an initialization period of 2 days and a batch size of 7 days that yields five batches. We calculate the average for each row in Table 6.20, which gives the five batch means as $\overline{c}_1 = 3017.9$, $\overline{c}_2 = 5332.3$, $\overline{c}_3 = 5381.4$, $\overline{c}_4 = 3148.3$, and $\overline{c}_5 = 1946.7$. The sum of the five values is 18,826.6, and the sum of their squares is 80,202,043.44. The confidence interval calculations are

$$\overline{x}_{\text{batch}} = \frac{18{,}826.6}{5} = 3765.32,$$

$$s^2_{\text{batch}} = \frac{80{,}202{,}043.44 - 5 \times 3765.32^2}{4} = 2{,}328{,}467.5,$$

which yields a 95% confidence interval of

$$\left(3765.3 - 2.776 \times \frac{1525.9}{\sqrt{5}},\quad 3765.3 + 2.776 \times \frac{1525.9}{\sqrt{5}}\right) = (1870.9, 5659.7).$$

Thus, we see that a run of this small size yields an estimate with a large degree of variability. With 95% confidence, we would say that the maximum error is 1894.4. Let us consider what would have happened if we had improperly used the individual 35 daily values (i.e., a batch size of one) to obtain the confidence interval. In that case, we would have estimated the maximum error to be 878.5, which greatly understates the inherent variation of the 3765.3 estimated value of the average daily cost. ■

▶ *Suggestion: Do Exercises 6.10–6.15.*

TABLE 6.20 Individual data points from a simulation run of length 37 days from the machine repair example.

	Daily Costs							Means
Week 0						2929	1441	—
Week 1	2558	3958	1772	6996	3451	1094	1296	3017.9
Week 2	6338	11105	7774	3710	1062	1931	5406	5332.3
Week 3	2270	2182	4365	3065	7111	7570	11107	5381.4
Week 4	4276	4024	2340	2345	1665	4042	3346	3148.3
Week 5	1728	2379	3050	1476	1490	1201	2303	1946.7

6.3 COMPARING SYSTEMS

Not only are simulations used to analyze existing or proposed systems, but they are also used as a decision tool in comparing two or more systems. In this section, we briefly discuss the analysis of simulation results when the goal is to compare two systems. For example, management may be considering adding a new workstation at a bottleneck point within a manufacturing facility. To help with that decision, a simulation of the facility with and without the new workstation could be run with the intent of comparing the throughput rate of the two options.

It is also quite common to compare several systems or to compare systems with their parameters set at multiple levels. The comparison of more than two systems should involve a careful design of the statistical experiments before any simulation runs are carried out. In the analysis of the simulations, a procedure called *analysis of variance* is used for comparing several systems. Such statistical issues are beyond the scope of this textbook; however, numerous textbooks are available that cover the statistical design and analysis for comparing multiple systems.

The context for this section is that two random samples have been obtained from a series of simulations describing two systems under consideration. If the study involves terminating simulations, the random samples come from replications. If the study involves steady-state simulations, the random sample may arise from a combination of replications and batching. The first random sample is assumed to describe system 1 and involves n_1 data points with their mean and variance being $\overline{X}_1$ and S_1^2, respectively. The second random sample (independent from the first) is assumed to describe system 2 and involves n_2 data points with their mean and variance being $\overline{X}_2$ and S_2^2.

Confidence Intervals for Means. When mean values are of interest, we generally would like to know the difference between the mean for system 1 and the mean for system 2. For large sample sizes, a confidence interval for the difference can be obtained by taking advantage of the fact that the difference in sample means can be approximated by a normal distribution. The variance for the difference is calculated by "pooling" the variations from the two samples; in other words, the estimate for the variance of the differences is given by

$$S^2_{\text{pooled}} = \frac{(n_1 - 1)S_1^2 + (n_2 - 1)S_2^2}{n_1 + n_2 - 2}.$$

A confidence interval can now be obtained by using the standard deviation as estimated from the combined data sets. In particular, the $1 - \alpha$ confidence interval for the difference in the mean values between system 1 and system 2 can be approximated by

$$(\overline{X}_1 - \overline{X}_2) \pm t_{(n_1+n_2-2,\frac{\alpha}{2})} S_{\text{pooled}} \sqrt{\frac{1}{n_1} + \frac{1}{n_2}}. \quad \textbf{(6.14)}$$

If the confidence interval does not include zero, we can say that we are at least $1 - \alpha$ confident that the means from system 1 and system 2 are different.

EXAMPLE 6.7 Table 6.20 contains simulation results from a simulation run of 37 days using one technician for the repairs. We run the simulation of the repair center again, except we use two technicians for repairs. The daily cost values include the cost of both technicians and the cost incurred due to machines being in the repair center. For the run involving two technicians, we use a start-up period of 2 days and a batch size of 7 days and run the simulation for 401 days; thus, we have 57 batches. These 57 data points yield a mean of \$2500 and a standard deviation of \$1200. Our goal is to use these data to determine whether the average cost of using two technicians is different from the average cost of using one technician. The estimate for the difference in the average cost of using one technician versus two technicians is \$1265. The pooled variance estimate is

$$S^2_{\text{pooled}} = \frac{4 \times 2{,}328{,}467 + 56 \times 1{,}440{,}000}{60} = 1{,}499{,}231.$$

For a 95% confidence interval, we use a t value of 2.0 (see Table C.3); therefore, the confidence interval is

$$1265 \pm 2 \times 1224.4 \sqrt{0.2 + 0.0175} = (123, 2407).$$

Because zero is not contained in the confidence interval, we can say that there is a statistical difference between the two systems, and that having two technicians will reduce the daily cost of the repair center. Similarly, we can be 95% confident that the reduction in daily costs resulting from having two technicians on duty is between \$123 and \$2407. (If this range in the estimate seems too large, then more simulation runs must be made.)

Confidence Intervals for Variances. Changes for systems are sometimes designed with the intent of decreasing variances. Comparisons involving the variances for two systems are based on the *ratio* of the variances. For example, if system 2 involves a proposed design change that claims to decrease the variance from system 1, then we would hope that the entire confidence interval centered around the ratio S_1^2/S_2^2 would be greater than one. The confidence interval involves critical values from the F distribution, and the F distribution has two parameters referred to as the degrees of freedom. Thus, the $1 - \alpha$ confidence interval for the ratio of S_1^2 to S_2^2 is given by

$$\left(\frac{S_1^2}{S_2^2} \frac{1}{F_{(n_1,n_2,\alpha/2)}}, \quad \frac{S_1^2}{S_2^2} F_{(n_2,n_1,\alpha/2)} \right). \quad \textbf{(6.15)}$$

If the confidence interval does not include 1, we can say that (with $1 - \alpha$ confidence) there is a statistical difference between the variances of system 1 and system 2. To illustrate this concept, we use the numbers from Example 6.7. The critical F values are taken from a book of statistical tables and are $F_{(5,57,0.025)} \approx 6.12$ and $F_{(57,5,0.025)} \approx 2.79$; therefore an approximate 95% confidence interval for the ratio of the variances is

$$\left(1.617\frac{1}{6.12}, 1.617 \times 2.79\right) = (0.26, 4.51).$$

Because one is included in the confidence interval for the ratio of the variances, we can only say that there is no statistical difference between the variances for the two systems. The variances that are being compared are the variances of mean daily costs averaged over a week. (Note that this is different from the variance for daily costs.) As with other examples based on this repair center system, the start-up periods and batch sizes should be much larger if actual estimates were to be made.

Confidence Intervals for Proportions. Sometimes design changes are considered because of their potential effect on tail probabilities. For example, after receiving complaints regarding long customer waiting times, management may try to institute changes to decrease the probability that queue times are longer than some specified level. Thus, we may want to develop a confidence interval for the difference between the tail probability for system 1 and the tail probability for system 2; denote these probabilities by $\hat{P}_1$ and $\hat{P}_2$, respectively. As long as n_1 and n_2 are large, the following can be used for the $1 - \alpha$ confidence interval for the difference

$$(\hat{P}_1 - \hat{P}_2) \pm z_{\alpha/2}\sqrt{\frac{\hat{P}_1(1 - \hat{P}_1)}{n_1} + \frac{\hat{P}_2(1 - \hat{P}_2)}{n_2}}. \quad \textbf{(6.16)}$$

We again use the numbers from Example 6.7 to illustrate the use of the confidence intervals for proportions. [As before, the value of n_1 is too small to appropriately use Eq. (6.16); however, the numerical values we use here should be sufficient to illustrate the proper calculations.] We are concerned with an average daily cost for any given week being larger than \$4500. So our goal is to determine the amount of reduction in the probability that a week's average daily cost would be greater than \$4500. From the first simulation run involving one technician (Table 6.20), there were 2 out of 5 weeks with daily averages greater than \$4500. In the second run, there were 6 out of 57 weeks with a daily average greater than \$4500. Thus, the 95% confidence interval is given by

$$0.4 - 0.105 \pm 1.96 \times \sqrt{\frac{0.4 \times 0.6}{5} + \frac{0.105 \times 0.895}{57}} = (-0.14, 0.73).$$

Thus, there is again too much statistical variation to draw any conclusions regarding a reduction in the probability that a week's average daily cost is greater than $4500.

▶ *Suggestion: Do Exercises 6.8 and 6.16.*

6.4 EXERCISES

6.1 Using three coins, simulate the production process of Example 6.1. Stop the simulation as soon as the simulated clock is greater than or equal to 100 and estimate the average number of widgets in the system. Compare your estimate with the estimate from Table 6.11.

6.2 Simulate the production process of Example 6.1, but assume that there are several machines available for the machining process. In other words, if a widget is acceptable (not destined for the scrap pile) and goes to the facility for machining, there are enough machines available so that a widget never has to wait to be machined. Thus, any number of widgets can be machined at the same time. Estimate the average number of widgets in the facility over a 100-minute period.

6.3 Simulate the production process of Example 6.1, but assume that arrivals are random. Specifically, the time between arriving widgets will equal 12 minutes half the time, 11 minutes one-third of the time, and 10 minutes one-sixth of the time. Use a die along with the coins to simulate this system. Estimate the number of widgets in the facility over a 100-minute period.

6.4 The manager of a service station next to a busy highway is concerned about congestion. The service station has two gas pumps and plenty of room for customers to wait if the pumps are busy. Data indicate that a car arrives every 3 minutes. Of the customers that arrive, 25% of them take 5 minutes 45 seconds in front of the pump and spend $10, 50% of them take 5 minutes 50 seconds in front of the pump and spend $10.50, and 25% of them take 5 minutes 55 seconds in front of the pump and spend $11. Simulate the arriving and servicing of cars and estimate the average receipts per hour and the average number of customers in the service station.

6.5 Consider the service station of Exercise 6.4, but assume that the time between arrivals is not fixed. Assume that the time between arriving cars is 2.5 minutes with probability 0.25, 3 minutes with probability 0.50, and 3.5 minutes with probability 0.25. Using simulation, estimate the average receipts per hour and the average number of customers in the service station.

6.6 A computer store sells two types of computers: the MicroSpecial and MicroSuperSpecial. People arrive at the computer store randomly throughout the day, such that the time between arriving customers is exponentially distributed with a mean time of 40 minutes. The store is small and there is only one clerk who helps in selection. (The clerk handles customers one at a time in a FIFO manner.) Twenty-five percent of the customers who enter the store end up buying nothing and using exactly 15 minutes of the clerk's time. Fifty percent of the customers who enter the store end up buying a MicroSpecial (yielding a profit of $225) and taking a random amount of the clerk's time, which is approximated by a continuous uniform distribution between 30 and 40 minutes. Twenty-five percent of the customers who enter the store end up buying a MicroSuperSpecial (yielding a profit of $700) and taking a random amount of the clerk's time, which is approximated by a continuous uniform distribution between 50 and 60 minutes. The store is open only 4 hours on Saturday. The policy is to close the door after the 4 hours but to continue to serve customers in the store when the doors close. Simulate the activities for a Saturday and answer the following questions.

(a) The store opens at 10:00 A.M. and closes the door at 2:00 P.M. Based on your simulation experiment, at what time is the clerk free to go home?

(b) What is your estimate for the total profit on Saturday?

(c) Simulate the process a second time, and determine how much your answers change.

6.7 Consider the simulation used to estimate the average sales receipts per house from Exercise 3.1. Since each house was independent from all other houses, the 25 data points formed a random sample. Use those 25 values to obtain a 95% confidence interval for the mean receipts per house. Calculate the theoretical value and compare it with your confidence interval.

6.8 Consider the selling of microwaves in Example 3.1. In that example, one-half of the customers who entered the store did not purchase anything. The store's management feels that by making certain changes in the store's appearance and by some additional training of the sales staff, the percent of customers who enter the store and do not purchase anything can be decreased to 40%. Furthermore, it has been estimated that the cost of these changes is equivalent to $1 per customer entering the store. Using simulation to justify your decision, do you feel that the extra expense of $1 per customer is worth the increase in purchases? Why or why not?

6.9. Write a computer simulation of a nonhomogenous Poisson process (see Example 3.8) with a mean rate function defined, for $t \geq 0$, by

$$\lambda(t) = 2 + \sin(2\pi t),$$

where the argument of the sine function is radians. Obtain 95% confidence intervals for the following.

(a) The mean number of arrivals during the interval $(0, 2)$.

(b) The standard deviation for the number of arrivals during the interval $(0, 2)$.

(c) The probability that there will be no arrivals during the interval $(0, 2)$.

6.10 The following 100 numbers are correlated data: 52.0, 27.3, 46.5, 43.0, 41.3, 44.2, 47.5, 56.8, 77.0, 50.9, 28.3, 30.9, 67.1, 50.3, 58.8, 60.0, 35.1, 40.5, 36.0, 41.1, 47.6, 63.6, 78.6, 59.6, 47.6, 44.5, 59.3, 61.0, 53.1, 26.7, 56.9, 52.4, 70.4,

58.0, 72.6, 58.1, 47.6, 37.7, 36.1, 59.3, 51.5, 40.2, 58.0, 77.9, 54.5, 57.5, 58.9, 44.6, 60.5, 57.5, 51.1, 60.2, 39.9, 37.5, 67.2, 61.5, 58.9, 58.3, 56.6, 44.9, 43.4, 44.5, 31.6, 20.1, 47.0, 58.0, 58.8, 54.6, 48.9, 58.9, 62.3, 62.8, 22.6, 66.3, 48.3, 56.8, 73.3, 47.5, 43.3, 42.5, 40.0, 54.2, 44.7, 51.7, 56.3, 59.7, 29.6, 49.0, 43.5, 33.2, 61.1, 80.1, 55.1, 76.8, 47.5, 56.5, 66.7, 65.8, 73.6, 62.6.

(a) Obtain the sample mean and sample variance of these data. (For comparison purposes, note that the true mean and variance of the underlying distribution are 50 and 200, respectively.)

(b) Estimate the lag 1 correlation coefficient. (For comparison purposes, note that the true lag 1 correlation coefficient is 0.2.)

(c) Using a batch size of 20, give a 95% confidence interval for the mean.

6.11 Correlated interarrival times can be generated using a Markov chain together with two interarrival distributions. Let φ_a and φ_b be two distributions, with means and variances given by θ_a, σ_a^2, θ_b, and σ_b^2, respectively. Furthermore, let $X_0, X_1, \cdots$, be a Markov chain with state space $E = \{a, b\}$ and with a transition matrix given by

$$\begin{bmatrix} p & 1-p \\ 1-p & p \end{bmatrix},$$

where $0 < p < 1$. The idea is that the Markov chain will be simulated with each time step of the chain corresponding to a new interarrival time. When the chain is in state i, interarrival times will be generated according to φ_i.

To start the simulation, we first generate a random number; if the random number is less than 0.5, set $X_0 = a$, otherwise set $X_0 = b$. The sequence of interarrival times can now be generated according to the following scheme:

1. Set $n = 0$.
2. If $X_n = a$, then generate T_n according to φ_a.
3. If $X_n = b$, then generate T_n according to φ_b.
4. Generate a random number. If the random number is less than p, let $X_{n+1} = X_n$; otherwise, set X_{n+1} to be the other state.
5. Increment n by one and return to step 2.

In this fashion, a sequence $T_0, T_1, \cdots$, of interarrival times is generated. These times are identically distributed, but not independent.

(a) Show that $E[T_0] = 0.5(\theta_a + \theta_b)$.

(b) Show that $\text{var}(T_0) = 0.5[\sigma_a^2 + \sigma_b^2 + 0.5(\theta_a - \theta_b)^2]$.

(c) Show that the lag 1 correlation coefficient is given by

$$\rho = \frac{0.5(2p-1)(\theta_a - \theta_b)^2}{\sigma_a^2 + \sigma_b^2 + 0.5(\theta_a - \theta_b)^2}.$$

6.12 Generate 100 correlated interarrival times according to the scheme described in Exercise 6.11. Let φ_a be exponential with mean 10, φ_b be exponential with mean 1, and $p = 0.85$. Based on your sample of size 100, estimate the mean, standard deviation, and lag 1 correlation coefficient. Compare your estimates with the actual values.

6.13 Generate 100 interarrival times such that the distributions of the times are normally distributed with mean 200, standard deviation 25, and lag 1 correlation 0.8. Based on your sample of size 100, estimate the mean, standard deviation, and lag 1 correlation coefficient. (*Hint:* Refer back to Exercise 1.25 and Example 3.4.)

The final three problems are for advanced programmers.

6.14 Write a computer simulation of an M/M/5/10 queueing system using the structure of the Future Events list. Set the mean arrival rate to be 14 per hour and the mean service time per machine to be 25 minutes. Use your simulation to build a confidence interval for the long-run overflow rate. Is the theoretical value within your confidence interval?

6.15 Write a computer simulation, using the structure of the Future Events list, for the manufacturing facility described in Example 5.12. The purpose of Example 5.12 was to illustrate an analytical approximation technique. Based on your simulation, was the approximation technique good?

6.16 A small service station next to an interstate highway has one pump and room for two other cars. Furthermore, we assume that the assumptions for an M/G/1/3 queueing system are appropriate. The mean arrival rate of cars is eight per hour and the service time distribution is approximated by a continuous uniform distribution varying from 3 to 10 minutes. The expected profit received from each car is $10. For an extra $75 per day, the owner of the station can increase the capacity for waiting cars by one (thus becoming an M/G/1/4 system). Is the extra $75 worthwhile? To answer this question, write a computer simulation with or without the Future Events list concept.

(a) Assume that the station is open 10 hours per day.

(b) Assume that the station is open 24 hours per day.

7

Inventory Theory

Applied probability can be taught as a collection of techniques useful for a wide variety of applications, or it can be taught as various application areas for which randomness plays an important role. The previous chapters have focused on particular techniques, with some applications being emphasized through examples and the homework problems. We now change tactics and present two chapters dealing with specific problem domains. The first is inventory, and the second is replacement.

Inventory theory is a useful stochastic concept that deals with the uncertainties associated with demand and supply of goods and services. It may seem advantageous from a production standpoint to have an infinite supply of raw material, but this approach may not be the most economical way of managing. Because storage rooms are limited, supply is limited, demand is stochastic, handling and storage of items have a cost, ordering often has a cost, being out of an item has a cost, and items may be perishable, the determination of an inventory policy that optimizes a cost or performance criterion is often needed. Given an objective to optimize, inventory theory uses probability principles to develop models that can answer questions about how much or how often to order.

It is important to offer one caveat with respect to these inventory models. The techniques and probabilistic reasoning used to solve the problems are more important than the problems themselves. The assumptions made in developing each inventory model tend to be restrictive, so the range of possible application is thus restricted. However, the general approach to formulating and solving the models is quite common. In actual application, approximations must often be employed to obtain solutions, and with an initial understanding of the basic modeling approach it becomes easier to develop good heuristics.

7.1 THE NEWS VENDOR PROBLEM

We begin with a classical example that is simplistic, yet illustrates some of the basic reasoning used in many inventory modeling situations. Because one of the main purposes of this chapter is to help develop the modeling process, we motivate each new type of model by a specific example. Then a general model is developed that can be used for the decision motivated by the example.

EXAMPLE 7.1 A news vendor buys newspapers at the start of every day to sell on the street corner. Each newspaper costs \$0.15 and is sold for \$0.25. Any unsold newspapers will be bought back by the supplier for \$0.10. After observing demand for a few months, the news vendor has determined that daily demand follows a uniform distribution between 21 and 40 newspapers. Thus, the question of interest is "how much inventory should be purchased at the start of each day in order to maximize the expected profits?" The decision involves balancing the risk of having extra papers at the end of the day with that of having too few papers and being unable to take advantage of potential sales. We shall first show how to solve the problem in general and then apply the above numbers to illustrate with a specific example. ■

The problem involves determining the inventory quantity to order, q, at the beginning of each day. The reason for the difficulty is that the daily demand, D, is not fixed and is known only through its cumulative distribution function, F (or, equivalently, through its probability mass function, f). In the ordering decision there is a trade-off between two "bad" possibilities: (1) The initial inventory order quantity may be too small (with probability $P\{D > q\}$), which results in some lost sales, and (2) the initial inventory order quantity may be too large (with probability $P\{D < q\}$), resulting in unused inventory.

Because demand is random, the actual daily receipts will vary; thus, we represent the daily profit (or loss) by a random variable C_q, which depends on the initial order quantity. In other words, the news vendor problem is to find the optimum q that maximizes $E[C_q]$. The relevant cost figures per newspaper are the wholesale purchase cost c_w, the retail purchase price c_r, and the salvage value c_s. For every newspaper that is sold, a profit of $c_1 = c_r - c_w$ is realized; for every newspaper initially purchased but not sold, a *loss* of $c_2 = c_w - c_s$ is incurred. If demand on a particular day is k and $k < q$, then k papers will be sold and $q - k$ papers will be returned. If $k > q$, then q papers will be sold and none returned. Thus, the expected profit per day is

$$E[C_q] = \sum_{k=0}^{q}[kc_1 - (q-k)c_2]f_k + qc_1 \sum_{k=q+1}^{\infty} f_k. \tag{7.1}$$

Because Eq.(7.1) is not continuous, we cannot maximize it by taking the derivative; therefore, if we are to use something other than brute force [determining the value of (7.1) for each possible q], we must discuss optimization procedures for discrete functions. The analogous expression for the derivative of continuous functions is the difference operator, denoted by Δ, for discrete functions (see Figure 7.1). In other words, if g is a discrete function, then its difference operator is defined by $\Delta g(k) = g(k+1) - g(k)$.

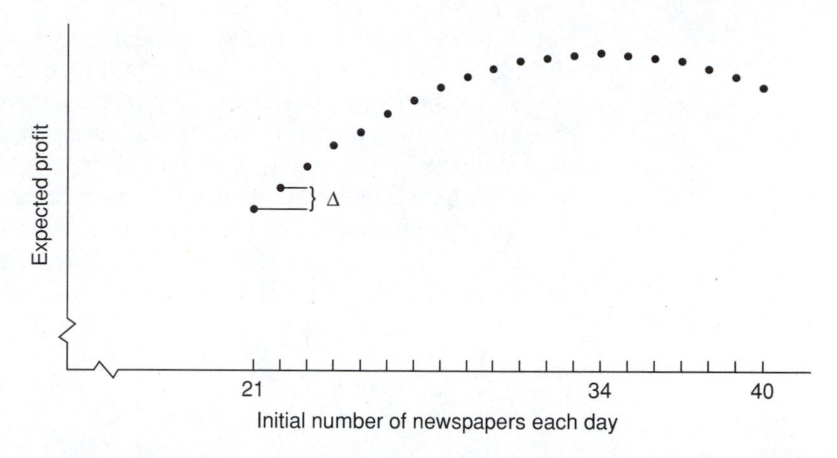

FIGURE 7.1 The expected profit function versus order quantity for Eq. (7.1).

Assume that g is a unimodal function with its maximum at k_0; that is, for $k < k_0$ the function g is increasing, and for $k > k_0$ the function is decreasing. To express this function using the Δ operator, $\Delta g(k) > 0$ for $k < k_0$, and $\Delta g(k) < 0$ for $k \geq qk_0$; thus,

$$k_0 = \min\{k : \Delta g(k) \leq 0\}.$$

The difficult part of the above equation is to obtain an expression for $\Delta g(k)$. Although not truly difficult, the procedure is tedious. For our problem, we have the following [starting with Eq. (7.1)]:

$$\begin{aligned}
\Delta E[C_q] &= E[C_{q+1}] - E[C_q] \\
&= \sum_{k=0}^{q+1} [kc_1 - (q+1-k)c_2] f_k + (q+1)c_1 \sum_{k=q+2}^{\infty} f_k \\
&\quad - \sum_{k=0}^{q} [kc_1 - (q-k)c_2] f_k - qc_1 \sum_{k=q+1}^{\infty} f_k \\
&= \sum_{k=0}^{q} [kc_1 - (q+1-k)c_2] f_k + (q+1)c_1 f_{q+1} \\
&\quad + (q+1)c_1 \sum_{k=q+1}^{\infty} f_k - (q+1)c_1 f_{q+1} \\
&\quad - \sum_{k=0}^{q} [kc_1 - (q-k)c_2] f_k - qc_1 \sum_{k=q+1}^{\infty} f_k \\
&= -\sum_{k=0}^{q} c_2 f_k + c_1 \sum_{k=q+1}^{\infty} f_k \\
&= -c_2 F(q) + c_1[1 - F(q)] \\
&= c_1 - (c_1 + c_2) F(q).
\end{aligned}$$

In the above expression, the third equality is a result of changing the limits of the first two summation terms so that they have the same limits as the last two summations, the fourth equality is a result of simplifying the terms of the third equality, the fifth equality is obtained by applying the definition of the cumulative distribution [see Eq. (1.3)], and the final equality results from rearranging a couple of terms from the preceding equation.

The maximum of the discrete function $E[C_q]$ is found by searching for the minimum value of q such that $\Delta E[C_q] \leq 0$. Thus, we have that the optimum order quantity, q^*, is the minimum q such that

$$\Delta E[C_q] = c_1 - (c_1 + c_2)F(q) \leq 0,$$

which is equivalent to

$$F(q) \geq \frac{c_1}{c_1 + c_2}. \tag{7.2}$$

We return to the numbers given at the start of this section and observe that the profit on each paper sold is $c_1 = \$0.10$ and the loss on each paper not sold is $c_2 = \$0.05$. Thus, we need to find the smallest q such that $F(q) \geq 0.667$. Because F is uniform, we have that $F(33) = 0.65$ and $F(34) = 0.70$, which implies that $q^* = 34$; therefore, at the start of each day, 34 newspapers should be ordered.

The news vendor problem can also be solved if demand is continuous. (Of course, continuous demand hardly fits this particular example, but it may easily fit other examples.) Let us suppose that the demand random variable is continuous, so that the objective function [analogous to Eq.(7.1)] becomes

$$E[C_q] = \int_{s=0}^{q} [sc_1 - (q - s)c_2]f(s)\, ds + qc_1 \int_{s=q}^{\infty} f(s)\, ds. \tag{7.3}$$

To minimize this expression with respect to the order quantity, we simply take the derivative[1] of Eq. (7.3) with respect to q and set it equal to zero. The derivative of the expected cost function is given as

$$\begin{aligned}\frac{d}{dq}E[C_q] &= qc_1 f(q) - \int_{s=0}^{q} c_2 f(s)\, ds - qc_1 f(q) + c_1 \int_{s=q}^{\infty} f(s)\, ds \\ &= -c_2 F(q) + c_1[1 - F(q)].\end{aligned}$$

Thus, the optimum order quantity, q^*, is given as that value that solves the following equation:

$$F(q^*) = \frac{c_1}{c_1 + c_2}.$$

► *Suggestion: Do Exercises 7.1–7.4.*

---

[1]Here we restate Leibniz's rule: If $G(y) = \int_{q(y)}^{p(y)} g(x, y)\, dx$, then $G'(y) = p'(y)g(p(y), y) - q'(y)g(q(y), y) + \int_{q(y)}^{p(y)} [\partial g(x, y)/\partial y]\, dx$.

7.2 SINGLE-PERIOD INVENTORY

The classical single-period inventory problem is very similar to the news vendor problem. In the news vendor problem, any inventory at the end of the period was returned to the supplier. In the single-period inventory problem, the ending inventory will be held for the next period and a cost will be incurred based on the amount of inventory held over. The modeling for this is carried out in two stages: First we solve the problem when there is no setup charge for placing the order; then it is solved with the inclusion of a setup charge. Based on the chapter regarding Markov chains, we should recognize that the inventory process can be modeled as a Markov chain. In this section, however, we first consider a myopic[2] optimum—that is, an optimum that only considers one period at a time without regard to future periods.

7.2.1 No Setup Costs

The trade-off in the single-period inventory model with no setup cost will be an inventory carrying cost versus a shortage cost. Again, we motivate the problem with a hypothetical example involving a discrete random variable for demand. After the discrete model has been formulated and solved, we then develop the model for the case when demand is continuous (or, at least, can be approximated by a continuous random variable).

EXAMPLE 7.2 A camera store specializes in a particularly popular and fancy camera. They guarantee that if they are out of stock, they will special-order the camera and promise delivery the next day. In fact, what the store does is purchase the camera retail from out of state and have it delivered through an express service. Thus, when the store is out of stock, they actually lose the sales price of the camera and the shipping charge, but they maintain their good reputation. The retail price of the camera is $680, and the special delivery charge adds another $25 to the cost. At the end of each month, there is an inventory holding cost of $20 for each camera in stock. The problem for the store is to know how many cameras to order at the beginning of each month so that costs will be minimized. Wholesale cost for the store to purchase the cameras is $400 each. (Assume that orders are delivered instantaneously, but only at the beginning of the month.) Demand follows a Poisson distribution with a mean of 7 cameras sold per month. ■

To begin formulating the inventory problem, we define three costs: c_h is the cost per item held in inventory at the end of the period, c_p is the penalty cost charged for each item short (i.e., for each item for which there is demand but no stock on hand to satisfy it), and c_w is the wholesale cost

[2]Myopia is a synonym for what is commonly called nearsightedness. Thus, a myopic policy is one that "sees" what is close but not what is far.

for each item. Let x be the inventory level at the beginning of the month immediately before an order is placed, let q be the order quantity, let y be the inventory level immediately after ordering (i.e., $y = x + q$), and let $C_{x,y}$ be the monthly cost. The decision variable has been changed from the order quantity (q) to the *order-up-to* quantity (y). Thus, y is the decision variable, and the cost equation[3] (see Figure 7.2) to be minimized is

$$E[C_{x,y}] = c_w(y - x) + \sum_{k=0}^{y} c_h(y - k)f_k + \sum_{k=y+1}^{\infty} c_p(k - y)f_k .$$

It is customary to separate the above equation into two parts: an ordering cost and a "loss function." The loss function, $L(y)$, is a function of the order-up-to quantity, y, and represents the cost of carrying inventory or being short. In other words, the loss function is defined for $y \geq 0$ as

$$L(y) = \sum_{k=0}^{y} c_h(y - k)f_k + \sum_{k=y+1}^{\infty} c_p(k - y)f_k, \tag{7.4}$$

which yields an objective function of

$$E[C_{x,y}] = c_w(y - x) + L(y). \tag{7.5}$$

Since the inventory involves a discrete random variable for demand, we proceed as in the previous section by working with differences instead of derivatives, namely,

$$\begin{aligned}
\Delta_y E[C_{x,y}] &= E[C_{x,y+1}] - E[C_{x,y}] \\
&= c_w(y + 1 - x) + \sum_{k=0}^{y+1} c_h(y + 1 - k)f_k + \sum_{k=y+2}^{\infty} c_p(k - y - 1)f_k \\
&\quad - c_w(y - x) - \sum_{k=0}^{y} c_h(y - k)f_k - \sum_{k=y+1}^{\infty} c_p(k - y)f_k \\
&= c_w + c_h F(y) - c_p[1 - F(y)].
\end{aligned}$$

Notice that several steps are missing between the second-to-last equality and the last equality; however, the steps are very similar to the analogous derivation for the news vendor problem.

Because we want to minimize cost, it follows that y^* is the smallest y such that $\Delta_y E[C_{x,y}] \geq 0$. This is equivalent to finding the smallest y such that

$$F(y) \geq \frac{c_p - c_w}{c_p + c_h}. \tag{7.6}$$

Notice that Eq. (7.6) does not depend upon the initial inventory level, x. (Of course, the equation does not depend upon x because we are only interested

[3]For a profit formulation of this problem, work Exercise 7.7.

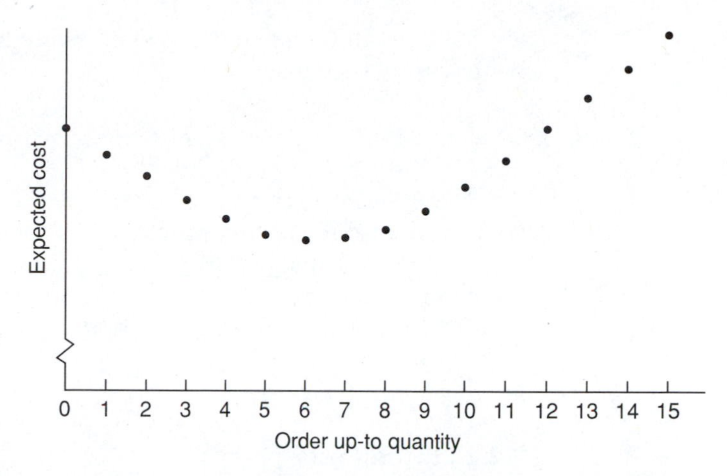

FIGURE 7.2 The expected cost function versus order-up-to quantity for Eq. (7.5).

in myopic policies.) Thus, at the end of each period, the policy is to order enough inventory to bring the stock level up to y^*.

We now return to the example that opened this section to illustrate the use of the inventory policy. The cost constants are $c_p = 705$, $c_w = 400$, and $c_h = 20$. Thus we need to find the value y such that

$$\sum_{k=0}^{y} \frac{7^k e^{-7}}{k!} \geq 0.4207$$

which implies that $y^* = 6$; that is, at the end of each month, order enough cameras so that the beginning inventory at the start of the following month will be 6 cameras.

If demand for the cameras were in the hundreds or thousands, we might choose to approximate the Poisson random variable with a normally distributed random variable—thus necessitating model development, assuming a continuous random variable. This will cause the loss function [Eq. (7.4)] to be redefined, for $y \geq 0$, as

$$L(y) = \int_{s=0}^{y} c_h(y - s)f(s)\, ds + \int_{s=y}^{\infty} c_p(s - y)f(s)\, ds. \tag{7.7}$$

We now take the derivative of Eq. (7.5) using the above continuous loss function to obtain

$$\begin{aligned} \frac{\partial}{\partial y} E[C_{x,y}] &= c_w + \int_{s=0}^{y} c_h f(s)\, ds - \int_{s=y}^{\infty} c_p f(s)\, ds \\ &= c_w + c_h F(y) - c_p[1 - F(y)]. \end{aligned}$$

The derivative is set to zero, which yields an inventory policy of always ordering enough material at the end of each period to bring the inventory level at the start of the next period up to the level y^*, where y^* is defined to be that value such that

$$F(y^*) = \frac{c_p - c_w}{c_p + c_h}. \tag{7.8}$$

There are at least two questions that should always be asked in the context of an optimization problem: (1) "Does an optimum exist?" and (2) "If it exists, is it unique?" For this particular problem, we know a unique optimum can always be found because the cost expression, $E[C_{x,y}]$ treated as a function of y, is convex.

▶ *Suggestion: Do Exercises 7.5–7.8.*

7.2.2 Setup Costs

The assumption of the previous section is that ordering ten items for inventory is simply ten times the cost of ordering one item for inventory; however, in many circumstances, there is a setup cost incurred whenever an order is placed.

EXAMPLE 7.3 Consider a manufacturing company that assembles an electronic instrument that includes a power supply. The power supply is the most expensive part of the assembly, so we are interested in the control of the inventory for the power supply unit. Orders are placed at the end of the week and delivery is made at the start of the following week. The number of units needed each week is random and can be approximated by a continuous uniform random variable that varies between 100 and 300. The cost of purchasing q power supply units is \$500 + 5q. In other words, there is a \$500 setup cost and a \$5 variable cost to purchase the power supplies when they are purchased in bulk (e.g., the purchase of 100 power supply units costs \$1000). Further, assume that there is a \$0.50 cost incurred for each item remaining in inventory at the end of the week and a \$25 cost for each item short during the week. (In other words, it costs \$25 for each unit special-ordered during the week.) ■

The cost model for this problem is identical to the previous problem except for the addition of a setup cost. We denote the setup cost by K; thus the cost model is

$$E[C_{x,y}] = \begin{cases} L(y) & \text{for } y = x, \\ K + c_w(y - x) + L(y) & \text{for } y > x. \end{cases} \tag{7.9}$$

An optimal policy based on Eq. (7.9) will lead to what is called an (s, S) inventory policy; that is, an order is placed if the ending period's inventory level is less than s, and enough inventory is ordered to bring the inventory level up to S. To understand how this model yields an (s, S) policy, consider the graphs illustrated in Figure 7.3.

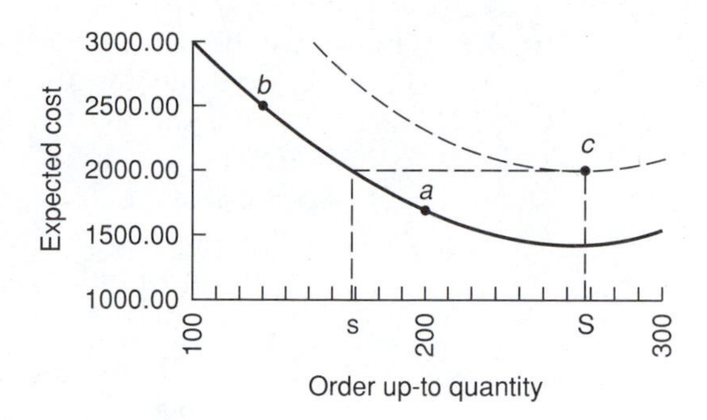

FIGURE 7.3 Cost functions used for the single-period inventory model with a setup cost.

The lower graph in Figure 7.3 is the cost function, ignoring the setup cost; the upper graph is the cost, including the setup cost. In other words, the lower graph gives the values of $c_w(y - x) + L(y)$ as a function of y and the upper graph is simply the same values added to the constant K. If an order up to S is placed, the top graph is relevant; therefore, the optimum order-up-to quantity, denoted by S, is the minimum of the graph. If no order is placed, the lower graph is relevant. Consider an initial inventory level (say $x = 200$ in Figure 7.3) that is between the values s and S. If no order is placed, the relevant cost is the point on the lower graph (a in Figure 7.3) that is above x; if an order up to S is placed, the relevant cost is the point on the upper graph (c in Figure 7.3) that is above S. It is thus seen that the optimum policy will be to not order, because the point (representing the expected cost) on the lower graph at x is less than the point (expected cost) on the upper graph at S. Now consider the value that is less than s (say $x = 125$ in Figure 7.3). If no order is placed, the relevant cost is the point on the lower graph (b in Figure 7.3) that is above the initial inventory x, which again must be compared to the point (c in Figure 7.3) on the upper graph that is above the order-up-to quantity S. It is thus seen that the optimum policy will be to place an order, because the point on the lower graph at x is greater than the point on the upper graph at S.

Since S is the minimum point on both the lower graph and the upper graph, we can find it by solving the "no setup cost" problem from the previous section. Thus, $S = y^*$, which is the solution from Eq. (7.9). The reorder point, s, is the value for which the no setup cost function equals the optimum setup cost function evaluated at the order-up-to quantity S. That is, the value of s is defined to be that value such that

$$c_w s + L(s) = K + c_w S + L(S). \qquad \textbf{(7.10)}$$

To illustrate the mechanics, consider the cost values given in the example at the start of this section. The value of S satisfies

$$F(S) = \frac{c_p - c_w}{c_p + c_h} = \frac{25 - 5}{25 + 0.5} = 0.7843.$$

In the example the demand density is defined to be $f(t) = 1/200$ for $100 < t < 300$ or, equivalently, the distribution is defined to be $F(t) = (t - 100)/200$ for $100 < t < 300$. Therefore, we need to solve the equation $(S - 100)/200 = 0.7843$, which yields $S = 100 + 0.7843 \times 200 \approx 257$. To obtain s we first determine the loss function:

$$\begin{aligned} L(y) &= 0.5 \int_{100}^{y} \frac{y - s}{200}\, ds + 25 \int_{y}^{300} \frac{s - y}{200}\, ds \\ &= \frac{51}{800} y^2 - 37.75y + 5637.5. \end{aligned}$$

The cost to order up to S is thus given as

$$\begin{aligned} K + c_w S + L(S) &= 500 + 5 \times 257 + L(257) \\ &= 1785 + 0.06375 \times 257^2 - 37.75 \times 257 + 5637.5 \\ &= 1931.37. \end{aligned}$$

The reorder point, s, is that value for which the inventory cost of not ordering is equal to the optimum ordering cost [Eq. (7.10)]. Thus, s is that value that satisfies the following

$$\begin{aligned} 1931.37 &= c_w s + L(s) \\ &= 5s + \frac{51}{800} s^2 - 37.75s + 5637.5 \\ &= 0.06375s^2 - 32.75s + 5637.5. \end{aligned}$$

The above quadratic equation yields two solutions—one larger and one smaller than S (see Figure 7.3). We want the smaller solution, which is $s = 168.3$. Therefore, the optimal inventory policy is to place an order whenever the ending inventory is equal to or less than 168, and, when an order is placed, to order enough to bring the next month's beginning inventory up to 257.

▶ *Suggestion: Do Exercises 7.9–7.11.*

7.3 MULTIPERIOD INVENTORY

It is an unusual circumstance when a single-period inventory model is truly appropriate. However, the previous section is worthwhile not only for its pedagogical value, but also because the problem formulation and solution structure is the beginning step in the multiperiod model. In this section we discuss inventory when the demand is a discrete random variable and the planning horizon involves several (but finite) periods.

As in the previous sections, the probability mass function giving the amount of demand during each period is denoted by $f(\cdot)$, and we assume a planning horizon containing $n_{\max}$ periods. For notational purposes, let the demand for period n be the random variable D_n; thus, $P\{D_n = k\} = f(k)$. The ending inventory for period n is denoted by the random variable X_n. After the inventory level is observed, a decision is made regarding an order; if an order is placed, the new inventory arrives immediately, so the beginning period's inventory is equal to the previous period's ending inventory plus the order quantity. If q items are ordered, there is an order cost given by $c_o(q)$. The inventory holding cost and shortage cost are given by c_h and c_p, respectively, and there are no back orders. Once an inventory ordering policy has been determined, the stochastic process $\{X_0, X_1, \cdots, X_{n_{max}}\}$ is a Markov chain.

The solution methodology we use is stochastic dynamic programming. Bellman, who is often considered the father of dynamic programming, defined[4] a concept called the "principle of optimality." Bellman's principle of optimality essentially holds that an optimal policy must have the property that, at any point in time, future decisions are optimal independent of the decisions previous to the current point in time. This is fundamentally a Markov property; therefore, stochastic dynamic programming is an ideal tool for identifying optimal decisions for processes satisfying the Markov property.

At the end of a period, the ending inventory is observed and a decision is made as to what the order-up-to quantity should be for the next period. Thus the optimum decision and the cost associated with the optimum decision made at the start of period n depend on the inventory at the end of period $n-1$. Denote the optimum decision made at the beginning of period n by $y_n(x)$ and total future cost (or value) of that decision by $v_n(x)$, where x is the ending inventory level of the previous period.

Dynamic programming is usually solved by a backward recursion. Thus, the first step towards a solution is to optimize the decision made in the final period; that is, we first determine $y_{n_{max}}$ and $v_{n_{max}}$. In other words, the first step is to solve the single-period problem,

$$v_{n_{max}}(x) = \min_{y \geq x} c_o(y - x) + L(y), \tag{7.11}$$

and $y_{n_{max}}$ is the value of y that minimizes the expression in the right-hand-side of the above equation. The question may arise as to how we solve Eq. (7.11) if we do not know the value of the ending inventory for $n_{max} - 1$. The answer is simple—find the optimal order-up-to quantity for every reasonable value of the ending inventory $x_{n_{max}-1}$. (Note that if c_o involves a setup cost plus a linear variable cost, then the term being minimized is $E[C_{x,y_n}]$ from Eq. (7.9).) Once $v_{n_{max}}$ is known, the remaining values of v_n can be calculated based on the typical dynamic programming recursive relationship:

$$v_{n-1}(x) = \min_{y \geq x}\{c_o(y - x) + L(y) + E_y[v_n(X_{n-1})]\}. \tag{7.12}$$

Notice that the expected value operator in Eq. (7.12) is written with y as a subscript. This is to indicate that the expected value depends on the decision

[4] R. Bellman, *Dynamic Programming* (Princeton, N.J.: Princeton University Press), 1957.

made for y, which is the beginning inventory level for period n. We illustrate these calculations through the following example.

EXAMPLE 7.4 A manufacturing company needs widgets in their manufacturing process; however, the exact number needed during the day is random and varies from zero to three with equal probability (i.e., f is discrete uniform). At the end of each day, an order is placed for widgets to be used during the next day, and delivery is made overnight. If not enough widgets are on hand for the day, there is a \$92 cost for each unit short. If widgets must be stored for the next day, there is a \$8 storage cost because of the need to refrigerate them overnight. Because of space restrictions, the most that can be kept overnight is five widgets. If an order for q widgets is placed, it costs $\$50 + 20q$ (i.e., a \$50 setup cost and a \$20 variable cost). It is now Tuesday evening, and one widget is on hand for tomorrow. We would like to determine the optimal ordering policy for the next three days. ■

We shall use the above example to illustrate the steps involved in using Eqs. (7.11) and (7.12). As is normal with dynamic programming problems, the computations become tedious, but those calculations are quick for a computer to handle as long as there is only a single state variable (there is only one state variable in this case—namely, the ending period's inventory).

We begin by calculating the single-period inventory costs. It should be clear that we would never order up to a number larger than three on Thursday night, because we need at most three widgets on Friday and the inventory costs beyond Friday do not affect the three-day planning horizon; however, it is not clear that we should be limited to just three items for either Wednesday or Thursday; therefore, we calculate the inventory ordering costs allowing for the possibility of ordering up to five widgets. The loss function calculation is according to Eq. (7.4); thus,

$$L(0) = 0 \times 0.25 + 92 \times 0.25 + 184 \times 0.25 + 276 \times 0.25 = 138$$

$$L(1) = 8 \times 0.25 + 0 \times 0.25 + 92 \times 0.25 + 184 \times 0.25 = 71$$

$$L(2) = 16 \times 0.25 + 8 \times 0.25 + 0 \times 0.25 + 92 \times 0.25 = 29$$

$$L(3) = 24 \times 0.25 + 16 \times 0.25 + 8 \times 0.25 + 0 \times 0.25 = 12$$

$$L(4) = 32 \times 0.25 + 24 \times 0.25 + 16 \times 0.25 + 8 \times 0.25 = 20$$

$$L(5) = 40 \times 0.25 + 32 \times 0.25 + 24 \times 0.25 + 16 \times 0.25 = 28$$

We add the inventory ordering costs to the above loss function values and obtain the values of the single-period inventory costs as shown in Table 7.1. (Because the order-up-to quantity, y, must be at least as large as the previous day's ending inventory, x, Table 7.1 contains no values in the lower portion.)

Our problem is to decide the beginning inventory levels for Wednesday through Friday. Friday's decision (actually the decision made Thursday evening) is given by Eq. (7.11), which is equivalent to taking the minimum value for each row of Table 7.1. These values are given in Table 7.2.

TABLE 7.1 Values for $E[C_{x,y}]$ used in Eq. 7.11.

x	Order-Up-To Quantity for Friday					
	$y = 0$	$y = 1$	$y = 2$	$y = 3$	$y = 4$	$y = 5$
0	0 + 138	70 + 71	90 + 29	110 + 12	130 + 20	150 + 28
1		0 + 71	70 + 29	90 + 12	110 + 20	130 + 28
2			0 + 29	70 + 12	90 + 20	110 + 28
3				0 + 12	70 + 20	90 + 28
4					0 + 20	70 + 28
5						0 + 28

Note that it is necessary in dynamic programming to determine the optimum decision for each state that might occur at the beginning of the period. The decision to be made Wednesday evening (i.e., for Thursday's beginning inventory level) is determined through the following restatement of Eq. (7.12):

$$v_2(x) = \min_{y \geq x}\{E[C_{x,y}] + E_y[v_3(X_2)]\}.$$

The values for $E[C_{x,y}]$ are given in Table 7.1, but the values for $E_y[v_3(X_2)]$ must be calculated using Table 7.2 as follows:

$$E_0[v_3(X_2)] = 119$$
$$E_1[v_3(X_2)] = 71 \times 0.25 + 119 \times 0.75 = 107$$
$$E_2[v_3(X_2)] = 29 \times 0.25 + 71 \times 0.25 + 119 \times 0.50 = 84.5$$
$$E_3[v_3(X_2)] = 12 \times 0.25 + 29 \times 0.25 + 71 \times 0.25 + 119 \times 0.25 = 57.75$$
$$E_4[v_3(X_2)] = 20 \times 0.25 + 12 \times 0.25 + 29 \times 0.25 + 71 \times 0.25 = 33$$
$$E_5[v_3(X_2)] = 28 \times 0.25 + 20 \times 0.25 + 12 \times 0.25 + 29 \times 0.25 = 22.25$$

To understand the above equations, remember that an expected value is simply the sum over all possibilities times the probability that the possibility will occur. Consider $E_0[v_3(X_2)]$. The "0" subscript indicates that Thursday begins with no inventory. If there is no inventory for Thursday, then, with probability one, there will be no inventory Thursday evening. If Thursday ends with no inventory, Friday's optimum cost [namely, $v_3(0)$ from Table 7.2] is 119. Consider $E_1[v_3(X_2)]$. If Thursday morning begins with an inventory level of 1, there is a 25% chance that Thursday evening will also have an inventory level of 1 (i.e., no demand for Thursday), in which case Friday's

TABLE 7.2 Optimum values for Thursday evening's decision.

x	0	1	2	3	4	5
v_3	119	71	29	12	20	28
y_3	2	1	2	3	4	5

TABLE 7.3 Values for $E[C_{x,y}] + E_y[v_3(X_2)]$.

	Order-Up-To Quantity for Thursday					
x	$y = 0$	$y = 1$	$y = 2$	$y = 3$	$y = 4$	$y = 5$
0	138 + 119	141 + 107	119 + 84.5	122 + 57.75	150 + 33	178 + 22.25
1		71 + 107	99 + 84.5	102 + 57.75	130 + 33	158 + 22.25
2			29 + 84.5	82 + 57.75	110 + 33	138 + 22.25
3				12 + 57.75	90 + 33	118 + 22.25
4					20 + 33	98 + 22.25
5						28 + 22.25

optimum cost [$v_3(1)$ from Table 7.2] is 71. Also, if Thursday morning begins with 1, there is a a 75% chance of no inventory being available Thursday evening, in which case the optimum cost for Friday is 119. Thus, the expected value of those two cost values equals 107.

The values for Thursday's costs (Table 7.1) are added to the expected future costs (i.e., $E_y[v_3(X_2)]$) to obtain the necessary values for Eq. (7.12), as in Table 7.3.

Wednesday evening's decision is now made by again minimizing each row within Table 7.3. These results are in Table 7.4.

The final decision to be made is for Tuesday evening. Because we already know Tuesday's ending inventory, we only have to compute one value; namely, we need to determine

$$v_1(1) = \min_{y \geq 1}\{E[C_{1,y}] + E_y[v_2(X_2)]\}.$$

The future costs (from Wednesday's point of view) are calculated as

$$\begin{aligned}
E_0[v_2(X_2)] &= 179.75 \\
E_1[v_2(X_2)] &= 159.75 \times 0.25 + 179.75 \times 0.75 = 174.75 \\
E_2[v_2(X_2)] &= 113.5 \times 0.25 + 159.75 \times 0.25 + 179.75 \times 0.50 = 158.1875 \\
E_3[v_2(X_2)] &= 69.75 \times 0.25 + 113.5 \times 0.25 + 159.75 \times 0.25 + 179.75 \times 0.25 \\
&= 130.6875 \\
E_4[v_2(X_2)] &= 53 \times 0.25 + 69.75 \times 0.25 + 113.5 \times 0.25 + 159.75 \times 0.25 = 99 \\
E_5[v_2(X_2)] &= 50.25 \times 0.25 + 53 \times 0.25 + 69.75 \times 0.25 + 113.5 \times 0.25 \\
&= 71.625
\end{aligned}$$

TABLE 7.4 Optimum values for Wednesday evening's decision.

x	0	1	2	3	4	5
v_2	179.75	159.75	113.5	69.75	53	50.25
y_2	3	3	2	3	4	5

TABLE 7.5 Values for $E[C_{x,y}] + E_y[v_2(X_2)]$.

	ORDER-UP-TO QUANTITY FOR WEDNESDAY					
x	$y = 0$	$y = 1$	$y = 2$	$y = 3$	$y = 4$	$y = 5$
1	—	71 + 174.75	99 + 158.1875	102 + 130.6875	130 + 99	158 + 71.625

Thus, the relevant costs for determining Tuesday evening's decision are shown in Table 7.5. Minimizing over the row gives $v_1(1) = 229$ and $y_1(1) = 4$. In other words, we order three additional widgets for Wednesday morning. The decision to be made Wednesday night depends on Wednesday's demand. If Wednesday evening ends with two or more widgets, no order is placed; if Wednesday evening ends with zero or one widget, an order is placed to bring Thursday's beginning inventory up to three widgets. For Thursday evening, if no widgets are present, we order two widgets to begin Friday's schedule; however, if any widgets remain Thursday evening, no additional widgets are ordered.

▶ *Suggestion: Do Exercises 7.12–7.17.*

7.4 EXERCISES

7.1 The following functions are defined only on the positive integers. Use the difference operator to solve the following.

(a) $\min g(n) = 2n^2 - 9n$.

(b) $\max g(n) = 1/(3n - 11)^2$.

(c) $\max g(n) = ne^{-0.3n}$.

7.2 Let $f(\cdot)$ be a probability mass function defined on the nonnegative integers, with $F(\cdot)$ its CDF and μ its mean. Define the function $g(n)$ for $n = 0, 1, \cdots$, by

$$g(n) = n \sum_{i=n}^{\infty} (i - n) f(i).$$

(a) Using the difference operator, show that $g(\cdot)$ is maximized at the smallest integer, n, such that

$$(2n + 1)[1 - F(n)] + \sum_{i=0}^{n} if(i) \geq \mu.$$

(b) Use the answer in (a) to obtain the optimum n when $f(\cdot)$ is the mass function for a Poisson random variable with mean 2.5.

7.3 A computer store is getting ready to order a batch of computers to sell during the coming summer. The wholesaler is giving the store a good price on the computers because they are being phased out and will not be available after the summer. Each computer will cost the store $550 to purchase and will sell for $995. The wholesaler has also offered to buy back any unsold equipment for $200. The manager of the computer store estimates that demand for the computer over the summer is a Poisson random variable. How many computers should be purchased by the store in preparation for the summer?

(a) The mean demand for the summer is 5.

(b) The mean demand for the summer is 50. (Note that the normal distribution is a good approximation for the Poisson.)

7.4 Consider the news vendor problem of Example 7.1. The newspaper company has established an incentive program. The company has grouped all news vendors in groups of five, and, at the end of each day, one person out of the five is randomly selected. The news vendor thus selected will receive a full refund of $0.15 for the unsold newspapers.

(a) Give the expression for the expected profit per day analogous to Eq. (7.1).

(b) Use difference equations to find an expression analogous to Eq. (7.2) that can be used to obtain the optimum order quantity.

(c) Find the optimum order quantity.

(d) Write a computer program to simulate this process. Based on the simulation, do you believe your answer to parts (b) and (c)? (Note: The simulations suggested in these exercises can be written without using a Future Events list.)

7.5 Using the single-period model, determine the optimum reorder point for the camera store in Example 7.2, except assume that the demand is Poisson with a mean of 70.

7.6 Next month's production at a manufacturing company will use widgets for part of its manufacturing process. Widgets cost $50 each, and there will be a $0.50 charge for each item held in inventory at the end of the month. If there is a shortage of widgets, the production process is seriously disrupted at a cost of $150 per widget short. What is the optimum ordering policy for the following scenarios?

(a) The demand is governed by the discrete uniform distribution varying between three and seven widgets (inclusive).

(b) The demand is governed by the continuous uniform distribution varying between 300 and 700 widgets.

(c) The cost structure as stated in the problem description is oversimplified. The penalty for being short is divided into two parts. There is a shortage cost due to a disruption to the manufacturing process of $1000, independent of the number of widgets short. In addition, there is a $60 cost per widget short. Derive a formula for the optimal ordering policy, assuming that the demand is a continuous uniform distribution.

(d) Apply the answer to part (c) assuming that demand varies between 300 and 700 widgets.

(e) Write a computer program to simulate this process. Based on the simulation, do you believe your answer to parts (c) and (d)?

7.7 The single-period inventory model without setup costs can be formulated to maximize profits instead of minimizing costs, as was done in Eq. (7.5). Show that the two formulations are equivalent.

7.8 The loss function for a single-period inventory is given by Eq. (7.7). The loss function includes a holding cost and a penalty cost. Assume that the goods are perishable and that any goods on hand at the beginning of the period must be sold during that period or they spoil and must be discarded at a cost of c_d per item. (That is, items can be held over for one period, but not for two periods.) Assume that all unsold goods (even those to be discarded) add to the inventory cost. How would Eq. (7.7) change?

(a) Use a FIFO policy—that is, assume that all items in inventory at the beginning of the period are sold first.

(b) Use a LIFO policy—that is, assume that all items in inventory at the beginning of the period are sold last so that customers get fresh goods if they are available.

7.9 Assume that there is a $170 handling cost each time an order is placed from the camera store of Example 7.2. Using a Poisson demand process with a mean of 70, determine the store's optimal ordering policy.

7.10 For the manufacturing problem described in Exercise 7.6, assume that there is an ordering cost of $750 incurred whenever an order is placed. Determine the optimal ordering policy when demand follows a continuous uniform distribution between 300 and 700.

(a) Assume that the penalty cost is $150 per widget short.

(b) Assume that the penalty cost is $60 per widget short plus an additional $1000 due to the process interruption.

7.11 In Example 7.3 the wholesaler of the power supplies has decided to offer a discount when large quantities are ordered. The current unit cost, excluding setup, is $5. The wholesaler has decided that orders of 100 or more will yield a reduction of cost to $4.50 per unit. Determine whether the availability of the discount changes the optimal ordering policy.

7.12 The optimal ordering policy for Example 7.4 was obtained for a planning horizon of three days. Determine the optimal ordering policy for a planning horizon of four days. (For simplicity, assume a seven-day work week.)

7.13 A small company produces printers on a weekly basis that are sold for $1700 each. If no printers are produced, there are no costs incurred. If one printer is produced, it costs $1400. If two printers are produced, the cost is $2000. If three printers are produced, the cost is $2500. The normal maximum production capacity for the company is three printers; however, by paying overtime, it is possible to produce one extra printer for an additional \$750. For each printer in inventory at the end of the week, there is a cost of \$15. Weekly demand can be approximated by a binomial random variable with $n = 4$ and $p = 0.4$. Any demand not met from existing inventory or from the week's production is lost. The current policy is to never inventory more than four units at the end of any week. It is now Friday, and one printer remains unsold for the week.

(a) What should be next week's production if a planning period of three weeks is used.

(b) What would be the cost savings, if any, if the current policy of allowing a maximum of four units stored at the end of the week was changed to five units?

(c) Part (a) can be solved using stochastic dynamic programming as discussed in Section 7.3. Assume that you do not know dynamic programming and answer (a) through a simulation analysis. Compare the amount of work required by the two approaches. After reaching a conclusion with dynamic programming and simulation, comment on your level of confidence in the two answers.

7.14 Derive a decision rule for a two-period, no-setup-cost inventory model. The basic cost structure should be the same as in Section 7.2.1, with continuous demand. Let the decision variables be y_1 and y_2, referring to the order-up-to quantities and periods 1 and 2, respectively.

(a) Show that y_1 can be obtained by minimizing the following,

$$E[C_{x,y_1}] = c_w(y_1 - x) + \int_{s=0}^{y_1} (c_h - c_w)(y_1 - s)f(s)\,ds$$

$$+ \int_{s=y_1}^{\infty} c_p(s - y_1)f(s)\,ds + c_w y_2^* + L(y_2^*),$$

where y_2^* is the optimum order-up-to quantity for period 2.

(b) Obtain expressions that can be used to find y_1^* and y_2^*.

(c) Find the optimum values for the order-up-to quantities using the data from Example 7.2, except with a mean demand of 70.

(d) Write a simulation for the situation of Example 7.2 and implement a two-period decision rule. Using your simulation, answer the following question: "How much difference is there between using a two-period rule and using a single-period rule?"

7.15 The multiperiod inventory problem can be modeled as a Markov chain. Consider an inventory policy with an order-up-to quantity given by y, where back orders are not allowed, and with a demand per period being given by the pmf $f(\cdot)$. Furthermore, let the costs c_o, c_h, and c_p be the ordering, holding, and shortage costs, respectively. Define a Markov chain with state space $E = \{0, 1, \cdots, y\}$, where X_n denotes the on-hand inventory at the end of each period, immediately before an order for additional inventory is placed. Finally, assume that orders are delivered instantaneously.

(a) Give the Markov matrix for the Markov chain.

(b) Give the Markov matrix using the numbers from Example 7.4, with $s = 1$ and $S = 5$.

(c) Determine the steady-state probabilities for the Markov chain and its associated long-run cost.

(d) What is the optimal ordering policy?

7.16 Develop a Markov chain formulation for the multiperiod inventory problem with setup costs.

(a) Give the Markov matrix for the Markov chain.

(b) Give the Markov matrix using the numbers from Example 7.2 with the addition of a $30 ordering cost. Use $s = 2$ and $S = 7$.

(c) Determine the steady-state probabilities for the Markov chain and its associated long-run cost.

(d) What is the optimal long-run ordering policy?

7.17 A manufacturer of specialty items has the following inventory policy. Three items are kept in inventory if there are no orders being processed. Whenever an order is received, the company immediately sends a request to begin the manufacturing of another item. If there is an item in inventory, the customer is shipped the item in inventory. If no items are in inventory, the customer's request is back-ordered and a discount of $600 is applied to the purchase price. The time to manufacture one item is exponentially distributed with a mean of two weeks. The items cost $12,000 to manufacture, and any item in inventory at the end of a week costs the company $75. The company averages 24 orders per year, and the arrival of orders appears to follow a Poisson process. The purpose of this exercise is to formulate the inventory problem as a queueing process. For this purpose view the ordering process as an M/M/∞ system. In such a case, notice that p_0 for the queueing system is equivalent to the company having an inventory level of three.

(a) What is the value of $r = \lambda/\mu$ for the M/M/∞ system?

(b) What is the expected number of items being manufactured within the facility? (Notice that this is L for the queueing system.)

(c) What is the probability that when an order is received, it must be back-ordered?

(d) Derive an expression for the long-run average weekly cost for this inventory problem.

(e) The current inventory policy is to maintain three items in stock when there are no outstanding orders. Is this optimum?

8

Replacement Theory

Almost all systems deteriorate with age or usage unless some corrective action is taken to maintain them. Repair and replacement policies are often implemented to reduce system deterioration and failure risk. Increasingly complex systems have generated interest in the research of replacement theory. Most of the models were originally developed for industrial and military systems; however, more recent applications of these models have extended to areas such as health, ecology, and the environment.

Determining repair and replacement policies for stochastically deteriorating systems involves a problem of decision making under uncertainty, because it is impossible to predict with certainty the timing and extent of the deterioration process. In repair and replacement models, these uncertainties are usually assumed to have known probability distribution functions, so that the deterioration process is stochastically predictable. In this chapter we present some of the basic models dealing with replacement and maintenance issues, focusing on the trade-off between maximizing the life of the equipment and minimizing increased costs due to early replacements and unforeseen failures.

In Section 8.1 we present the classical age replacement maintenance policy, and in Section 8.2 an extension of the age replacement policy that includes repairs that do not change the failure distribution is developed. Finally, Section 8.3 presents the block replacement maintenance policy.

8.1 AGE REPLACEMENT

An age replacement policy is one in which the equipment is replaced as soon as it fails or when it reaches a prespecified age—whichever occurs first. As in Chapter 7 we shall consider both a discrete case and a continuous case. The basic concept for these models is the consideration of two types of actions that may be taken at any point in time: (1) Replace the equipment with (probabilistically) identical equipment, or (2) leave the equipment alone.

8.1.1 Discrete Life Times

We begin by assuming that the random variable describing the time until failure for the equipment is a discrete random variable. We also assume that the equipment being replaced is only one component. In order to provide a proper framework for the replacement problem, an example is used for motivation.

EXAMPLE 8.1 Consider a specially designed switch used to initiate production processing each day. At the start of each day, the switch is used to begin processing; however, it is not fully reliable and must be replaced periodically. Experience indicates that the switch will fail somewhere between the 20th day and the 30th day according to the following probability mass function:

Day	20	21	22	23	24	25	26	27	28	29	30
Prob.	0.02	0.03	0.05	0.1	0.3	0.2	0.15	0.08	0.04	0.025	0.005

The switch costs \$300 to replace; however, an additional cost of \$200 in lost production is incurred if the switch is replaced during the morning at failure time. Because production only occurs during the day, it may be worthwhile to replace the switch as a preventive measure during some evening before it fails. The disadvantage of the preventive action is that we do not get full utilization from the switch. Thus our problem is to determine a replacement strategy that optimally balances the trade-off between avoiding the additional replacement cost and fully utilizing the life of the switch. ■

A typical replacement strategy is to determine a replacement age (t_0) such that, if the switch fails before t_0, it is replaced; if the switch works on day t_0, replace it that evening. (A schematic of this replacement policy is given in Figure 8.1.) With this strategy, we can now formulate the optimization problem with an objective of minimizing the expected long-run average daily cost. To state the problem generally, let T denote the random variable representing the failure time, with its CDF being $F(t)$. Furthermore, let c_r denote the cost of replacement, c_f represent the additional cost associated with failure, and $z(t_0)$ denote the long-run average daily cost associated with

FIGURE 8.1 Schematic representation for an age replacement policy.

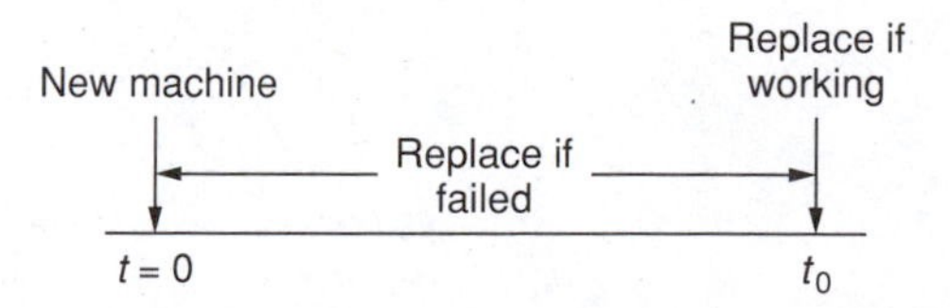

the replacement age of t_0. Because some switches will fail and some will be replaced before failure, different switches will have different replacement costs. Let C_{t_0} be the random variable indicating the replacement costs for a switch while using a replacement age of t_0. With this notation, the objective function is given as

$$z(t_0) = \frac{E[C_{t_0}]}{E[\min(T, t_0)]}.$$

In obtaining the numerator, we make two observations. First, the replacement cost, c_r, is always incurred, and, second, the failure cost, c_f, is incurred only when the switch fails on or before the replacement day, t_0. Thus,

$$E[C_{t_0}] = c_r + c_f \Pr\{T \le t_0\}.$$

To obtain the denominator, let the random variable S denote the actual replacement time, that is, $S = \min(T, t_0)$. The CDF of the random variable S is thus

$$G(s) = P\{S \le s\} = \begin{cases} F(s) & \text{if } s < t_0 \\ 1 & \text{if } s \ge t_0 \end{cases},$$

where $P\{T \le s\} = F(s)$, i.e., F is the CDF of T. (See Figure 8.2 for a representation of G.) We now use the third statement of Property 1.11 to obtain the denominator, namely,

$$E[\min(T, t_0)] = E[S] = \int_0^\infty [1 - G(s)]\,ds = \int_0^{t_0} [1 - F(s)]\,ds.$$

Therefore, the expected long-run average cost associated with a particular replacement policy can be written as

$$z(t_0) = \frac{c_r + c_f F(t_0)}{\int_0^{t_0} [1 - F(s)]\,ds}. \tag{8.1}$$

Because our problem deals with a discrete random variable, we introduce some additional notation so that the computation of Eq. (8.1) will be more

FIGURE 8.2 The CDF $G(\cdot)$ of S derived from the CDF $F(\cdot)$ of T.

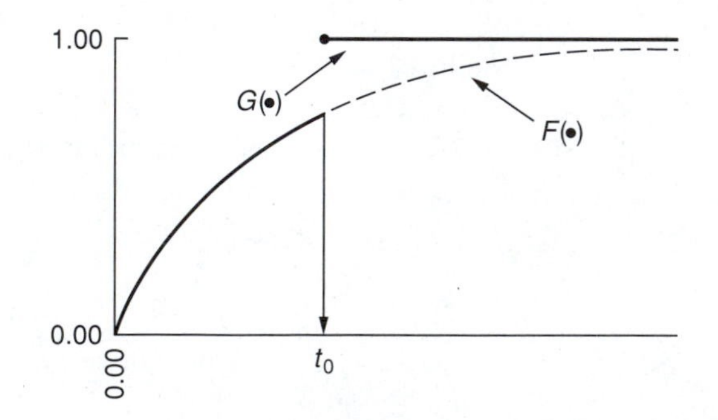

straightforward. Specifically, let $F_k = F(k) = \Pr\{T \leq k\}$; that is, F_k is the probability that the switch fails on or before day k. Also, let $\overline{F_k} = 1 - F(k)$; that is, $\overline{F_k}$ is the probability that the switch fails *after* day k. The optimal replacement time is given as t_0^*, where t_0^* is the k that solves the following,

$$\min_k z(k) = \frac{c_r + c_f F_k}{\sum_{i=0}^{k-1} \overline{F_i}}, \tag{8.2}$$

where $\overline{F_0} = 1$. It should also be observed that if replacement occurs on day t_0, then the sum in the denominator begins at zero and goes until $t_0 - 1$.

Unfortunately, Eq. (8.2) does not yield itself to a nice single equation from which the optimum can be determined; thus, we find the optimum by "brute force." The easiest way to do this is to make a table and continually evaluate the long-run average cost, $z(k)$, for successive values of k until the first optimum is obtained. The following table illustrates these computations:

k	f_k	F_k	$c_r + c_f F_k$	$\overline{F_k}$	$\sum_{i=0}^{k-1} \overline{F_i}$	$z(k)$
19	0.00	0.00	300	1.00	19.00	15.79
20	0.02	0.02	304	0.98	20.00	15.20
21	0.03	0.05	310	0.95	20.98	14.78
22	0.05	0.10	320	0.90	21.93	14.59
23	0.10	0.20	340	0.80	22.83	14.89

We stop the above computations when the first local minimum is identified and therefore set the optimum replacement time as $t_0^* = 22$. In this approach we make some assumptions about the shape of the long-run average cost function—namely, that the first local minimum is the global minimum. It is not appropriate at the level of this textbook to go into all the details about such an assumption; however, one concept that is useful to know when hoping to make such an assumption has to do with the hazard rate function. An important characteristic of random variables that relate to failure times is its hazard rate.

DEFINITION 8.1 *The function $h(\cdot)$ defined for $t \geq 0$ by*

$$h(t) = \frac{f(t)}{1 - F(t)}$$

is called the HAZARD RATE FUNCTION, *where F is a CDF*[1] *and f is either a pdf or pmf, depending on whether the random variable of interest is discrete or continuous.*

[1]It is also possible to obtain the distribution from the hazard rate through the relationship $F(t) = 1 - \exp\left\{-\int_0^t h(s)\,ds\right\}$.

Intuitively, the hazard rate evaluated at t gives the *rate* of failure at time t given that the item has not yet failed at t. A random variable is said to have an *increasing failure rate* (IFR) distribution if its hazard rate is increasing for all $t \geq 0$. Thus, a random variable that has an IFR distribution refers to an item that in some sense "gets worse" as it ages. As a general rule, replacement problems that deal with IFR distributions have costs that increase with age. Thus, cost functions associated with IFR distributions are usually well behaved, and Eq. (8.1) taken as a function of t_0 will have a unique minimum. Because the above example problem has an IFR distribution, we know that the local minimum is a global minimum, and the optimal policy is to replace the switch at the end of day 22 if failure does not occur during the first 22 days of operation.

▶ *Suggestion: Do Exercises 8.1–8.4.*

8.1.2 Continuous Life Times

When the lifetime of the system being considered for a replacement policy can be described by a continuous random variable, it is possible to obtain a closed-form expression for the optimum replacement time. The basic optimization equation [Eq. (8.1)] is the same for both continuous and discrete cases. Therefore, we begin by taking the derivative of Eq. (8.1) with respect to t_0. It is not difficult to show that

$$\frac{dz(t_0)}{dt_0} = 0$$

yields the equation

$$c_f f(t_0) \int_0^{t_0} [1 - F(s)]\, ds = [c_r + c_f F(t_0)][1 - F(t_0)]. \qquad \textbf{(8.3)}$$

Before continuing with the derivative, we return to Definition 8.1. The hazard rate function for a distribution is an important descriptor and is often given instead of the distribution function to describe a random variable. In general, replacement policies are usually developed for systems that have IFR distributions, so we shall assume an increasing hazard rate function and convert Eq. (8.3) into an expression using its hazard rate. We divide both sides of Eq. (8.3) by c_f and the factor $1 - F(t_0)$ to obtain

$$[1 - F(t_0)] + h(t_0) \int_0^{t_0} [1 - F(s)]\, ds = 1 + \frac{c_r}{c_f}. \qquad \textbf{(8.4)}$$

Before we can make definite statements regarding a solution to Eq. (8.4), we need to know how many solutions it has. We first observe that the right-hand side is a constant; therefore, if the left-hand side is increasing, there can be at most one solution. The derivative of the left-hand side with respect to t_0 yields

$$-f(t_0) + [1 - F(t_0)]h(t_0) + \frac{dh(t_0)}{dt_0} \int_0^{t_0} [1 - F(s)]\, ds,$$

which can be simplified to

$$\frac{dh(t_0)}{dt_0} \int_0^{t_0} [1 - F(s)]\, ds.$$

Because $h(\cdot)$ was assumed to be increasing, the above expression is positive for all $t_0 \geq 0$, which implies that the left-hand side of Eq. (8.4) is increasing. Therefore, we know that there is at most one solution to Eq. (8.4). (Figure 8.3 presents an example graph of the left-hand side of Eq. (8.4) as a function of the replacement age.)

The optimal age replacement policy can now be stated as follows: Let t_0 satisfy Eq. (8.4). If a failure occurs before t_0, replace at the failure time; if no failure occurs before t_0, replace at t_0.

EXAMPLE 8.2 A smelting process involves steel pots that are lined with a ceramic material. The ceramic lining is used because the electrolytic solution within the pots is corrosive; however, the lining will develop cracks with age, which allows the electrolytic solution within the pot to find a path to the steel shell. If the lining fails (i.e., develops cracks) while the operation is in process, the solution will eat through the steel pot and spill on the floor, which involves extra cleanup costs and extra repair time for the steel shells. Upon failure, or at a scheduled replacement, the steel shell is repaired and a new ceramic lining is placed in the shell. The lifetime of the lining is described by a Weibull distribution with scale parameter $\lambda = 0.0392$ per month and shape parameter $\alpha = 16.4$. Thus the expected lifetime of the lining is 24.7 months, with a standard deviation of 1.84 months. Replacement cost is \$30,000, with

FIGURE 8.3 A graphical solution of Eq. (8.4) for Example 8.2.

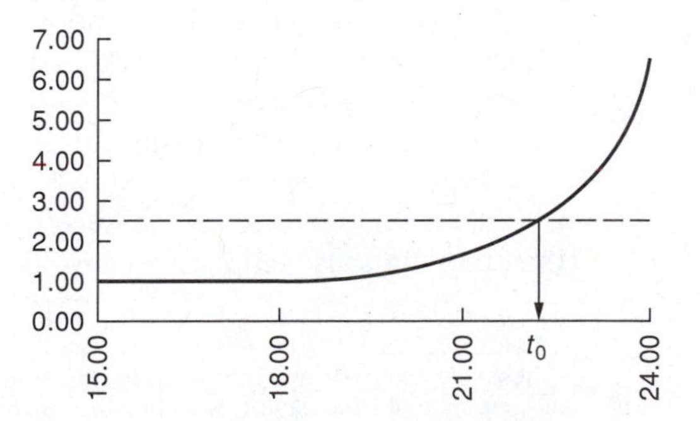

an additional \$20,000 incurred if a failure occurs;[2] thus, $c_r/c_f = 1.5$. The hazard rate function for the Weibull distribution is $h(t) = \alpha\lambda(\lambda t)^{\alpha-1}$ so that the Weibull has an IFR if $\alpha > 1$. Therefore, we need to find the value of t such that

$$e^{-(\lambda t)^\alpha} + \alpha\lambda(\lambda t)^{\alpha-1} \int_0^t e^{-(\lambda s)^\alpha}\, ds = 2.5.$$

The above equation cannot be solved analytically, but it can be easily evaluated numerically. (Appendix B contains some guidelines for numerical integration.) Evaluating the above numerically yields $t_0 = 22.14$; therefore, any pot lining that fails with an age less than 22.14 months will be immediately replaced, and the lining of any pot still working with an age of 22.14 months will be replaced even though no problems with the lining had yet become evident. The expected long-run cost for this policy is obtained by substituting the value of t_0 back into Eq. (8.1) to yield \$1447.24 per month. ■

■ **EXAMPLE 8.3** **Simulation.** To ensure a good understanding of the replacement process, we simulate the failure of the ceramic lining of steel pots for Example 8.2. Because the Weibull distribution governs failure time, our first step is to obtain the inverse of the Weibull CDF. Using the inverse mapping method, the transformation from a random number to a Weibull random variate with scale and shape parameters λ and α, respectively, is

$$T = \frac{1}{\lambda}[-\ln(R)]^{1/\alpha}.$$

For the simulation, we generate a random number (starting at line 13 of Table C.6) and obtain a failure time. If the failure time is before the replacement time, t_0, then a failure cost is incurred; if the random failure time is larger than the replacement time, the failure cost is not incurred. After a replacement of the lining (either at the failure time or replacement time), the pot is instantaneously placed in service. We maintain the cumulative cost and the cumulative time. Table 8.1 shows the results of ten replacements using an age replacement policy of $t_0 = 22.14$. Based on this simulation, the estimated average cost for the replacement policy is \$1452 per month.

For comparison, Table 8.2 shows the results of ten replacements if a replacement policy of $t_0 = 24.70$ had been used. The simulation would then yield an estimated average of \$1488 per month. Of course, no conclusions can be drawn from these simulations, but they do illustrate the dynamics of the process. Several replicates of longer runs would need to be made if conclusions were to be drawn as to the preference of one policy over the other. This also emphasizes a weakness of simulation: A single calculation

[2]These numbers are not very realistic, but they are given for an easy comparison with the discrete case of Example 8.1. Note that the means and standard deviations of the two examples are similar.

TABLE 8.1 Simulation of Example 8.2 with $t_0 = 22.14$.

Time New Machine Starts	Random Number	Failure Time	Replacement Time	Cost	Cumulative Cost
0	0.7400	23.710	22.14	30,000	30,000
22.14	0.4135	25.318	22.14	30,000	60,000
44.28	0.1927	26.298	22.14	30,000	90,000
66.42	0.6486	24.241	22.14	30,000	120,000
88.56	0.0130	27.900	22.14	30,000	150,000
110.70	0.2967	25.815	22.14	30,000	180,000
132.84	0.0997	26.843	22.14	30,000	210,000
154.98	0.9583	21.044	21.04	50,000	260,000
176.02	0.0827	26.971	22.14	30,000	290,000
198.16	0.2605	25.976	22.14	30,000	320,000
220.30					

from the analytical approach is needed to obtain the optimal policy, whereas many simulation runs must be made simply to obtain an estimate of the optimum. ■

▶ *Suggestion: Do Exercises 8.5, 8.6 and 8.8.*

8.2 MINIMAL REPAIR

Some types of failure do not always require replacement; it is sometimes possible to repair the failed equipment instead of completely replacing it. If a repair simply returns the equipment to service without any truly corrective work, it is called *minimal repair;* that is, a minimal repair is a repair that returns the equipment to a working condition with the remaining life of the

TABLE 8.2 Simulation of Example 8.2 with $t_0 = 24.70$.

Time New Machine Starts	Random Number	Failure Time	Replacement Time	Cost	Cumulative Cost
0	0.7400	23.710	23.71	50,000	50,000
23.71	0.4135	25.318	24.70	30,000	80,000
48.41	0.1927	26.298	24.70	30,000	110,000
73.11	0.6486	24.241	24.24	50,000	160,000
97.35	0.0130	27.900	24.70	30,000	190,000
122.05	0.2967	25.815	24.70	30,000	220,000
146.75	0.0997	26.843	24.70	30,000	250,000
171.45	0.9583	21.044	21.04	50,000	300,000
192.49	0.0827	26.971	24.70	30,000	330,000
217.19	0.2605	25.976	24.70	30,000	360,000
241.89					

equipment being the same as if it had never failed. For example, suppose you have a car and the tire has just blown out. You could buy a new car or you could replace the tire. If you choose the latter approach and buy a new tire, the life of the car remains unaffected by the repair, which is therefore a minimal repair. An example of a nonminimal repair would be a blown engine. If you overhaul the engine, that would most likely affect the life of the car and is therefore not a minimal repair.

There are many variations possible when minimal repair is considered. In this section we consider two scenarios. The two cases are presented to illustrate the modeling methodology and should not be viewed as the only models to use whenever minimal repair is appropriate. In most situations a new model would need to be developed that is designed uniquely for the given problem.

8.2.1 Minimal Repairs without Early Replacements

Once again we consider an age replacement policy, except that now, whenever a failure occurs before the replacement age, a minimal repair is performed. Let t_0 designate the age limit at which point a working system will be replaced, let c_m designate the cost incurred for each minimal repair, and let c_r designate the replacement cost. As in the previous section, our first task is to determine the expected cost per system and the expected age per system. The objective function for the optimal policy is then the ratio between the two expected values.

The expected cost per system is the replacement cost, c_r, plus the expected cost for minimal repairs, c_m, times the expected number of failures before time t_0. It can be shown that the expected number of minimal repairs during the interval $(0, t]$ is equal to $\int_0^t h(s)\,ds$, where $h(\cdot)$ is the hazard rate for the system. Because replacement occurs only at the replacement time, the age of the system is t_0. Thus, the value of the age limit for replacement is given by

$$\min_{t_0} z(t_0) = \frac{c_r + c_m \int_0^{t_0} h(s)\,ds}{t_0}. \tag{8.5}$$

To find the value of t_0 for which the above function is minimized, we take the derivative of the right-hand side of Eq. (8.5). Setting the derivative equal to zero yields the equation

$$t_0 h(t_0) - \int_0^{t_0} h(s)\,ds = \frac{c_r}{c_m}. \tag{8.6}$$

Again the question must be asked as to how many solutions might exist for Eq. (8.6). And, as before, as long as the model deals with a system having an IFR distribution, there will be at most one solution. The optimal minimal-repair policy is, therefore, to determine the value of t_0 that satisfies Eq. (8.6)—minimally repair any failures that occur before t_0 and replace at t_0.

EXAMPLE 8.4 We return to Example 8.2 and assume that a procedure has been developed that can patch the cracks in the lining that caused the pot to fail. When a failure occurs, there is still the $20,000 failure cost incurred, but the patch costs an additional $4000; thus, $c_m = \$24,000$. The integrals of Eqs. (8.5) and (8.6) involve the hazard rate function, which is easily integrated for the Weibull distribution; in particular, $\int_0^t h(s)\,ds = (\lambda t)^\alpha$. Therefore, the average long-run cost for the minimal-repair model using a Weibull distribution is

$$z(t_0) = \frac{c_r + c_m(\lambda t_0)^\alpha}{t_0},$$

and Eq. (8.6) becomes

$$t_0^* = \frac{1}{\lambda}\left(\frac{c_r}{(\alpha - 1)c_m}\right)^{1/\alpha}.$$

Because $\lambda = 0.0392$ and $\alpha = 16.4$, the optimum replacement time is $t_0^* = 21.89$, and the expected long-run average cost is $1459.59 per month. ■

8.2.2 Minimal Repairs with Early Replacements

Another scenario when minimal repairs are a reasonable option is to establish two age-control points, t_1 and t_2. (See the schematic in Figure 8.4.) Any failure before age t_1 is corrected by a minimal repair, a replacement is made for a system that fails between the ages of t_1 and t_2, and a preventative replacement is made at the age of t_2 if the system is still operating. There are three relevant costs: c_m is the cost for each minimal repair, c_r is the cost to replace the system, and c_f is an additional cost that must be added to c_r for a replacement due to a failure.

The expected cost per system is given by

$$c_r + c_m \int_0^{t_1} h(s)\,ds + c_f \frac{F(t_2) - F(t_1)}{1 - F(t_1)}.$$

The first term is simply the cost of replacement, which must always be incurred; the second term is the cost of a minimal repair times the expected number of minimal repairs during the interval $(0, t_1]$, and the third term is the cost of failure times the probability that the system fails in the interval $(t_1, t_2]$. Notice that the probability of failure is a conditional probability because we know that the system must be working at time t_1.

FIGURE 8.4 Schematic representation for minimal-repair age replacement policy.

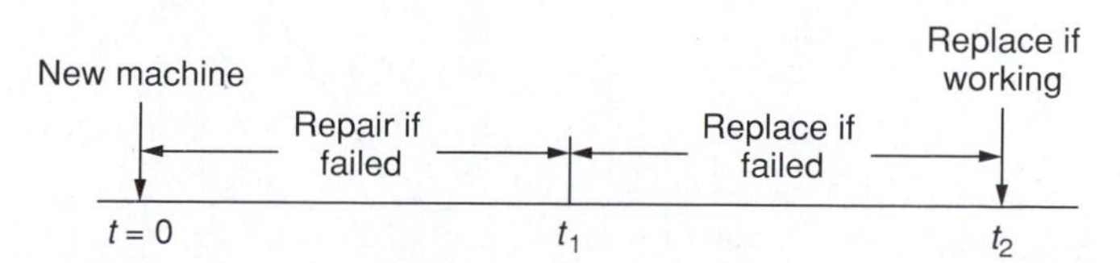

To obtain the expected life length of the system, let T be the random variable denoting the age at which the system is replaced, and, as before, let $F(\cdot)$ denote the system's failure distribution. Then,

$$\Pr\{T > t\} = \begin{cases} 1 & \text{if } t < t_1, \\ [1 - F(t)]/[1 - F(t_1)] & \text{if } t_1 \leq t < t_2, \\ 0 & \text{if } t \geq t_2. \end{cases} \tag{8.7}$$

The expected life of the system is thus given by

$$t_1 + \frac{1}{1 - F(t_1)} \int_{t_1}^{t_2} [1 - F(s)]\, ds.$$

(See the graph of the distribution functions in Figure 8.5.)

The optimization problem for the (t_1, t_2) policy is again the expected cost per system divided by the expected life length per system. This yields a two-dimensional optimization problem as

$$\min_{t_1, t_2} z(t_1, t_2) = \frac{c_r + c_m \int_0^{t_1} h(s)\, ds + c_f[F(t_2) - F(t_1)]/[1 - F(t_1)]}{t_1 + \int_{t_1}^{t_2} [1 - F(s)]/[1 - F(t_1)]\, ds}. \tag{8.8}$$

Because this is a two-dimensional unconstrained optimization problem, we need to take the partial derivative of the right-hand side of Eq. (8.8) with respect to t_1 and t_2. (The partials are not difficult, just tedious.) One of the standard nonlinear search techniques can then be used to find the optimum (t_1, t_2) policy.

FIGURE 8.5 Illustration of the replacement time distribution from Eq. (8.7).

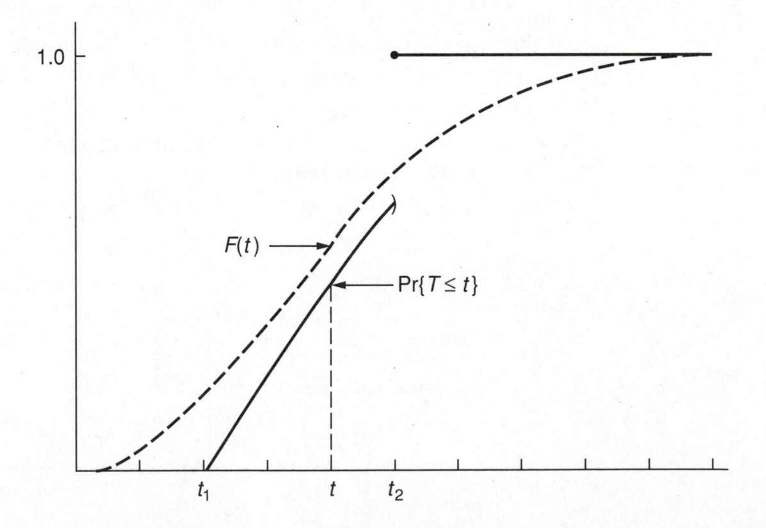

EXAMPLE 8.5 So that we might compare the three different age-control policies mentioned in this chapter, we return to Example 8.4 and calculate the optimum values for t_1 and t_2. The function to be minimized [Eq. (8.8)] becomes

$$z(t_1, t_2) = \frac{50{,}000 + 24{,}000(0.0392t_1)^{16.4} - 20{,}000e^{(0.0392t_1)^{16.4}}e^{-(0.0392t_2)^{16.4}}}{t_1 + e^{(0.0392t_1)^{16.4}}\int_{t_1}^{t_2} e^{-(0.0392s)^{16.4}}\,ds}.$$

Using a nonlinear search routine, the optimum values turn out to be $t_1^* = 19.50$ and $t_2^* = 22.14$, yielding an expected long-run average cost of \$1446.39. It is interesting to compare these numbers with those of Examples 8.2 and 8.4. The optimal age replacement problem without minimal repairs (Example 8.2) is equivalent to the problem in Eq. (8.8) where t_1 is restricted to be zero, and the minimal-repair problem without early replacement (Example 8.4) is equivalent to Eq. (8.8) where t_2 is restricted to be equal to t_1. There is an almost insignificant difference between the optimal no minimal repair policy (Example 8.2) and the optimal (t_1, t_2) policy, but that does not imply minimal repair is never worthwhile. Because the optimal $t_1^* = 19.5$, the probability that a minimal repair is performed equals 0.012; thus, the *expected* cost savings is small. However, given the occurrence of an early failure, the actual cost savings may be significant. ■

▶ *Suggestion: Do Exercises 8.7 and 8.11.*

8.3 BLOCK REPLACEMENT

One disadvantage of an age replacement policy is that such a policy demands good record keeping. In a situation in which there are many items subject to failure, it may be advantageous to use a block replacement policy—that is, a replacement policy that replaces all items at fixed intervals of time independent of the age of the individual items (see Figure 8.6 for a schematic representing a block replacement policy). We will further assume that the cost of the items is relatively low compared to a setup cost needed for the replacement. Consider, for example, a city that must maintain its street lights. The major cost for replacement is the sending out of the crew. There is little difference in replacing one or two lights.

For modeling purposes we shall return to the discrete period case and let f_m, for $m = 1, 2, \cdots$, denote the failure probability mass function for an item (the items' failure times are assumed to be probabilistically identical), let

FIGURE 8.6 Schematic representation for a block replacement policy.

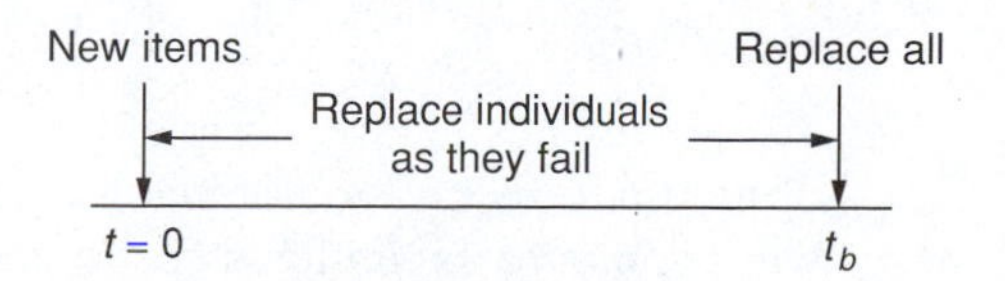

c_r be the individual replacement costs, and let K be the setup cost. The policy under consideration is to establish a block replacement time, t_b. That is, at time 0 all items are assumed new—when an item fails it must be immediately replaced with a new part; then at time t_b all items are replaced—even those that had recently been replaced. Finally, we shall let $z(t_b)$ denote the expected long-run average cost for a replacement policy using a replacement time of t_b.

The difficult aspect of obtaining an expression for $z(t_b)$ is the determination of the number of individual replacements made. We will not be able to obtain a closed form expression for that number, but it can be easily obtained recursively. Let $\overline{n}_0$ denote the number of items in the system, and let $\overline{n}_k$ denote the expected number of items replaced during period k. These can be determined by the following recursive scheme:

$$\begin{aligned}
\overline{n}_1 &= f_1\overline{n}_0 \\
\overline{n}_2 &= f_2\overline{n}_0 + f_1\overline{n}_1 \\
\overline{n}_3 &= f_3\overline{n}_0 + f_2\overline{n}_1 + f_1\overline{n}_2 \\
&\vdots \\
\overline{n}_m &= f_m\overline{n}_0 + f_{m-1}\overline{n}_1 + \cdots + f_1\overline{n}_{m-1}.
\end{aligned} \tag{8.9}$$

Notice, for example, that $\overline{n}_2$ involves the factor $f_1\overline{n}_1$ because those items that were new in time period 1 may fail in the following time period.

Utilizing the above definition, the expected long-run average cost associated with a block replacement time of t_b is given by

$$z(t_b) = \frac{(K + \overline{n}_0 c_r) + (K + c_r)\sum_{m=1}^{t_b}\overline{n}_m}{t_b}.$$

The first expression in the numerator on the right-hand side of the above equation (namely, $K + \overline{n}_0 c_r$) reflects the cost incurred during each block replacement time, and the second term reflects the costs of all the individual replacements made. In the development of the second term, we assumed that as soon as a failure occurs, it is replaced immediately; thus, if three failures occur within a single period, the three failures would occur at separate times and would call for three individual replacements. Likewise, if failures occur in the last period (between times $t_b - 1$ and t_b), the failed items are replaced at their failure time and then again when all items are replaced at time t_b.

Therefore, the optimal block replacement time is given as t_b^*, where t_b^* is the i that solves the following

$$\min_i z(i) = \frac{1}{i}\left[K + \overline{n}_0 c_r + (K + c_r)\sum_{m=1}^{i}\overline{n}_m\right]. \tag{8.10}$$

Sometimes it is possible for the costs to be such that it is never optimum to replace any items before failure. Therefore, it is important to always compare the minimal cost of the block replacement policy (i.e., the cost at the first local minimum) against the policy that never replaces ahead of time (i.e., the policy that replaces an item when it fails and never replaces working items). Such a policy has an average cost given by

$$z(\infty) = \frac{\overline{n}_0(K + c_r)}{E[T]}, \tag{8.11}$$

where $E[T] = \sum_{m=0}^{\infty} mf(m) = \sum_{m=0}^{\infty} \overline{F_m}$.

EXAMPLE 8.6 A manufacturing firm produces aluminum powder. One of the uses of the powder is as an explosive, so care is taken during a major portion of the manufacturing process to ensure that the powder is produced in an oxygen-free environment. Therefore, the firm has installed 50 sensors at various locations to record the presence of oxygen. As soon as a sensor stops working, it must be immediately replaced; however, the process must be shut down to replace a sensor, costing approximately $300 just for the shutdown and start-up period. The variable cost to replace a sensor is $40, which includes the sensor itself as well as labor costs and a slight additional cost of lost production. The probability mass function describing the failure time of the sensors is as follows:

Week	2	3	4	5	6	7	8	9	10	11	12
Prob.	0.02	0.03	0.05	0.1	0.3	0.2	0.15	0.08	0.04	0.025	0.005

Thus, we have the following calculations:

For $i = 1$

$$\overline{n}_0 = 50$$
$$\overline{n}_1 = 50 \times 0 = 0$$
$$z(i) = (300 + 50 \times 40) = 2300$$

For $i = 2$

$$\overline{n}_2 = 50 \times 0.02 + 0 \times 0 = 1$$
$$z(i) = \frac{1}{2}(300 + 50 \times 40 + 340 \times 1) = 1320$$

For $i = 3$

$$\overline{n}_3 = 50 \times 0.03 + 0 \times 0.02 + 1 \times 0 = 1.5$$
$$z(i) = \frac{1}{3}[300 + 50 \times 40 + 340 \times (1 + 1.5)] = 1050$$

For $i = 4$

$$\overline{n}_4 = 50 \times 0.05 + 0 \times 0.03 + 1 \times 0.02 + 1.5 \times 0 = 2.52$$
$$z(i) = \frac{1}{4}[300 + 50 \times 40 + 340 \times (1 + 1.5 + 2.52)] = 1002$$

For $i = 5$

$$\overline{n}_5 = 50 \times 0.1 + 0 \times 0.05 + 1 \times 0.03 + 1.5 \times 0.02 + 2.52 \times 0 = 5.06$$

$$z(i) = \frac{1}{5}[300 + 50 \times 40 + 340 \times (1 + 1.5 + 2.52 + 5.06)] = 1145$$

Thus, the optimal block replacement time is at the end of the fourth week. We need now to compare this to the cost of only individual replacement [Eq. (8.11)], which yields a weekly cost of $2543. Therefore, the optimal block replacement cost of $1002 is the best policy. ■

▶ *Suggestion: Do Exercises 8.9, 8.10 and 8.12.*

8.4 EXERCISES

8.1 This problem[3] involves considering the importance of keeping track of history when discussing the reliability of a machine. Let T be a random variable that indicates the time until failure for the machine. Assume that T has a uniform distribution from zero to two years and answer the question, "What is the probability that the machine will continue to work for at least three more months?"

(a) Assume that the machine is new.

(b) Assume that the machine is one year old and has not yet failed.

(c) Now assume that T has an exponential distribution with a mean of one year and answer parts (a) and (b) again.

(d) Is it important to know how old the machine is in order to answer the question, "What is the probability that the machine will continue to work for at least three more months"?

8.2 In its production process an old factory uses a conveyor belt that seems to be always in need of repair. It costs $500 to have a repair crew "fix" the belt; however, after it is fixed it still remains as unreliable as it was immediately after the last time it was fixed. After the belt is fixed, the probability that it will work for exactly k days is $0.7 \times 0.3^{k-1}$. When the belt fails, production costs an additional $600 because of the interruption to the production process. If the belt works during the day, the repair crew can be called in during the evening off-shift and make adjustments to the belt, so that the probability law governing the next failure is the same as if the belt had failed during the day and the repair crew had fixed it that day. (When the repair crew fixes the belt during the evening, the $600 process interruption cost is saved.) Explain why the optimum replacement policy for this belt is to never replace before failure.

[3]This problem is from the first chapter and is included here in case it was skipped.

8.3 A specialized battery is a necessary component to a production machine. Purchase and installation of the battery costs $4000. Although the battery is advertised as an "18-month" battery, the actual probability mass function governing its life length is

Month	Prob.	Month	Prob.	Month	Prob.
1	0.05	8	0.00	15	0.25
2	0.03	9	0.00	16	0.15
3	0.02	10	0.02	17	0.08
4	0.0	11	0.03	18	0.04
5	0.0	12	0.05	19	0.025
6	0.0	13	0.10	20	0.005
7	0.0	14	0.15		

For purposes of this problem, assume that failure always occurs just before the end of the month. Replacement can be made at the end of a month without any disruption to the production schedule; however, if the battery fails before replacement, an additional cost of $1000 is incurred. (Notice that the above distribution is not IFR because of the initial probability of failure. This is quite common, because initial failures are usually due to manufacturing defects. After the "break-in" period, the distribution is IFR.)

(a) What is the optimal replacement time?

(b) Assume that the cost of record keeping is $10 per month. In other words, in order to follow the replacement policy, records need to be maintained regarding the life of the battery. A policy that ignores the life of the battery and simply replaces at failure has no monthly "record-keeping" costs. What is the optimal replacement policy?

8.4 The purpose of this problem is to consider warranties and to provide practice in model development. You will need to adjust the numerator in Eq. (8.2) slightly in order to answer the questions. We continue with the above problem dealing with the specialized battery; that is, you should use the data from Exercise 8.3 assuming that there are no record-keeping costs.

(a) The battery comes with a 6-month warranty; that is, if the battery fails in the first 6 months, a new one is installed free of charge. Does this change the optimal policy? (Assume that if battery failure occurs in the first 6 months, the warranty is renewed when the new battery is installed. Of course, the failure cost of $1000 is still incurred.)

(b) For an additional $450 in the purchase price of the battery, the 6-month warranty can be extended to a 12-month warranty. Is it worthwhile? (Assume that if battery failure occurs in the first 12 months, a new 12-month warranty free of charge comes with the new battery.)

8.5 Simulate the situation described in Exercise 8.4 (a). Does your simulation agree with your analytical results?

8.6 Simulate the process described in Example 8.5. The theoretical optimum average value is approximately \$1446. Is this within a 95% confidence interval derived from your simulation runs?

8.7 Show that a solution to Eq. (8.6) yields the unique minimum for Eq. (8.5) when the system has an increasing failure rate distribution.

8.8 The temperature control in an oven used within a manufacturing process must be periodically reset. Once the control has been reset, it will be accurate for a random length of time, which is described by the Weibull distribution with scale parameter $\lambda = 0.01$ per day and shape parameter $\alpha = 2$. To reset the control, the process must be stopped at a cost of \$750. If the control is not reset, defective parts will be produced after the control loses its accuracy. As soon as the defective parts are identified, the process is stopped and the control reset. The expected cost incurred due to the production of the defective parts is \$550. What is the optimal age replacement policy for the resetting of the control?

8.9 Part of the manufacturing process for powdered aluminum must be carried out in an oxygen-free environment. There are 25 sensors that are used to ensure that the environment stays free of oxygen. To replace one or more sensors, the process must be shut down at a cost of \$900 (including lost production). In addition, the sensors themselves cost \$75 to replace (including labor). The life of the sensors can be approximated by a Poisson distribution with a mean of 10 months. What is the optimum block replacement policy?

8.10 Verify the optimum block replacement policy derived in Exercise 8.9 by a simulation analysis.

8.11 Consider an alternative minimal-repair policy defined by the age t_1. Any failure before age t_1 is corrected by a minimal repair. A replacement is made at the first failure that occurs after t_1 (i.e., no early replacement.) Let c_m be the cost for each minimal repair, c_r the cost of replacement at failure, and $F(\cdot)$ the system's failure distribution.

(a) Derive the long-run expected cost per unit time.

(b) Under what general circumstances will the policy in (a) be better than the minimal repair policy without early replacement given by Eq. (8.5)?

8.12 Equation (8.10) gives the optimum block replacement time, but to use it directly (as in Example 8.6) requires repetitive calculations. Show that a recursive approach can be taken where the long-run average costs are given by

$$z(1) = K(1 + \overline{n}_1) + c_r(\overline{n}_0 + \overline{n}_1)$$

$$z(i) = z(i-1)\frac{i-1}{i} + \frac{(K + c_r)\overline{n}_i}{i} \quad \text{for } i = 2, 3, \cdots,$$

where $\overline{n}_i$ is given by Eq. (8.9).

9 Markov Decision Processes[1]

Markov chains provide a useful modeling tool for determining expected profits or costs associated with certain types of systems. The key characteristic that allows for a Markov model is a probability law in which the future behavior of the system is independent of the past behavior given the present condition of the system. When this Markov property is present, the dynamics of the process can be described by a matrix containing the one-step transition probabilities and a vector of initial conditions. In some circumstances the transition probabilities may depend on decisions made just before the transition time. Furthermore, not only the transition probabilities, but also associated costs or profits per transition, may depend on decisions made at the transition times. For example, consider a slight variation of Exercise 2.1. This problem featured Joe and Pete playing a game of matching pennies (the pennies were biased), and the Markov model for the game used a state space representing the number of pennies in Joe's pocket. We shall generalize the previous homework problem by having two rules for the game instead of one: Rule 1 states that Joe wins when the coins match and Pete wins when the coins do not match; rule 2 is the opposite—namely, that Pete wins when the coins match and Joe wins when they do not match. Now, before each play of the game, the previous winner gets to decide which rule is to be used. The dynamics of this game can no longer be modeled as a simple Markov chain because we need to know how Joe and Pete will make their decisions before transition probabilities can be determined.

[1]This chapter would normally be skipped in a one-semester undergraduate course.

Processes that involve decisions affecting the transition probabilities often yield models in which optimization questions naturally arise. When the basic structure of Markov chains is combined with decision processes and optimization questions, a new model called a *Markov decision process* is formed. In Markov decision processes (in contrast to Markov chains) future outcomes depend not only on the current state but also on the decisions made. The purpose of this chapter is to present some of the basic concepts and techniques of Markov decision theory and to indicate the types of problems that are amenable to modeling as such processes.

9.1 BASIC DEFINITIONS

The basic structure and dynamics of a Markov decision process will be introduced by way of an example. In this example we introduce the basic elements of a Markov decision process—namely, a stochastic process with a state space denoted by E and a decision process with an action space denoted by A.

EXAMPLE 9.1 Let $X = \{X_0, X_1, \cdots\}$ be a stochastic process with a four-state state space $E = \{a, b, c, d\}$. This process will represent a machine that can be in one of four operating conditions, denoted by the states a through d, indicating increasing levels of deterioration. As the machine deteriorates, not only is it more expensive to operate, but also production is lost. Standard maintenance activities are always carried out in states b through d so that the machine may improve due to maintenance; however, improvement is not guaranteed. In addition to the state space, there is an *action space* that gives the decisions possible at each step. (We sometimes use the words "decisions" and "actions" interchangeably when referring to the elements of the action space.) In this example we shall assume the action space is $A = \{1, 2\}$; that is, at each step there are two possible actions: use an inexperienced operator (action 1) or use an experienced operator (action 2). To complete the description of a Markov decision problem, we need a cost vector and a transition matrix for each possible action in the action space. For our example, define the two cost vectors[2] and two Markov matrices as

$$\boldsymbol{f}_1 = (100, 125, 150, 500)^T,$$
$$\boldsymbol{f}_2 = (300, 325, 350, 600)^T,$$
$$\boldsymbol{P}_1 = \begin{bmatrix} 0.1 & 0.3 & 0.6 & 0.0 \\ 0.0 & 0.2 & 0.5 & 0.3 \\ 0.0 & 0.1 & 0.2 & 0.7 \\ 0.8 & 0.1 & 0.0 & 0.1 \end{bmatrix},$$

[2]A superscript T denotes transpose.

$$P_2 = \begin{bmatrix} 0.6 & 0.3 & 0.1 & 0.0 \\ 0.75 & 0.1 & 0.1 & 0.05 \\ 0.8 & 0.2 & 0.0 & 0.0 \\ 0.9 & 0.1 & 0.0 & 0.0 \end{bmatrix}.$$

The dynamics of the process are illustrated in Figure 9.1 and are as follows: If, at time n, the process is in state i and the decision k is made, then a cost of $f_k(i)$ is incurred and the probability that the next state will be j is given by $P_k(i, j)$. To illustrate, if $X_n = a$ and decision 1 is made, then a cost of \$100 is incurred (representing the operator cost, lost production cost, and machine operation cost) and $\Pr\{X_{n+1} = a\} = 0.1$; or, if $X_n = d$ and decision 2 is made, then a cost of \$600 is incurred (representing the operator cost, machine operation cost, major maintenance cost, and lost-production cost) and $\Pr\{X_{n+1} = a\} = 0.9$. ■

In general, the decision made at a given point in time is a random variable, which we shall denote by D_n. Thus, there are two stochastic processes defined—the system description process, $X = \{X_0, X_1, \cdots\}$, and the decision process, $D = \{D_0, D_1, \cdots\}$. We can now give explicitly the form for the processes considered in this chapter.

DEFINITION 9.1 *Let X be a system description process with state space E and let D be a decision process with action space A. The process (X, D) is a* MARKOV DECISION PROCESS *if, for $j \in E$ and $n = 0, 1, \cdots$, the following holds:*

$$\Pr\{X_{n+1} = j \mid X_0, D_0, \cdots, X_n, D_n\} = \Pr\{X_{n+1} = j \mid X_n, D_n\}.$$

Furthermore, for each $k \in A$, let $\boldsymbol{f}_k$ be a cost vector and $\boldsymbol{P}_k$ be a Markov matrix. Then

$$\Pr\{X_{n+1} = j \mid X_n = i, D_n = k\} = P_k(i, j)$$

and the cost $f_k(i)$ is incurred whenever $X_n = i$ and $D_n = k$.

An obvious question is, "How can decisions be made to minimize costs?" A secondary question is, "What does 'minimize' mean?" We first discuss the different ways decisions can be made.

DEFINITION 9.2 *A* POLICY *is any rule, using current information, past information, and/or randomization that specifies which action to take at each point in time. The set of all (decision) policies is denoted by $\mathfrak{D}$.*

FIGURE 9.1 Sequence of events in a Markov decision process.

Observe state		Take action		Incur cost		Transition to next state
$X_n = i$	→	$D_n = k$	→	$f_k(i)$	→	$P_k(i, j)$

The following are some legitimate policies for the above problem:

Policy 1 Always choose action 1, independent of the state for X (i.e., let $D_n \equiv 1$ for all n).

Policy 2 If X_n is in state a or b, let $D_n = 1$; if X_n is in state c or d, let $D_n = 2$.

Policy 3 If X_n is in state a or b, let $D_n = 1$; if X_n is in state c, toss a (fair) coin and let $D_n = 1$ if the toss results in a head and let $D_n = 2$ if the toss results in a tail; if X_n is in state d, let $D_n = 2$.

Policy 4 Let $D_n \equiv 1$ for $n = 0$ and 1. For $n \geq 2$, if $X_n > X_{n-1}$ and $X_{n-2} = a$, let $D_n = 1$; if $X_n > X_{n-1}, X_{n-2} = b$, and $D_{n-1} = 2$, let $D_n = 1$; otherwise, let $D_n = 2$.

As you look over these example policies, observe the wide range of possibilities. Policy 3 involves a randomization rule, and policy 4 uses history. Once a policy is selected, the probability law governing the evolution of the process is determined. However, for an arbitrary policy, the Markov decision process is *not* necessarily a Markov chain because we allow decisions to depend upon history. For example, if policy 4 is used, the decision maker needs X_n, X_{n-1}, and X_{n-2} in order to know which decision is to be made.

We are now ready to answer the question regarding the meaning of the term *minimize*. There are two common criteria used: (1) expected total discounted cost, and (2) average long-run cost.

Expected Total Discounted Cost Criterion. The total discounted cost problem is equivalent to using a present worth calculation for the basis of decision making. Specifically, let α be a discount factor such that one dollar obtained at time $n = 1$ has a present value of α at time $n = 0$. [In traditional economic terms, if r is a rate of return (interest rate) specified by the management of a particular company, then $\alpha = 1/(1 + r)$.] The total discounted cost for a particular Markov decision process is thus given by $E[\sum_{n=0}^{\infty} \alpha^n f_{D_n}(X_n)]$. For example, assume that, in the above example, policy 1 is chosen (i.e., the inexperienced operator is always used), and a discount factor of $\alpha = 0.95$ (equivalent to a rate of return of approximately 5.3% per period) is used. In that case the example reduces to the computation of the total discounted cost of a standard Markov chain as was discussed in Chapter 2. The total expected discounted cost is calculated according to Property 2.21 and is given by $(\boldsymbol{I} - \alpha \boldsymbol{P}_1)^{-1}\boldsymbol{f}_1$, which yields the vector $\boldsymbol{v} = (4502, 4591, 4676, 4815)^T$. In other words, if the process starts in state a, the expected present value of all future costs is 4502. (It should be observed that when using a discount factor, the total discounted cost depends on the initial state.)

This example illustrates the fact that a specific policy needs to be selected before expectations can be taken. To designate this dependence on the policy, a subscript will be used with the expectation operator. Thus, $E_d[\cdot]$ denotes an expectation under the probability law specified by the policy

$d \in \mathfrak{D}$. The total discounted value of a Markov decision process under a discount factor of α using the policy $d \in \mathfrak{D}$ will be denoted by v_d^α; that is,

$$v_d^\alpha(i) = E_d\left[\sum_{n=0}^{\infty} \alpha^n f_{D_n}(X_n) \mid X_0 = i\right]$$

for $i \in E$ and $0 < \alpha < 1$. Thus, the discounted cost optimization problem can be stated as follows: Find $d^\alpha \in \mathfrak{D}$ such that $v_{d^\alpha}^\alpha(i) = v^\alpha(i)$ where the vector $\boldsymbol{v}^\alpha$ is defined, for $i \in E$, by

$$v^\alpha(i) = \min_{d \in \mathfrak{D}} v_d^\alpha(i). \tag{9.1}$$

It should be pointed out that the question of the existence of an optimal policy can be a difficult one when the state space is infinite; however, for the purposes of this text, we shall only consider problems in which its existence is ensured by assuming that both the state space and action space are finite.

Average Long-Run Cost Criterion. Using an infinite horizon planning period, the total (undiscounted) cost may be infinite for all possible decisions, so that total cost cannot be used to distinguish between alternative policies. However, if cost per transition is compared, then alternatives may be evaluated. Thus a commonly used criterion is $\lim_{m\to\infty} \frac{1}{m}\sum_{n=0}^{m-1} f_{D_n}(X_n)$. For example, we again assume that policy 1 is used; thus, action 1 is always chosen. Using the Markov chain results from the previous chapter, the long-run cost can be calculated according to Property 2.20. In other words, we first calculate the steady-state probabilities using the matrix $\boldsymbol{P}_1$, which yields $\boldsymbol{\pi} = (0.253, 0.167, 0.295, 0.285)$; then, the vector $\boldsymbol{\pi}$ is multiplied by the vector $\boldsymbol{f}_1$ yielding a long-run average cost of 232.925.

For a fixed policy $d \in \mathfrak{D}$, the average long-run cost for the Markov decision process will be denoted by φ_d; in other words,

$$\varphi_d = \lim_{m\to\infty} \frac{f_{D_0}(X_0) + \cdots + f_{D_{m-1}}(X_{m-1})}{m}.$$

Thus, the optimization problem can be stated as follows: Find $d^* \in \mathfrak{D}$ such that $\varphi_{d^*} = \varphi^*$, where φ^* is defined by

$$\varphi^* = \min_{d \in \mathfrak{D}} \varphi_d. \tag{9.2}$$

As before, the existence question can be a difficult one for infinite state spaces. We leave such questions to the advanced textbooks.

9.2 STATIONARY POLICIES

The Markov decision problem as stated in Definitions 9.1 and 9.2 appears difficult because of the generality permitted by the policies. However, it turns out that, under fairly general conditions, the optimum policy always has a very nice structure so that the search for an optimum can be limited to a much smaller set of policies. In particular, policies of the type exemplified by policy 3 and policy 4 can be excluded in the search for an optimum. Consider the following two definitions.

DEFINITION 9.3 *An* ACTION FUNCTION *is a vector that maps the state space into the action space; that is, an action function assigns an action to each state.*

In other words, if $\boldsymbol{a}$ is an action function, then $a(i) \in A$ for each $i \in E$. In the example policies given immediately after Definition 9.2, policy 2 is equivalent to the action function $a = (1, 1, 2, 2)$, where the action space is $A = \{1, 2\}$.

DEFINITION 9.4 *A* STATIONARY POLICY *is a policy that can be defined by an action function. The stationary policy defined by the function* $\boldsymbol{a}$ *takes action* $a(i)$ *at time* n *if* $X_n = i$*, independent of previous states, previous actions, and time* n*.*

The key idea of a stationary policy is that it is independent of time and is a nonrandomized policy that depends only on the current state of the process and, therefore, ignores history. Computationally, a stationary policy is convenient in that the Markov decision process under a stationary policy is always a Markov chain. To see this, let the transition matrix $\boldsymbol{P}^a$ be defined, for $i, j \in E$, by

$$P^a(i, j) = P_{a(i)}(i, j) \tag{9.3}$$

and let the cost vector $\boldsymbol{f}^a$ be defined, for $i \in E$, by

$$f^a(i) = f_{a(i)}(i). \tag{9.4}$$

Using the example from the previous section, we define the policy that uses the inexperienced operator whenever the machine is in state a or b and uses the experienced operator whenever the machine is in state c or d. The Markov decision process thus forms a chain with the Markov matrix,

$$\boldsymbol{P}^a = \begin{bmatrix} 0.1 & 0.3 & 0.6 & 0.0 \\ 0.0 & 0.2 & 0.5 & 0.3 \\ 0.8 & 0.2 & 0.0 & 0.0 \\ 0.9 & 0.1 & 0.0 & 0.0 \end{bmatrix},$$

and cost vector given by

$$\boldsymbol{f}^a = (100, 125, 350, 600)^T.$$

Any stationary policy can be evaluated by forming the Markov matrix and cost vector associated with it according to Eqs. (9.3) and (9.4). The reason that this is important is that the search for the optimum can always be restricted to stationary policies as is given in the following property.

Property 9.5 If the state space E is finite, there exists a stationary policy that solves the problem given in Eq. (9.1). Furthermore, if every stationary policy yields an irreducible Markov chain, there exists a stationary policy that solves the problem given in Eq. (9.2). [The optimum policy may depend on the discount factor and may be different for Eqs. (9.1) and (9.2).]

For those familiar with linear programming, this property is analogous to the result that permits the simplex algorithm to be useful. Linear programming starts with an uncountably infinite set of feasible solutions; then, by taking advantage of convexity, the property is established that allows the analyst to focus only on a finite set of solutions, namely, the set of extreme points. The simplex algorithm is a procedure that starts at an easily defined extreme point and moves in a logical fashion to another extreme point in such a way that the solution is always improved. When no more improvement is possible, the optimum has been found. In Markov decision theory we start with an extremely large class of possible policies, many of which produce processes that are not Markovian. Property 9.5 allows the analyst to focus on a much smaller set of policies, each one of which produces a Markov chain. In the remaining sections, algorithms will be developed that start with an easily defined stationary policy and move to another stationary policy in such a way as to always improve until no more improvement is possible, in which case an optimum has been found.

9.3 DISCOUNTED COST ALGORITHMS

This section presents three different procedures for finding the optimal policy under an expected total discounted cost criterion. These procedures are based on a *fixed-point* property that holds for the optimal value function. In this section we first discuss this key property.

In mathematics a function is called *invariant* with respect to an operation if the operation does not vary the function. For example, the steady-state vector, $\boldsymbol{\pi}$, is also called the invariant vector for the Markov matrix $\boldsymbol{P}$ because the operation $\boldsymbol{\pi P}$ does not vary the vector $\boldsymbol{\pi}$. For Markov decision processes the operation will be more complicated than simple matrix multiplication, but the basic idea of an invariant function will still hold. If the invariant function is unique, then it is called a *fixed point* for the operation.

For those familiar with dynamic programming, the invariant property as applied to Markov decision processes will be recognized as the standard dynamic programming recursive relationship. The operation involves minimizing current costs plus all future costs, where the future costs must be discounted to the present in order to produce a total present value.

Property 9.6 ***Fixed-Point Theorem for Markov Decision Processes*** Let $\boldsymbol{v}^{\alpha}$ be the optimal value function as defined by Eq. (9.1) with $0 < \alpha < 1$. The function $\boldsymbol{v}^{\alpha}$ satisfies, for each $i \in E$, the following:

$$\boldsymbol{v}^{\alpha}(i) = \min_{k \in A}\{f_k(i) + \alpha \sum_{j \in E} P_k(i,j) v^{\alpha}(j)\}. \tag{9.5}$$

Furthermore, it is the only function satisfying this property.

Property 9.6 provides a means to determine if a given function happens to be the optimal function. If we are given a function that turns out to be the optimal value function, it is easy to obtain the optimal policy through the next property.[3]

Property 9.7 Let $\boldsymbol{v}^{\alpha}$ be the optimal value function as defined by Eq. (9.1) with $0 < \alpha < 1$. Define an action function, for each $i \in E$, as follows:

$$a(i) = \operatorname{argmin}_{k \in A}\{f_k(i) + \alpha \sum_{j \in E} P_k(i,j) v^{\alpha}(j)\}.$$

The stationary policy defined by the action function $\boldsymbol{a}$ is an optimal policy.

At the moment, we do not know how to obtain $\boldsymbol{v}^{\alpha}$, but Property 9.7 tells how to obtain the optimal policy once $\boldsymbol{v}^{\alpha}$ is known. To illustrate, let us assert that the optimal value function for the machine problem given in Example 9.1 is $\boldsymbol{v}^{\alpha} = (4287, 4382, 4441, 4613)$ for $\alpha = 0.95$. Our first task is to verify the assertion through the fixed-point theorem for Markov decision processes. The calculations needed to verify that the given vector is the optimal value function are as follows. (Note that the calculations are needed for each state in the state space.)

$$v^{\alpha}(a) = \min\left\{ 100 + 0.95(0.1, 0.3, 0.6, 0.0)\begin{pmatrix}4287\\4382\\4441\\4613\end{pmatrix}; \quad 300 + 0.95(0.6, 0.3, 0.1, 0.0)\begin{pmatrix}4287\\4382\\4441\\4613\end{pmatrix}\right\}$$
$$= \min\{4287; 4414\} = 4287.$$

$$v^{\alpha}(b) = \min\left\{ 125 + 0.95(0.0, 0.2, 0.5, 0.3)\begin{pmatrix}4287\\4382\\4441\\4613\end{pmatrix}; \quad 325 + 0.95(0.75, 0.1, 0.1, 0.05)\begin{pmatrix}4287\\4382\\4441\\4613\end{pmatrix}\right\}$$
$$= \min\{4382; 4437\} = 4382.$$

[3]The term *argmin* used in the property refers to the *argument* that yields the minimum value.

$$v^\alpha(c) = \min\left\{ 150 + 0.95(0.0, 0.1, 0.2, 0.7)\begin{pmatrix}4287\\4382\\4441\\4613\end{pmatrix}; \right.$$

$$\left. 350 + 0.95(0.8, 0.2, 0.0, 0.0)\begin{pmatrix}4287\\4382\\4441\\4613\end{pmatrix} \right\}$$

$$= \min\{4478; 4441\} = 4441.$$

$$v^\alpha(d) = \min\left\{ 500 + 0.95(0.8, 0.1, 0.0, 0.1)\begin{pmatrix}4287\\4382\\4441\\4613\end{pmatrix}; \right.$$

$$\left. 600 + 0.95(0.9, 0.1, 0.0, 0.0)\begin{pmatrix}4287\\4382\\4441\\4613\end{pmatrix} \right\}$$

$$= \min\{4613; 4682\} = 4613.$$

Because, for each $i \in E$, the minimum of the two values yielded the asserted value of $v^\alpha(i)$, we know that it is optimum by Property 9.6. Looking back over the above calculations, we can also pick out the argument (i.e., action) that resulted in the minimum value. For state $i = a$, the first action yielded the minimum; for $i = b$, the first action yielded the minimum; for $i = c$, the second action yielded the minimum; and for $i = d$, the first action yielded the minimum. Therefore, from Property 9.7, the stationary optimal policy is defined by the action function $\boldsymbol{a} = (1, 1, 2, 1)$.

9.3.1 Value Improvement for Discounted Costs

The fixed-point theorem for Markov decision processes (Property 9.6) allows for an easy iteration procedure that will limit to the optimal value function. The procedure starts with a guess for the optimal value function. That guess is then used as the value for $\boldsymbol{v}^\alpha$ in the right-hand side of Eq. (9.5), and another value for $\boldsymbol{v}^\alpha$ is obtained. If the second value obtained is the same as the initial guess, we have an optimum; otherwise, the second value for $\boldsymbol{v}^\alpha$ is used in the right-hand side of (9.5) to obtain a third, and so forth. The concept of a fixed point is that by repeating such a successive substitution scheme, the fixed point will be obtained.

Algorithm 9.8 **Value Improvement Algorithm** The following iteration procedure will yield an approximation to the optimal value function as defined by Eq. (9.1).

Step 1 Make sure that $\alpha < 1$, choose a small positive value for ϵ, set $n = 0$, and let $v_0(i) = 0$ for each $i \in E$. (We set $\boldsymbol{v}_0 = \boldsymbol{0}$ for convenience; any initial solution is sufficient.)

Step 2 For each $i \in E$, define $v_{n+1}(i)$ by

$$v_{n+1}(i) = \min_{k \in A}\{f_k(i) + \alpha \sum_{j \in E} P_k(i,j) v_n(j)\}.$$

Step 3 Define δ by

$$\delta = \max_{i \in E}\{|v_{n+1}(i) - v_n(i)|\}.$$

Step 4 If $\delta < \epsilon$, let $\boldsymbol{v}^\alpha = v_{n+1}$ and stop; otherwise, increment n by one and return to step 2. ■

There are two major problems with the value improvement algorithm: (1) It can be slow to converge and (2) there is no simple rule for establishing a convergence criterion (i.e., setting a value for ϵ). In theory, it is true that, as the number of iterations approaches infinity, the value function becomes the optimum. However, in practice we need to stop short of infinity, and therefore the rule that says to stop when the change in the value function becomes negligible is commonly used. Exactly what qualifies as negligible is a judgment call that is sometimes difficult to make.

Another aspect of the value improvement algorithm is that the intermediate values produced by the algorithm do not give any indication of what the optimal policy is; in other words, the values for $\boldsymbol{v}_n$ are not helpful for determining the optimal (long-run) policy unless the algorithm has converged. (However, $\boldsymbol{v}_n$ does represent the optimal value associated with a finite horizon problem when $\boldsymbol{v}_0 = 0$; that is, $\boldsymbol{v}_n$ gives the minimal discounted cost, assuming that there are n transitions remaining in the life of the process.) The optimal long-run policy is obtained by taking the final value function from the algorithm and using Property 9.7. The major advantage of the value improvement algorithm is its computational simplicity.

When the value improvement algorithm is applied to the example machine problem defined in the previous section (using $\alpha = 0.95$), the following sequence of values is obtained:

$$\begin{aligned} \boldsymbol{v}_0 &= (0, 0, 0, 0) \\ \boldsymbol{v}_1 &= (100, 125, 150, 500) \\ \boldsymbol{v}_2 &= (230.62, 362.50, 449.75, 635.38) \\ \boldsymbol{v}_3 &= (481.58, 588.59, 594.15, 770.07) \\ &\vdots \end{aligned}$$

9.3.2 Policy Improvement for Discounted Costs

The algorithm of the previous section focused on the value function. In this section we present an algorithm that focuses on the policy and then calculates the value associated with that particular policy. The result is that convergence is significantly faster, but there are more calculations for each iteration. Specifically, the policy improvement algorithm involves an inverse routine (see step 3 below) that can be time-consuming for large problems and subject to round-off errors. However, if the problem is such that an accurate inverse is possible, then policy improvement is preferred over value improvement.

Algorithm 9.9 **Policy Improvement Algorithm** The following iteration procedure will yield the optimal value function as defined by Eq. (9.1) and its associated optimal stationary policy.

Step 1 Make sure that $\alpha < 1$, set $n = 0$, and define the action function $\boldsymbol{a}_0$ by

$$a_0(i) = \text{argmin}_{k \in A} f_k(i)$$

for each $i \in E$.

Step 2 Define the matrix $\boldsymbol{P}$ and the vector $\boldsymbol{f}$ by

$$f(i) = f_{a_n(i)}(i)$$
$$P(i,j) = P_{a_n(i)}(i,j)$$

for each $i, j \in E$.

Step 3 Define the value function $\boldsymbol{v}$ by

$$\boldsymbol{v} = (\boldsymbol{I} - \alpha \boldsymbol{P})^{-1} \boldsymbol{f}.$$

Step 4 Define the action function $\boldsymbol{a}_{n+1}$ by

$$a_{n+1}(i) = \text{argmin}_{k \in A} \{f_k(i) + \alpha \sum_{j \in E} P_k(i,j) v(j)\}$$

for each $i \in E$.

Step 5 If $\boldsymbol{a}_{n+1} = \boldsymbol{a}_n$, let $\boldsymbol{v}^\alpha = \boldsymbol{v}$ and $\boldsymbol{a}^\alpha = \boldsymbol{a}_n$, and stop; otherwise, increment n by one and return to step 2.

The basic concept of the policy improvement algorithm involves taking a stationary policy, calculating the cost vector and transition matrix associated with that policy [step 2, see Eqs. (9.3) and (9.4)], then determining the expected total discounted cost associated with that cost vector and transition matrix (step 3, see Property 2.21). If that policy was an optimal policy, then we would get that policy again through the use of Property 9.7 and the value function just calculated (step 4). This algorithm works because, if the policy is not the optimum, we are guaranteed that the policy formed in step 4 will have a value associated with it that is better (not worse) than the previous one. To illustrate this algorithm we outline the results obtained from applying the algorithm to our example problem:

ITERATION I

Step 1

$$\boldsymbol{a}_0 = (1, 1, 1, 1)$$

Step 2

$$\boldsymbol{f} = (100, 125, 150, 500)^T$$

$$\boldsymbol{P} = \begin{bmatrix} 0.1 & 0.3 & 0.6 & 0.0 \\ 0.0 & 0.2 & 0.5 & 0.3 \\ 0.0 & 0.1 & 0.2 & 0.7 \\ 0.8 & 0.1 & 0.0 & 0.1 \end{bmatrix}$$

Step 3

$$\boldsymbol{v} = (4502, 4591, 4676, 4815)^T$$

Step 4

$$\begin{aligned} a_1(a) &= \operatorname{argmin}\left\{ 100 + 0.95(0.1, 0.3, 0.6, 0.0)\begin{pmatrix} 4502 \\ 4591 \\ 4676 \\ 4815 \end{pmatrix};\right. \\ &\qquad\left. 300 + 0.95(0.6, 0.3, 0.1, 0.0)\begin{pmatrix} 4502 \\ 4591 \\ 4676 \\ 4815 \end{pmatrix}\right\} \\ &= \operatorname{argmin}\{4502; 4619\} = 1. \end{aligned}$$

$$\begin{aligned} a_1(b) &= \operatorname{argmin}\left\{ 125 + 0.95(0.0, 0.2, 0.5, 0.3)\begin{pmatrix} 4502 \\ 4591 \\ 4676 \\ 4815 \end{pmatrix};\right. \\ &\qquad\left. 325 + 0.95(0.75, 0.1, 0.1, 0.05)\begin{pmatrix} 4502 \\ 4591 \\ 4676 \\ 4815 \end{pmatrix}\right\} \\ &= \operatorname{argmin}\{4591; 4642\} = 1. \end{aligned}$$

$$\begin{aligned} a_1(c) &= \operatorname{argmin}\left\{ 150 + 0.95(0.0, 0.1, 0.2, 0.7)\begin{pmatrix} 4502 \\ 4591 \\ 4676 \\ 4815 \end{pmatrix};\right. \\ &\qquad\left. 350 + 0.95(0.8, 0.2, 0.0, 0.0)\begin{pmatrix} 4502 \\ 4591 \\ 4676 \\ 4815 \end{pmatrix}\right\} \\ &= \operatorname{argmin}\{4676; 4644\} = 2. \end{aligned}$$

$$a_1(d) = \operatorname{argmin}\left\{ 500 + 0.95(0.8, 0.1, 0.0, 0.1)\begin{pmatrix}4502\\4591\\4676\\4815\end{pmatrix};\right.$$
$$\left. 600 + 0.95(0.9, 0.1, 0.0, 0.0)\begin{pmatrix}4502\\4591\\4676\\4815\end{pmatrix}\right\}$$
$$= \operatorname{argmin}\{4815; 4885\} = 1.$$

Thus, $\boldsymbol{a}_1 = (1, 1, 2, 1)$.

Step 5 Because $\boldsymbol{a}_0 \neq \boldsymbol{a}_1$, repeat the above steps 2 through 5 using the stationary policy defined by $\boldsymbol{a}_1$.

ITERATION II

Step 2

$$\boldsymbol{f} = (100, 125, 350, 500)^T$$

$$\boldsymbol{P} = \begin{bmatrix} 0.1 & 0.3 & 0.6 & 0.0 \\ 0.0 & 0.2 & 0.5 & 0.3 \\ 0.8 & 0.2 & 0.0 & 0.0 \\ 0.8 & 0.1 & 0.0 & 0.1 \end{bmatrix}$$

Step 3

$$\boldsymbol{v} = (4287, 4382, 4441, 4613)^T$$

Step 4 The calculations for step 4 will be a repeat of those calculations immediately following Property 9.7 because the value function is the same. Therefore, the result will be

$$\boldsymbol{a}_2 = (1, 1, 2, 1).$$

Step 5 The algorithm is finished because $\boldsymbol{a}_1 = \boldsymbol{a}_2$; therefore, the results from the most recent steps 3 and 4 are optimum.

As a final computational note, it should be pointed out that there are more efficient numerical routines for obtaining the value of $\boldsymbol{v}$ than through the inverse as it is written in step 3. Numerical procedures for solving a system of linear equations are usually written without finding the inverse; thus, if the policy improvement algorithm is written into a computer program, a numerical routine to solve

$$(\boldsymbol{I} - \alpha\boldsymbol{P})\boldsymbol{v} = \boldsymbol{f}$$

as a linear system of equations should be used. ■

9.3.3 Linear Programming for Discounted Costs

It is possible to solve Markov decision processes using linear programming, although it is more limited than the other approaches. For ease of presentation, we have restricted our discussion to finite states and action spaces; however, the other two algorithms work for infinite state spaces, whereas linear programming can only be used for finite dimensional problems. The key to the linear programming formulation is the following property.

Property 9.10 ***Lemma for Linear Programming*** Let $\boldsymbol{v}^\alpha$ be the optimal value function as defined by Eq. (9.1) with $0 < \alpha < 1$, and let $\boldsymbol{u}$ be another real-valued function on the (finite) state space E. If $\boldsymbol{u}$ is such that

$$u(i) \leq \min_{k \in A}\{f_k(i) + \alpha \sum_{j \in E} P_k(i,j)u(j)\}$$

for all $i \in E$, then $\boldsymbol{u} \leq \boldsymbol{v}^\alpha$.

Consider a set made up of all functions that satisfy the inequality of Property 9.10. From Eq. (9.5) we know that the optimal value function $\boldsymbol{v}^\alpha$ would be included in that set. The force of this lemma for linear programming is that the largest of all functions in that set is $\boldsymbol{v}^\alpha$ (see Property 9.6). In other words, $\boldsymbol{v}^\alpha$ is the solution to the following mathematical programming problem (stated in matrix notation):

max $\boldsymbol{u}$

subject to:

$$\boldsymbol{u} \leq \min_{k \in A}\{\boldsymbol{f}_k + \alpha \boldsymbol{P}_k \boldsymbol{u}\}$$

The maximization in the above problem is a component by component operation; therefore, max $\boldsymbol{u}$ is equivalent to $\max \sum_{i \in E} u(i)$. Furthermore, the inequality being true for the minimum over all $k \in A$ is equivalent to it holding for each $k \in A$. Therefore, we have the following algorithm.

Algorithm 9.11 **Linear Programming for Discounted Costs** The optimal solution to the following linear program gives the minimum value function, $\boldsymbol{v}^\alpha$, with $0 < \alpha < 1$, for the problem defined by Eq. (9.1).

$\max \sum_{i \in E} u(i)$

subject to:

$$u(i) \leq f_k(i) + \alpha \sum_{j \in E} P_k(i,j)u(j) \quad \text{for each } i \in E \text{ and } k \in A.$$

The optimal policy is to choose an action k for state i such that $s_{i,k} = 0$, where $s_{i,k}$ is the slack variable associated with the equation corresponding to state i and action k.

Notice that the variables in the linear programming formulation are unrestricted as to sign. Many of the software packages available for linear programming assume that all variables are restricted to be nonnegative; however, this is easy to remedy without doubling the size of the problem by the standard technique of letting unrestricted variables be the difference of two nonnegative variables. If $f_k(i) \geq 0$ for all $k \in A$ and $i \in E$, the optimal solution will be nonnegative, so that nothing will be lost by a nonnegative restriction. If some of the $\boldsymbol{f}_k$ components are negative, then let δ be the absolute value of the most negative and add δ to all values of $\boldsymbol{f}_k$. Then the linear program using the new values for $\boldsymbol{f}_k$ with the variables restricted will yield the proper optimum, and the value of the objective function will be too high by the amount $\delta/(1 - \alpha)$.

To illustrate the linear programming formulation, we again use the machine problem from the previous section, yielding

$$\max z = u_a + u_b + u_c + u_d$$

subject to:

$$\begin{aligned}
u_a &\leq 100 + 0.095u_a + 0.285u_b + 0.57u_c \\
u_a &\leq 300 + 0.57u_a + 0.285u_b + 0.095u_c \\
u_b &\leq 125 + 0.19u_b + 0.475u_c + 0.285u_d \\
u_b &\leq 325 + 0.7125u_a + 0.095u_b + 0.095u_c + 0.0475u_d \\
u_c &\leq 150 + 0.095u_b + 0.19u_c + 0.665u_d \\
u_c &\leq 350 + 0.76u_a + 0.19u_b \\
u_d &\leq 500 + 0.76u_a + 0.095u_b + 0.095u_d \\
u_d &\leq 600 + 0.855u_a + 0.095u_b
\end{aligned}$$

Before solving this system, we moved the variables on the right-hand side to the left and added slack variables. Then, the solution to this problem yields $u_a = 4287$, $u_b = 4382$, $u_c = 4441$, and $u_d = 4613$, and the slacks associated with the first, third, sixth, and seventh equations are zero.

The linear programming algorithm used a maximizing objective function for the minimum cost problem; it would use a minimizing objective function for the maximum profit problem, and, in addition, the inequalities in the constraints would be reversed. In other words, the programming formulation for a discounted maximum *profit* problem would be to *minimize* $\boldsymbol{u}$ subject to $\boldsymbol{u} \geq \boldsymbol{f}_k + \alpha \boldsymbol{P}_k \boldsymbol{u}$ for each $k \in A$. Again, the decision variables are unrestricted as to sign.

9.4 AVERAGE COST ALGORITHMS

The average cost criterion problem is slightly more difficult than the discounted cost criterion problem because it was the discount factor that produced a fixed point. However, there is a recursive equation that is analogous to Property 9.6, as follows.

Property 9.12 Assume that every stationary policy yields a Markov chain with only one irreducible set. There exists a scalar φ^* and a vector $\boldsymbol{h}$ such that, for all $i \in E$,

$$\varphi^* + h(i) = \min_{k \in A}\{f_k(i) + \sum_{j \in E} P_k(i,j)h(j)\}. \tag{9.6}$$

The scalar φ^* is the optimal cost as defined by Eq. (9.2), and the optimal action function is defined by

$$a(i) = \operatorname{argmin}_{k \in A}\{f_k(i) + \sum_{j \in E} P(i,j)h(j)\}.$$

The vector $\boldsymbol{h}$ is unique up to an additive constant.

Let us use this property to determine if the optimal policy previously determined for a discount factor of 0.95 is also the optimal policy under the average cost criterion. In other words, we would like to determine if the stationary policy defined by $\boldsymbol{a} = (1, 1, 2, 1)$ is optimal using the long-run average cost criterion. The first step is to solve the following system of equations:

$$\begin{aligned}
\varphi^* + h_a &= 100 + 0.1h_a + 0.3h_b + 0.6h_c \\
\varphi^* + h_b &= 125 \qquad\qquad + 0.2h_b + 0.5h_c + 0.3h_d \\
\varphi^* + h_c &= 350 + 0.8h_a + 0.2h_b \\
\varphi^* + h_d &= 500 + 0.8h_a + 0.1h_b \qquad\qquad + 0.1h_d.
\end{aligned}$$

An immediate difficulty in the above system might be seen in that there are four equations and five unknowns; however, this fact should be expected because the $\boldsymbol{h}$ vector is not unique. This will be discussed in more detail later; for now, simply set h_a equal to zero and solve for the remaining four unknowns. The solution should yield $\varphi^* = 219.24$ and $\boldsymbol{h} = (0.0, 97.10, 150.18, 322.75)$. In other words, these values have been determined so that, for each $k = a(i)$, the following equality holds:

$$\varphi^* + h(i) = f_k(i) + \boldsymbol{P}_k\boldsymbol{h}(i)$$

Now, to determine optimality, we must verify that for each $k \neq a(i)$ the following inequality holds:

$$\varphi^* + h(i) \leq f_k(i) + \boldsymbol{P}_k\boldsymbol{h}(i).$$

That is, the following must hold for optimality to be true:

$$\begin{aligned}
\varphi^* + h_a &\leq 300 + 0.6h_a \;\; + 0.3h_b + 0.1h_c \\
\varphi^* + h_b &\leq 325 + 0.75h_a + 0.1h_b + 0.1h_c + 0.05h_d
\end{aligned}$$

$$\varphi^* + h_c \le 150 \qquad\quad + 0.1h_b + 0.2h_c + 0.7h_d$$
$$\varphi^* + h_d \le 600 + 0.9h_a + 0.1h_b.$$

Because the above holds, we have optimality.

It should also be observed that once the optimal policy is known, the value for φ^* can be obtained using Property 2.20 from Chapter 2. First, obtain the long-run probabilities associated with the Markov matrix given by

$$\mathbf{P}^a = \begin{bmatrix} 0.1 & 0.3 & 0.6 & 0.0 \\ 0.0 & 0.2 & 0.5 & 0.3 \\ 0.8 & 0.2 & 0.0 & 0.0 \\ 0.8 & 0.1 & 0.0 & 0.1 \end{bmatrix}.$$

These probabilities are $\boldsymbol{\pi} = (0.3630, 0.2287, 0.3321, 0.0762)$. Secondly, multiply those probabilities by the cost $\boldsymbol{f}^a = (100, 125, 350, 500)^T$, which yields $\varphi^* = 219.24$.

There is a close connection between the discounted cost problem and the long-run average cost problem. To illustrate some of these relationships, we present Table 9.1, which contains some information derived from the value function associated with the optimal policy for $\alpha = 0.95, \alpha = 0.99$, and $\alpha = 0.999$.

The value for each component in the vector $(1 - \alpha)\boldsymbol{v}^\alpha$ approaches φ^* as the discount factor approaches one. To understand this, remember that the geometric series is $\sum_{n=0}^{\infty} \alpha^n = 1/(1 - \alpha)$; thus, if a cost of c is incurred every period with a discount factor of α, its total value will be $v = c/(1-\alpha)$. Or, conversely, a total cost of v is equivalent to an average per period cost of $c = (1 - \alpha)v$. This gives the intuitive justification behind the following property.

Property 9.13 Let $\boldsymbol{v}^\alpha$ be the optimal value function defined by Eq. (9.1), let φ^* be the optimal cost defined by Eq. (9.2), and assume that every stationary policy yields a Markov chain with only one irreducible set. Then,

$$\lim_{\alpha \to 1} (1 - \alpha)v^\alpha(i) = \varphi^*$$

for any $i \in E$.

Referring to Table 9.1, notice that the third row under each discount factor gives the relative difference in total cost for starting in the different states. In other words, there is a \$97 advantage in starting in state a instead of b under a discount factor of $\alpha = 0.99$. Thus, it follows that $h(b) = 97$ because the $\boldsymbol{h}$ vector gives these relative differences—namely, if $h(a) = 0$, then $h(i)$ for $i \in E$ gives the additional cost of starting in state i instead of state a. When we initially solved the above system of equations to determine the values for φ^* and $\boldsymbol{h}$, we set $h(a)$ equal to zero. If we had started by setting $h(b)$ equal to zero, we would have obtained $h(a) = -97.10, h(c) = 53.08$, and $h(d) = 225.65$. This is stated explicitly in the following property.

TABLE 9.1 Discounted cost and long-run average cost relationships.

DISCOUNT	VECTOR	$i = a$	$i = b$	$i = c$	$i = d$
	$\boldsymbol{v}^\alpha$	4287	4382	4441	4613
$\alpha = 0.95$	$(1-\alpha)\boldsymbol{v}^\alpha$	214.35	219.10	222.05	230.65
	$\boldsymbol{v}^\alpha - v^\alpha(a)$	0	94	154	326
	$\boldsymbol{v}^\alpha$	21827	21923	21978	22150
$\alpha = 0.99$	$(1-\alpha)\boldsymbol{v}^\alpha$	218.27	219.23	219.78	221.50
	$\boldsymbol{v}^\alpha - v^\alpha(a)$	0	97	151	323
	$\boldsymbol{v}^\alpha$	219136	219233	219286	219459
$\alpha = 0.999$	$(1-\alpha)\boldsymbol{v}^\alpha$	219.13	219.23	219.28	219.45
	$\boldsymbol{v}^\alpha - v^\alpha(a)$	0	97	150	323

Property 9.14 Let $\boldsymbol{v}^\alpha$ be the optimal value function defined by Eq. (9.1), let φ^* be the optimal cost defined by Eq. (9.2), and let $\boldsymbol{h}$ be the vector defined by Property 9.12. Then

$$\lim_{\alpha \to 1} v^\alpha(i) - v^\alpha(j) = h(i) - h(j)$$

for $i, j \in E$.

This property is the justification for arbitrarily setting $h(a)$ equal to zero when we solve for φ^* and $\boldsymbol{h}$. In fact, it is legitimate to pick any single state i and set its $h(i)$ value equal to any given number.

9.4.1 Policy Improvement for Average Costs

Property 9.12 enables the design of an algorithm similar to Algorithm 9.9. The procedure begins with an arbitrary policy, determines the φ^* and $\boldsymbol{h}$ values associated with it, and then either establishes that the policy is optimal or produces a better policy. The specifics are as follows:

■ **Algorithm 9.15** **Policy Improvement Algorithm** The following iteration procedure will yield the optimal value function as defined by Eq. (9.2) and its associated optimal stationary policy.

Step 1 Set $n = 0$, let the first state in the state space be denoted by the number 1, and define the action function $\boldsymbol{a}_0$ by

$$a_0(i) = \text{argmin}_{k \in A} f_k(i)$$

for each $i \in E$.

Step 2 Define the matrix $\boldsymbol{P}$ and the vector $\boldsymbol{f}$ by

$$f(i) = f_{a_n(i)}(i)$$
$$P(i,j) = P_{a_n(i)}(i,j)$$

for each $i, j \in E$.

Step 3 Determine values for φ and $\boldsymbol{h}$ by solving the system of equations given by

$$\varphi + \boldsymbol{h} = \boldsymbol{f} + \boldsymbol{P}\boldsymbol{h},$$

where $h(1) = 0$.

Step 4 Define the action function $\boldsymbol{a}_{n+1}$ by

$$a_{n+1}(i) = \operatorname{argmin}_{k\in A}\{f_k(i) + \sum_{j\in E} P_k(i,j)h(j)\}$$

for each $i \in E$.

Step 5 If $\boldsymbol{a}_n = \boldsymbol{a}_{n+1}$, let $\varphi^* = \varphi$, and $\boldsymbol{a}^* = \boldsymbol{a}_n$, and stop; otherwise, increment n by one and return to step 2. ■

To illustrate this algorithm we outline the results obtained from applying the algorithm to our example problem.

ITERATION I

Step 1

$$\boldsymbol{a}_0 = (1, 1, 1, 1)$$

Step 2

$$\boldsymbol{f} = (100, 125, 150, 500)^T$$

$$\boldsymbol{P} = \begin{bmatrix} 0.1 & 0.3 & 0.6 & 0.0 \\ 0.0 & 0.2 & 0.5 & 0.3 \\ 0.0 & 0.1 & 0.2 & 0.7 \\ 0.8 & 0.1 & 0.0 & 0.1 \end{bmatrix}$$

Step 3 Solve the following (where h_a has been set to zero):

$$\begin{aligned}
\varphi \quad &= 100 + 0.3h_b + 0.6h_c \\
\varphi + h_b &= 125 + 0.2h_b + 0.5h_c + 0.3h_d \\
\varphi + h_c &= 150 + 0.1h_b + 0.2h_c + 0.7h_d \\
\varphi + h_d &= 500 + 0.1h_b \qquad\qquad + 0.1h_d
\end{aligned}$$

to obtain $\varphi = 232.86$ and $\boldsymbol{h} = (0, 90.40, 176.23, 306.87)^T$.

Step 4

$$a_1(a) = \operatorname{argmin}\left\{ 100 + (0.1, 0.3, 0.6, 0.0)\begin{pmatrix} 0 \\ 90.40 \\ 176.23 \\ 306.87 \end{pmatrix}; \right.$$

$$\left. 300 + (0.6, 0.3, 0.1, 0.0)\begin{pmatrix} 0 \\ 90.40 \\ 176.23 \\ 306.87 \end{pmatrix} \right\}$$

$$= \operatorname{argmin}\{232.86; 344.74\} = 1.$$

$$a_1(b) = \operatorname{argmin}\left\{ 125 + (0.0, 0.2, 0.5, 0.3)\begin{pmatrix} 0 \\ 90.40 \\ 176.23 \\ 306.87 \end{pmatrix}; \quad 325 + (0.75, 0.1, 0.1, 0.05)\begin{pmatrix} 0 \\ 90.40 \\ 176.23 \\ 306.87 \end{pmatrix} \right\}$$

$$= \operatorname{argmin}\{323.26; 367.01\} = 1.$$

$$a_1(c) = \operatorname{argmin}\left\{ 150 + (0.0, 0.1, 0.2, 0.7)\begin{pmatrix} 0 \\ 90.40 \\ 176.23 \\ 306.87 \end{pmatrix}; \quad 350 + (0.8, 0.2, 0.0, 0.0)\begin{pmatrix} 0 \\ 90.40 \\ 176.23 \\ 306.87 \end{pmatrix} \right\}$$

$$= \operatorname{argmin}\{409.09; 368.08\} = 2.$$

$$a_1(d) = \operatorname{argmin}\left\{ 500 + (0.8, 0.1, 0.0, 0.1)\begin{pmatrix} 0 \\ 90.40 \\ 176.23 \\ 306.87 \end{pmatrix}; \quad 600 + (0.9, 0.1, 0.0, 0.0)\begin{pmatrix} 0 \\ 90.40 \\ 176.23 \\ 306.87 \end{pmatrix} \right\}$$

$$= \operatorname{argmin}\{539.73; 609.04\} = 1.$$

Thus, $\boldsymbol{a}_1 = (1, 1, 2, 1)$.

Step 5 Because $\boldsymbol{a}_0 \neq \boldsymbol{a}_1$, repeat the above using the stationary policy defined by $\boldsymbol{a}_1$.

ITERATION II

Step 2

$$\boldsymbol{f} = (100, 125, 350, 500)^T$$

$$\boldsymbol{P} = \begin{bmatrix} 0.1 & 0.3 & 0.6 & 0.0 \\ 0.0 & 0.2 & 0.5 & 0.3 \\ 0.8 & 0.2 & 0.0 & 0.0 \\ 0.8 & 0.1 & 0.0 & 0.1 \end{bmatrix}$$

Step 3 Solve the system of equations on page 240 by first setting $h(a)$ equal to zero to obtain $\varphi = 219.24$ and $\boldsymbol{h} = (0.0, 97.10, 150.18, 322.75)^T$.

Step 4

$$
\begin{aligned}
a_2(a) &= \operatorname{argmin}\left\{ 100 + (0.1, 0.3, 0.6, 0.0)\begin{pmatrix} 0 \\ 97.10 \\ 150.18 \\ 322.75 \end{pmatrix};\right. \\
&\qquad\qquad \left. 300 + (0.6, 0.3, 0.1, 0.0)\begin{pmatrix} 0 \\ 97.10 \\ 150.18 \\ 322.75 \end{pmatrix}\right\} \\
&= \operatorname{argmin}\{219.24; 344.15\} = 1.
\end{aligned}
$$

$$
\begin{aligned}
a_2(b) &= \operatorname{argmin}\left\{ 125 + (0.0, 0.2, 0.5, 0.3)\begin{pmatrix} 0 \\ 97.10 \\ 150.18 \\ 322.75 \end{pmatrix};\right. \\
&\qquad\qquad \left. 325 + (0.75, 0.1, 0.1, 0.05)\begin{pmatrix} 0 \\ 97.10 \\ 150.18 \\ 322.75 \end{pmatrix}\right\} \\
&= \operatorname{argmin}\{316.33; 415.67\} = 1.
\end{aligned}
$$

$$
\begin{aligned}
a_2(c) &= \operatorname{argmin}\left\{ 150 + (0.0, 0.1, 0.2, 0.7)\begin{pmatrix} 0 \\ 97.10 \\ 150.18 \\ 322.75 \end{pmatrix};\right. \\
&\qquad\qquad \left. 350 + (0.8, 0.2, 0.0, 0.0)\begin{pmatrix} 0 \\ 97.10 \\ 150.18 \\ 322.75 \end{pmatrix}\right\} \\
&= \operatorname{argmin}\{415.67; 369.42\} = 2.
\end{aligned}
$$

$$
\begin{aligned}
a_2(d) &= \operatorname{argmin}\left\{ 500 + (0.8, 0.1, 0.0, 0.1)\begin{pmatrix} 0 \\ 97.10 \\ 150.18 \\ 322.75 \end{pmatrix};\right. \\
&\qquad\qquad \left. 600 + (0.9, 0.1, 0.0, 0.0)\begin{pmatrix} 0 \\ 97.10 \\ 150.18 \\ 322.75 \end{pmatrix}\right\} \\
&= \operatorname{argmin}\{541.98; 609.71\} = 1.
\end{aligned}
$$

Thus, $\boldsymbol{a}_2 = (1, 1, 2, 1)$.

Step 5 The algorithm is finished because $\boldsymbol{a}_1 = \boldsymbol{a}_2$; therefore, the results from the most recent steps 3 and 4 are optimum.

As a final computational note, a matrix form for the equations to be solved in step 3 can be given. Let $\boldsymbol{S}$ be a matrix equal to $\boldsymbol{I} - \boldsymbol{P}$ except that the first column is all ones; that is,

$$S(i,j) = \begin{cases} 1 & \text{if } j = 1, \\ 1 - P(i,j) & \text{if } i = j \text{ and } j \neq 1, \\ -P(i,j) & \text{if } i \neq j \text{ and } j \neq 1, \end{cases}$$

where we have identified the first state in the state space as state 1. As mentioned at the end of Section 9.3.2, using numerical procedures that specifically solve linear systems of equations is more efficient than first determining the inverse and then multiplying the inverse times the right-hand-side vector. The system that must be solved for the average cost criterion problem is

$$\boldsymbol{Sx} = \boldsymbol{f}.$$

The values for the quantities needed in the algorithm are then obtained as $h(1) = 0$, $\varphi^* = x(1)$, and $h(i) = x(i)$ for $i > 1$. ■

9.4.2 Linear Programming for Average Costs

The linear programming formulation for the long-run average cost problem takes a completely different approach than the formulation for the discounted cost problem, because there is no property analogous to Property 9.10 for average costs. The decision variables for the discounted cost linear program are components of the value function with a maximizing objective function. The decision variables for the long-run average cost linear program are components of the policy function with a minimizing objective function.

In order for the feasible set of possible policies to be convex, the linear program for the long-run average cost problem considers not only stationary policies but also randomized policies. A randomized policy (see the example policy 3 on page 228) is a policy that assigns a probability mass function over the action space for each state in the state space. Specifically, let $\nu_i(k)$ for $k \in A$ be a probability mass function for each $i \in E$; that is,

$$\nu_i(k) = \Pr\{D_n = k \mid X_n = i\}.$$

Of course, a stationary policy defined by the action function $\boldsymbol{a}$ is a subset of these randomized policies, where $\nu_i(k)$ is equal to one if $k = a(i)$ and is equal to zero otherwise.

Each randomized policy will yield a steady-state probability vector, $\boldsymbol{\pi}$. Multiplying the steady-state probabilities with the randomized policy produces a joint probability mass function of the state and action taken in the long run, which we shall denote as $x(i,k)$ for $i \in E$ and $k \in A$; that is,

$$x(i,k) = \nu_i(k)\pi(i) = \lim_{n\to\infty} \Pr\{X_n = i, D_n = k\}. \tag{9.7}$$

For a fixed $i \in E$, the function $\boldsymbol{\nu}_i$ is a probability mass function, so it sums to one; therefore,

$$\sum_{k \in A} x(i,k) = \sum_{k \in A} \nu_i(k)\pi(i) = \pi(i). \tag{9.8}$$

The decision variables for the linear program will be the joint probabilities designated by $x(i,k)$ for $i \in E$ and $k \in A$. For a fixed sequence of the joint probabilities, the expected value of the average cost for the associated randomized policy is

$$\varphi = \sum_{i \in E} \sum_{k \in A} x(i,k) f_k(i),$$

and the three conditions that must be fulfilled are

1. As a mass function, the probabilities must sum to one.
2. As a mass function, the probabilities must be nonnegative.
3. As steady-state probabilities, the equation $\boldsymbol{\pi P} = \boldsymbol{\pi}$ must hold.

Conditions 1 and 2 are straightforward, but the third condition needs some discussion. Specifically, we need to obtain the vector $\boldsymbol{\pi}$ and the matrix $\boldsymbol{P}$ from the given joint probabilities. It should not be difficult to show (see the chapter exercises) that the appropriate matrix for a fixed randomized policy is given by

$$P(i,j) = \sum_{k \in A} \nu_i(k) P_k(i,j)$$

for $i \in E$ and $k \in A$. Therefore, the equation $\boldsymbol{\pi P} = \boldsymbol{\pi}$ becomes

$$\pi(j) = \sum_{i \in E} \pi(i) \sum_{k \in A} \nu_i(k) P_k(i,j)$$

for $j \in E$. We can now combine Eqs. (9.7) and (9.8) with the above equation to write $\boldsymbol{\pi P} = \boldsymbol{\pi}$ in terms of the joint probabilities:

$$\sum_{k \in A} x(j,k) = \sum_{i \in E} \sum_{k \in A} x(i,k) P_k(i,j). \tag{9.9}$$

With the above background, these relationships can be structured as the following linear program, the solution of which yields the optimal policy for the long-run average cost criterion problem.

Algorithm 9.16 **Linear Programming for Average Costs** The optimal solution to the following linear program gives the minimum value for the problem defined by Eq. (9.2).

$$\min \varphi = \sum_{i \in E} \sum_{k \in A} x(i,k) f_k(i)$$

subject to:

$$\sum_{k \in A} x(j,k) = \sum_{i \in E} \sum_{k \in A} x(i,k) P_k(i,j) \quad \text{for each } j \in E$$

$$\sum_{i \in E} \sum_{k \in A} x(i,k) = 1$$

$$x(i,k) \geq 0 \qquad \text{for each } i \in E \text{ and } k \in A,$$

where the first equation in the constraint set is redundant and may be deleted. The optimal policy is to choose an action k for state i such that $x(i, k) > 0$. ■

Because we know that a (nonrandomized) stationary policy is optimal, there will only be one positive value of $x(i, k)$ for each $i \in E$; therefore, the optimal action function is to let $a(i)$ be the value of k for which $x(i, k) > 0$. It should also be observed that the steady-state probabilities under the optimal policy are given as $\pi(i) = x(i, a(i))$ for $i \in E$.

To illustrate the linear programming formulation, we again return to our machine problem example:

$$\begin{aligned} \min \varphi = {} & 100x_{a1} + 125x_{b1} + 150x_{c1} + 500x_{d1} \\ & + 300x_{a2} + 325x_{b2} + 350x_{c2} + 600x_{d2} \end{aligned}$$

subject to:

$$\begin{aligned} x_{a1} + x_{a2} &= 0.1x_{a1} + 0.6x_{a2} + 0.75x_{b2} + 0.8x_{c2} + 0.8x_{d1} + 0.9x_{d2} \\ x_{b1} + x_{b2} &= 0.3x_{a1} + 0.3x_{a2} + 0.2x_{b1} + 0.1x_{b2} + 0.1x_{c1} + 0.2x_{c2} \\ &\quad + 0.1x_{d1} + 0.1x_{d2} \\ x_{c1} + x_{c2} &= 0.6x_{a1} + 0.1x_{a2} + 0.5x_{b1} + 0.1x_{b2} + 0.2x_{c1} \\ x_{d1} + x_{d2} &= 0.3x_{b1} + 0.05x_{b2} + 0.7x_{c1} + 0.1x_{d1} \\ x_{a1} + x_{b1} + x_{c1} + x_{d1} + x_{a2} + x_{b2} + x_{c2} + x_{d2} &= 1 \\ x_{ik} \geq 0 \qquad & \text{for all } i \text{ and } k. \end{aligned}$$

The solution to the linear program yields an objective function value of 219.24, with the decision variables being $x_{a1} = 0.363, x_{b1} = 0.229, x_{c2} = 0.332, x_{d1} = 0.076$, and all other variables zero. Thus, the optimal action function is $\boldsymbol{a} = (1, 1, 2, 1)$, its average per period cost is $\varphi^* = 219.24$, and the steady-state probabilities for the resulting Markov chain are given by $\boldsymbol{\pi} = (0.363, 0.229, 0.332, 0.076)$.

9.5 THE OPTIMAL STOPPING PROBLEM

The final section of this chapter deals with a special case of the previous material; however, because of its special structure it is more efficient to solve the problem using its own formulation. The context for the optimal stopping problem is that we have a Markov chain $X_0, X_1, \cdots$, with state space E, Markov matrix $\boldsymbol{P}$, and *profit* function $\boldsymbol{f}$. Instead of a profit occurring at each visit, the profit will be realized only once—when the process is stopped. The process works as follows. The Markov chain will start in an initial state. The decision maker considers the state of the process and decides whether to continue the process or to stop. If the decision to continue is made, then no cost or profit is made and a transition of the chain occurs according to the Markov matrix $\boldsymbol{P}$, and again the decision maker decides whether to continue or to stop after the new state is observed. If the decision is made to stop, then a profit is made according to the profit function $\boldsymbol{f}$ and the process terminates.

In order to state the problem mathematically, we first define a type of stopping rule. Let S be a subset of the state space E, and define the random variable N_S by

$$N_S = \min\{n \geq 0 : X_n \in S\}. \tag{9.10}$$

The random variable N_S is thus the first time the Markov chain enters the set of states designated by S, and we shall call the set S a *stopping set*. Once a stopping set has been identified, its associated stopping rule random variable is defined [through Eq. (9.10)], and the stopping problem using that particular stopping rule has a value given by

$$w_S(i) = E[f(X_{N_S}) \mid X_0 = i].$$

In other words, $w_S(i)$ is the expected profit made if the process starts in state i, and it is continued until the first time that the process reaches a state in the set S. Thus, the optimal stopping problem is to find the set $S^* \subseteq E$ such that $w_{S^*}(i) = w(i)$, where the vector $\boldsymbol{w}$ is defined, for $i \in E$, by

$$w(i) = \max_{S \subseteq E} w_S(i). \tag{9.11}$$

EXAMPLE 9.2 For example, consider the Markov chain (Figure 9.2) with state space $E = \{a, b, c, d, e\}$, Markov matrix

$$\boldsymbol{P} = \begin{bmatrix} 1.0 & 0.0 & 0.0 & 0.0 & 0.0 \\ 0.5 & 0.0 & 0.3 & 0.2 & 0.0 \\ 0.0 & 0.2 & 0.6 & 0.2 & 0.0 \\ 0.0 & 0.1 & 0.5 & 0.2 & 0.2 \\ 0.0 & 0.0 & 0.0 & 0.0 & 1.0 \end{bmatrix},$$

and profit function

$$\boldsymbol{f} = (1, 2, 5, 8, 10)^T.$$

First, observe the structure for this matrix and corresponding state diagram given in Figure 9.2. States a and e are absorbing, and states $b, c,$ and d are transient. If a stopping set does not contain states a and e, there is a positive probability that the process will never stop, yielding a profit of zero; therefore,

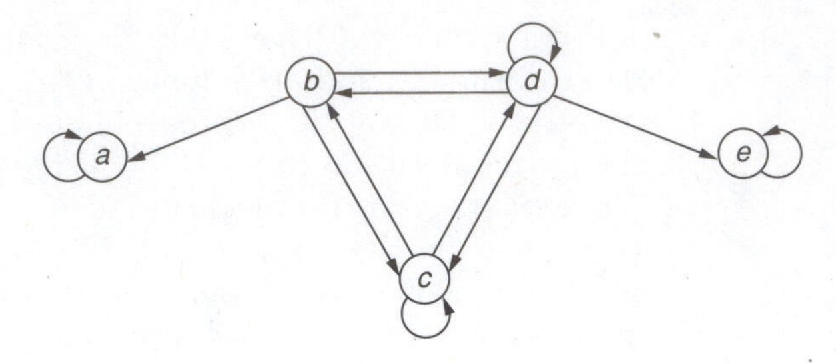

FIGURE 9.2 State diagram for the Markov chain of Example 9.2.

we only consider stopping sets containing the absorbing states. However, the inclusion of states $b, c,$ and d in the stopping set forms the crux of the optimal stopping problem. Consider the stopping set $S = \{a, c, d, e\}$. If the process starts in any of the states within S, it stops immediately; therefore, $w(a) = 1, w(c) = 5, w(d) = 8$, and $w(e) = 10$. Before calculating $w(b)$, observe that, at the next transition for the Markov chain, the stopping rule designated by S will force the chain to stop; therefore we simply need to take the expected value over the possible paths of this chain to determine $w(b)$. Thus, $w(b) = 0.5w(a) + 0.3w(c) + 0.2w(d) = 3.6$. (Of course, in general it is harder to determine the values for $\boldsymbol{w}$, but this example should serve to illustrate the dynamics of the optimal stopping problem.) Note that, although $w(b) = 3.6$, sometimes the process will actually yield a profit of 1, sometimes it will yield a profit of 5, and sometimes it will yield a profit of 8 under the assumption that $X_0 = b$. In general, $w_S(i)$ represents the expected profit if the Markov chain starts in state i and is stopped on the first visit to the set S. If the Markov chain starts in an irreducible set, the maximum profit that could be made is equal to the maximum profit in the irreducible set; thus, absorbing states always belong to optimal stopping sets. ■

It is also easy to incorporate a discount factor into the optimal stopping problem by discounting the final profit back to a present value. Thus we can state the discounted problem similarly as finding the set $S^* \subset E$ such that $w^\alpha_{S^*}(i) = w^\alpha(i)$ where the vector $\boldsymbol{w}^\alpha$ is defined, for $i \in E$, by

$$w^\alpha(i) = \max_{S \subseteq E} E\left[\alpha^{N_S} f(X_{N_S}) \mid X_0 = i\right], \tag{9.12}$$

for $0 < \alpha \leq 1$.

The basis for the linear programming formulation for the discounted cost Markov decision process was the lemma given in Property 9.10. A similar lemma allows the formulation of the optimal stopping problem as a linear program.

Property 9.17 ***Lemma for Linear Programming*** Let $\boldsymbol{w}^\alpha$ be the optimal value function as defined by Eq. (9.12) with $0 < \alpha \leq 1$, and let $\boldsymbol{u}$ be another real-valued function on the (finite) state space E. If $\boldsymbol{u}$ is such that $\boldsymbol{u} \geq \boldsymbol{f}$

and

$$u(i) \geq \alpha \sum_{j \in E} P(i,j)u(j)$$

for all $i \in E$, then $\boldsymbol{u} \geq \boldsymbol{w}^{\alpha}$.

It can also be shown that the optimum function $\boldsymbol{w}^{\alpha}$ has the properties listed in the lemma; therefore, $\boldsymbol{w}^{\alpha}$ is the minimum such function. In order for $w^{\alpha}(i)$ to be optimal, the expected profit for state i [namely, $w^{\alpha}(i)$] must also be no smaller than the payoff, $f(i)$, if the process stops on state i. This results in the formulation that follows.

Algorithm 9.18 **Linear Programming for Optimal Stopping** The optimal solution to the following linear program gives the maximal value function, $\boldsymbol{w}^{\alpha}$, with $0 < \alpha \leq 1$, for the problem defined by Eq. (9.12).

$\min \sum_{i \in E} u(i)$

subject to:

$$u(i) \geq \alpha \sum_{j \in E} P(i,j)u(j) \quad \text{for each } i \in E$$

$$u(i) \geq f(i) \quad \text{for each } i \in E$$

The optimal stopping set is given by $S^* = \{i \in E : w^{\alpha}(i) = f(i)\}$.

Taking advantage of this algorithm to solve the (undiscounted) example problem of this section yields the following:

$\min z = u_a + u_b + u_c + u_d + u_d$

subject to:

$$\begin{aligned}
u_b &\geq 0.5u_a + 0.3u_c + 0.2u_d \\
u_c &\geq + 0.2u_b + 0.6u_c + 0.2u_d \\
u_d &\geq + 0.1u_b + 0.5u_c + 0.2u_d + 0.2u_e \\
u_a &\geq 1,\ u_b \geq 2,\ u_c \geq 5,\ u_d \geq 8,\ u_e \geq 10,
\end{aligned}$$

where the trivial restrictions $u_a \geq u_a$ and $u_e \geq u_e$ have been excluded.

The solution to the above linear program yields $w(a) = 1.0$, $w(b) = 3.882$, $w(c) = 5.941$, $w(d) = 8.0$, and $w(e) = 10.0$, which gives the optimal stopping set as $S^* = \{a, d, e\}$.

9.6 EXERCISES

To obtain the maximum benefit from these homework problems, it would be best to have a computer available. As a starting point, write a program implementing the policy improvement algorithm for both the discounted cost and average cost criteria. The only difficult part of those codes is the matrix inversion procedure. If you do not have a subroutine available that inverts a matrix, consult the appendix for a description of an algorithm for matrix inversion.

9.1 Returning to the maintenance problem given in Example 9.1, we wish to maximize profits. Profits are determined by revenue minus costs, where revenue is a function of the state but not of the decision and is given by the function $\boldsymbol{r} = (900, 400, 450, 750)^T$. The costs and transition probabilities are as presented in the example.

(a) Compute the profit function $\boldsymbol{g}_1$ associated with the stationary policy that uses action 1 in every state.

(b) Using a discount factor $\alpha = 0.95$, verify that the vector $\boldsymbol{v}^\alpha = (8651.88, 8199.73, 8233.37, 8402.65)^T$ is the optimal value that maximizes the total discounted profit. (You need to use Property 9.6 with a minor modification because this is a maximizing problem.)

(c) Using a discount factor of $\alpha = 0.7$, use the value iteration algorithm to find the optimal value function that maximizes the total discounted profit.

(d) Using the policy improvement algorithm, find the policy that maximizes the total discounted profit if the discount factor is such that \$1 today is worth \$1.12 after one time period.

(e) Set up the linear programming formulation that solves the problem in part (d).

9.2 Use the data from Exercise 9.1 to answer the following questions:

(a) Find the average (undiscounted) profit per transition for the policy $\boldsymbol{a} = (2, 1, 2, 1)$.

(b) Show that the policy $\boldsymbol{a} = (2, 1, 2, 1)$ is not the policy that maximizes the profit per transition. (Use Property 9.12 after replacing the "min" operator with the "max" operator.)

(c) Using the policy improvement algorithm, find the policy that maximizes the average profit per transition.

(d) Find the policy that maximizes the average profit per transition using the linear programming formulation.

9.3 Joe recently graduated with a degree in operations research emphasizing stochastic processes. He wants to use his knowledge to advise people about presidential candidates. Joe has collected data on the past presidents according to their party (the two major parties[4] are the Labor Party and the Worker's Choice Party) and has determined that if the economy is good, fair,

---

[4]We use fictitious (in the United States) party names to emphasize that this problem is for illustrative purposes only. We do not mean to imply that such decisions should be made on purely short-term economic issues removed from moral issues.

or bad the selection of a president from a specific party will have various effects on the economy after his or her term. The data collected show that if the state of the economy is good and a Labor Party candidate is elected president, 3.2 million new jobs are created and a total of 4.5 trillion dollars are spent during the presidential term, whereas a Worker's Choice candidate only creates 3.0 million jobs and spends 4 trillion dollars. However, a candidate from the Labor Party who receives the country in good economic conditions will leave the country after his or her term in good, fair, or bad economic condition with probability of 0.75, 0.20, and 0.05, respectively, whereas the Worker's Choice counterpart will leave the country in those conditions with probabilities 0.80, 0.15, and 0.05, respectively.

If the initial state of the economy is fair a Labor Party president creates 2 million jobs, spends $3.5 trillion, and the probabilities of leaving a good, fair, or bad economy are 0.3, 0.5, and 0.2, respectively. A Worker's Choice president creates 2.3 million jobs, spends $4.5 trillion, and the probabilities of leaving a good, fair, or bad economy are 0.2, 0.6, and 0.2, respectively.

If the initial state of the economy is bad a Labor president creates 1 million jobs, spends $2.5 trillion, and the probabilities of leaving a good, fair, or bad economy are 0.2, 0.2, and 0.6, respectively. A Worker's Choice president creates 1.2 million jobs, spends $3 trillion, and the probabilities of leaving a good, fair, or bad economy are 0.1, 0.3, and 0.6, respectively.

When the state of the economy is bad, independent candidates try to capitalize on the bad economy and run for president. If an independent candidate is elected president 1.5 million jobs are created, $3.3 trillion are spent, and the economy is left in good, fair, or poor condition with probabilities 0.05, 0.40, and 0.55, respectively.

(a) Use Markov decision processes to determine the optimal voting strategy for presidential elections if the average number of new jobs per presidential term are to be maximized.

(b) Use Markov decision processes to determine how to vote if the objective is to minimize the total discounted spending if the discount factor is $\alpha = 0.9$ per term.

(c) Find the policy that minimizes the total discounted spending per job created if the discount factor is $\alpha = 0.9$ per term.

9.4 Consider a gambling machine that has five light bulbs labeled "a" through "e." When a person starts the game one of the five bulbs light up. At this point, the player has two options: (1) The game can be continued or (2) the game can be stopped. If the game is continued, then another bulb will be lit according to the Markov matrix below. When the game is stopped, a payoff is received corresponding to the current bulb that is lit.

Let reward function (in pennies) and one-step transition probabilities of the Markov chain with state space $E = \{a, b, c, d, e\}$ be given by

$$\boldsymbol{f} = (35, 5, 10, 20, 25)^T,$$

$$\boldsymbol{P} = \begin{bmatrix} 0.3 & 0 & 0 & 0.7 & 0 \\ 0 & 1 & 0 & 0 & 0 \\ 0.2 & 0.5 & 0.3 & 0 & 0 \\ 0.2 & 0 & 0 & 0.8 & 0 \\ 0.1 & 0.3 & 0.2 & 0.1 & 0.3 \end{bmatrix}.$$

(a) Determine the optimal strategy.

(b) Formulate the problem as a Markov decision process specifying the state space E, action space A, the $P_k(i,j)$ for each $i,j \in E$ and $k \in A$, and the reward functions f_k.

(c) Now assume that the game is started by inserting some nickels, and, as long as the game is continued, no additional coins are needed. After the coins are inserted, the first bulb is lit according to a uniform distribution. What is the minimum amount that should be charged per game if the machine is to make a profit in the long run? (For simplicity assume that operating and maintenance costs are negligible.)

9.5 Formulate the optimal stopping time problem as a Markov decision process using linear programming and show that the formulation reduces to the linear programming problem formulation of the optimal stopping time problem.

9.6 Consider the following optimal stopping time problem with payoff function $\boldsymbol{f}$, transition probabilities $\boldsymbol{P}$, state space $E = \{a, b, c, d, e\}$, and discount factor $\alpha = 0.5$, where

$$\boldsymbol{f} = (50, 40, 10, 20, 30)^T,$$

$$\boldsymbol{P} = \begin{bmatrix} 0.3 & 0 & 0 & 0.7 & 0 \\ 0 & 1 & 0 & 0 & 0 \\ 0.5 & 0.2 & 0.3 & 0 & 0 \\ 0.6 & 0 & 0 & 0.4 & 0 \\ 0.2 & 0.1 & 0.1 & 0.3 & 0.3 \end{bmatrix}.$$

Besides the payoff, there is a continuation fee given by the vector

$$\boldsymbol{h} = (1, 1, 2, 2, 3)^T.$$

In other words, if the Markov chain is in state i at time n and the decision to continue is made, then a fee of $h(i)$ is paid, regardless of the time n.

(a) Use a Markov decision process to find the optimal strategy that a gambler should follow to maximize his profit. That is, find the set of states where his decision should be to stop.

(b) Show that the linear programming formulation of this problem can be reduced to an optimal stopping time problem.

9.7 The custodian of an inventory system for a store counts the number of items on hand at the end of each week and places an order for additional items. Orders can be made only on Friday night and arrive early the following Monday morning. (The store is closed during the weekends.)

The ordering cost for j additional units is given by

$$c(j) = \begin{cases} 0 & \text{if } j = 0 \\ 100 + 500j & \text{if } j > 0. \end{cases}$$

The demand is 0, 1, 2, 3, 4, or 5 items per week, with probabilities 0.05, 0.15, 0.22, 0.22, 0.18, and 0.18, respectively. Items are sold for $750 each. If a customer asks for an item after all the inventory has been sold, then the manager of the store will buy the item from a competitor for a cost of $800 and sell it to the customer—thus taking a loss but maintaining good will. The company's cost of inventory is affected by a discount factor of 0.90.

(a) Set up the problem as a Markov decision process. (Let the action space be $A = \{0, 1, 2, 3, 4, 5\}$, where $k \in A$ refers to an order-up-to quantity of size k, so that if the inventory level is i and $i < k$, then the quantity ordered is $k - i$.)

(b) Find the optimal number of items that should be ordered at the end of the week to minimize the total discounted cost.

(c) Assume that there is an additional holding cost of $10 per item remaining at the end of the week. Find the optimal number of items that should be ordered at the end of the week to minimize the total discounted cost.

(d) Assume that instead of buying the item from a competitor, the store manager gives the customer a 10% reduction in the purchase price and places the item on back order (i.e., promises to deliver it to the customer the following Monday). What is the optimum ordering policy now? [Assume that the $10 holding cost from part (c) is still relevant.]

(e) How does the above answer change using an average cost criterion (i.e., no discount factor)?

9.8 Consider the following discrete time queueing system. One, two, or three customers arrive to the system at the beginning of each day with probabilities 0.25, 0.60, or 0.15, respectively. The system can hold at most five customers, so any customers in excess of five leave the system, never to return. (In other words, if, at the start of the day, the system contains four customers and two customers arrive, then only one will join the system and the other will disappear.) At the start of the day, after the arrival of the day's customers, a decision must be made as to whether one or two workers will be used for the day. If one worker is used, there is a 50% chance that only one customer will be served during the day and a 50% chance that two customers will be served during the day. If two workers are used, there is a 60% chance that two customers will be served, a 30% chance that three customers will be served, and a 10% chance that four customers will be served. (If there are only two customers present when two workers are used, then obviously there is a 100% chance that two customers will be served. A similar adjustment in the percentages needs to be made when three customers are present.) Each worker used costs $75 per day. At the end of the day, any customers that must wait until the next day will cost the system $30. Using Markov decision theory with the average cost criterion, determine how many workers should be used at the start of each day.

9.9 A machine produces items that are always inspected. The inspection is perfect in that the inspector always makes a correct determination as to whether the item is good or bad. The machine can be in one of two states: good or bad. If the machine is in the good state, it will produce a good item with a probability of 95%; if the machine is in the bad state, it will always produce bad items. If the machine is in the good state, there is a 2% probability that it will make a transition to the bad state for the next produced item. If the machine is in the bad state, it will stay in the bad state. The raw material and production costs are $500 per item; items are sold for $1000 per item if good and scrapped with a salvage value of $25 per item if bad. It costs $2500 to overhaul the machine (i.e., place it instantaneously in the good state). The state of the machine is not known directly, but only through the inspection of the produced items. Each time a part is produced, that part is inspected and a decision is made either to continue with the machine as is or to overhaul the machine.

(a) Consider the Markov chain $X = \{X_0, X_1, \cdots\}$, which represents the state of the machine. Because the state of X is not observable, we must work with the Markov chain $Z = \{Z_0, Z_1, \cdots\}$, where Z is the probability that the machine is bad after the item is inspected; that is, $Z_n = P\{X_n =$ "bad"$\}$, which has the state space given by the continuous interval [0,1]. The process Z depends on the inspection process $I = \{I_0, I_1, \cdots\}$, where $I_n = 0$ if the nth part was inspected and found to be bad and $I_n = 1$ if the nth part was inspected and found to be good. (Thus the dynamics is as follows: A decision is made based on Z_n and I_n, the X process makes a transition, an observation is made, and the cycle repeats.) If $Z_n = p$, what is the probability that $I_{n+1} = 0$?

(b) If $Z_n = p$, what is the probability that $I_{n+1} = 1$?

(c) If $Z_n = p$, what are the possible values of Z_{n+1}?

(d) The goal of this problem is to determine when to fix the machine. Formulate this problem as a Markov decision process with state space [0,1] and with a discount factor of 0.9. In other words, write out the relationship [Eq. (9.5)] that the optimal function must satisfy.

(e) The optimal policy can be expressed as a control limit policy. That is, there exists a value p^* such that if $Z_n = p < p^*$, the decision is to leave the machine alone; if $p \geq p^*$, the decision is to fix the machine. Using this fact and your expression from part (d), find the optimal policy. (*Hint:* Observe that there is only a discrete set of possible values that p can take.)

10

Advanced Queues[1]

The previous chapter introduced the concept of modeling processes using queueing theory and indicated the wide variety of applications possible through the use of queues. However, a major drawback to the application of the queueing theory discussed so far is its dependence on the Poisson and exponential assumptions regarding the probability laws for the arrival and service processes. One of the properties of the exponential distribution is that its mean equals its standard deviation, which indicates a considerable amount of variation. Therefore, for many service processes, the exponential distribution is not a good approximation. However, the memorylessness property of the exponential distribution is the key to building mathematically tractable models. Without the memorylessness property, the Markovian property is lost and the ability to build tractable models is significantly reduced.

There are two separate (although related) problems for the analyst who uses queueing processes for modeling: (1) to develop a model that adequately approximates the physical system being studied, and (2) to analyze the model and obtain the desired measures of effectiveness.

The first part of this chapter introduces methods for analyzing queueing systems more complicated than the simple systems discussed so far. Over the past 15 or 20 years, there has been an increasing emphasis on practical queueing models and the development of computationally tractable methods for these models; much of this emphasis has been due to extensive work by M. F. Neuts.[2] The main feature of these models is that matrix methods, which are easily written into computer routines, are used in the queueing analysis.

[1]This chapter would normally be skipped in a one-semester undergraduate course.

[2]We recommend the text, M. F. Neuts, *Matrix-Geometric Solutions in Stochastic Models* (Baltimore, Md.: The Johns Hopkins University Press, 1981), as a place to begin your further reading in this area.

This should be contrasted with the older methods of queueing analysis using transform techniques, which are not easily written into computer code.

After introducing these matrix geometric techniques, we shall present a class of distributions called *phase-type distributions,* which will allow the modeling of nonexponential processes while at the same time maintaining a Markovian structure. Methods to compute measures of effectiveness for queueing models with phase-type distributions are also presented.

10.1 DIFFERENCE EQUATIONS

Before presenting the matrix geometric technique for solving queueing models, we return once again to the simple M/M/1 queue. We shall present an alternative method for obtaining the steady-state probabilities, and then expand on that method to obtain a matrix geometric procedure. As you recall, the M/M/1 model yields the system of equations given by Eq. (5.1) that must be solved to obtain the steady-state probabilities. The solution method used in Chapter 5 to solve that system of equations was a successive substitution procedure. In this section we study the characteristic equation method for linear difference equations. For further reading, almost any introductory textbook on numerical analysis will contain these procedures.

We first present the general theory for solving difference equations of order two and then apply this knowledge to our queueing problem. Suppose we wish to find expressions for $x_0, x_1, \cdots$ that satisfy the following system of difference equations:

$$\begin{aligned} a_0x_0 + a_1x_1 + a_2x_2 &= 0 \\ a_0x_1 + a_1x_2 + a_2x_3 &= 0 \\ &\vdots \\ a_0x_{n-1} + a_1x_n + a_2x_{n+1} &= 0 \\ &\vdots \end{aligned}$$

where a_0, a_1, and a_2 are constants. The first step in obtaining an expression for x_n is to form the *characteristic function* defined by

$$f(z) = a_0 + a_1z + a_2z^2.$$

As long as $a_0 \neq 0$, $a_2 \neq 0$, and the roots of $f(\cdot)$ are real, the solutions to the system of difference equations are easy to represent. Suppose the characteristic function has two distinct roots,[3] called z_1 and z_2. Then

$$x_n = c_1z_1^n + c_2z_2^n$$

[3]The value z is a root if $f(z) = 0$.

is a solution to the system of difference equations, where c_1 and c_2 are constants to be determined from some other conditions (boundary conditions) that fully specify the system of equations. Suppose the characteristic function has a single root (of multiplicity 2), called z. Then

$$x_n = c_1 z^n + c_2 n z^{n-1}$$

is a solution to the system of difference equations.

One issue that is sometimes confusing involves knowing the limits on the index when the general expression is written. If the recursive equation is $x_{n+1} = x_n + x_{n-1}$ for $n = n_0, n_0 + 1, \cdots$, then the general expression must be true for all x_n, with n starting at $n_0 - 1$ because that is the first term that appears in the recursive equations.

EXAMPLE 10.1 **Fibonacci Sequence.** We wish to find values of x_n that satisfy the following:

$$\begin{aligned} x_2 &= x_0 + x_1 \\ x_3 &= x_1 + x_2 \\ x_4 &= x_2 + x_3 \\ &\vdots \end{aligned}$$

The characteristic function is $f(z) = z^2 - z - 1$, and the quadratic equation gives the roots as $z = (1 \pm \sqrt{5})/2$; therefore, a general solution for these difference equations is $x_n = c_1[(1+\sqrt{5})/2]^n + c_2[(1-\sqrt{5})/2]^n$. The Fibonacci numbers are those numbers that satisfy the recursion $x_{n+1} = x_n + x_{n-1}$ for $n = 1, 2, \cdots$ with $x_0 = 0$ and $x_1 = 1$. These boundary conditions constitute the information needed to obtain specific values for the constants; namely, c_1 and c_2 are found by rewriting the boundary conditions using the general form of x_n:

$$c_1\left(\frac{1+\sqrt{5}}{2}\right)^0 + c_2\left(\frac{1-\sqrt{5}}{2}\right)^0 = 0$$

$$c_1\left(\frac{1+\sqrt{5}}{2}\right)^1 + c_2\left(\frac{1-\sqrt{5}}{2}\right)^1 = 1.$$

This yields

$$c_1 + c_2 = 0$$

$$c_1(1+\sqrt{5}) + c_2(1-\sqrt{5}) = 2;$$

thus, $c_1 = -c_2 = 1/\sqrt{5}$ and the general expression for the Fibonacci numbers is

$$x_n = \frac{1}{\sqrt{5}}\left[\left(\frac{1+\sqrt{5}}{2}\right)^n - \left(\frac{1-\sqrt{5}}{2}\right)^n\right],$$

for $n = 0, 1, \cdots$. The fact that these are integers for all values of n is surprising, but if you check out some of these values, the sequence $0, 1, 1, 2, 3, 5, 8, 13, \cdots$ should be obtained. ■

■ **EXAMPLE 10.2** **The M/M/1 Queue.** The M/M/1 queue produces a difference equation [from Eq. (5.1)] of the form

$$\lambda p_{n-1} - (\lambda + \mu)p_n + \mu p_{n+1} = 0$$

for $n = 1, 2, \cdots$; thus the characteristic function for the M/M/1 queue is $f(z) = \lambda - (\lambda + \mu)z + \mu z^2$. From the quadratic equation we easily find that the roots[4] of the characteristic function are $z_1 = \rho$ and $z_2 = 1$, where $\rho = \lambda/\mu$; therefore, the general form of the solution is

$$p_n = c_1 \rho^n + c_2$$

for $n = 0, 1, \cdots$. To obtain specific values for c_1 and c_2, we look next at the norming equation: $\sum_{n=0}^{\infty} p_n = 1$. The first observation from the norming equation is that c_2 must be set equal to zero; otherwise, the infinite sum would always diverge. Once $c_2 = 0$ is fixed, the steps for finding c_1 are identical to Eq. (5.5), where c_1 takes the place of p_0, yielding $c_1 = 1 - \rho$, as long as $\lambda < \mu$. ■

10.2 BATCH ARRIVALS

We begin the discussion of the matrix geometric approach by investigating a queueing system in which *batches* of customers arrive together; we designate such a system as an $M^{[X]}/M/1$ queueing system. For example, customers might arrive to the queueing system in cars, and the number of people in each car is a random variable denoted by X. Or, in a manufacturing setting, items might arrive at a processing center on a pallet, but then the items must be processed individually; this would also lead to a batch arrival queueing system. As usual, we let the mean arrival rate for the Poisson process (i.e., arrival rate of batches) be λ; then the rate at which the process makes a transition from state n to $n + i$ is

$$\lambda_i = \lambda P\{X = i\},$$

where X denotes the size of a batch. Finally, we assume that the server treats customers individually with a mean (exponential) service rate of μ.

To illustrate the structure of this process, we develop the equations for the $M^{[X]}/M/1$ queueing system assuming a maximum batch size of two. Thus, $\lambda = \lambda_1 + \lambda_2$ and the generator matrix, $\boldsymbol{Q}$, is given by

---

[4]Note that $(\lambda + \mu)^2 - 4\lambda\mu = (\lambda - \mu)^2$.

$$
\begin{array}{c|cc|cc|cc|c}
0 & 1 & 2 & 3 & 4 & 5 & 6 & \cdots \\
-\lambda & \lambda_1 & \lambda_2 & & & & & \\
\hline
\mu & -(\lambda+\mu) & \lambda_1 & \lambda_2 & 0 & & & \\
0 & \mu & -(\lambda+\mu) & \lambda_1 & \lambda_2 & & & \\
\hline
 & 0 & \mu & -(\lambda+\mu) & \lambda_1 & \lambda_2 & 0 & \\
 & 0 & 0 & \mu & -(\lambda+\mu) & \lambda_1 & \lambda_2 & \\
\hline
 & & & 0 & \mu & -(\lambda+\mu) & \lambda_1 & \cdots \\
 & & & 0 & 0 & \mu & -(\lambda+\mu) & \cdots \\
\hline
 & & & & & \ddots & \ddots & \ddots
\end{array}
$$

where blanks within the matrix are interpreted as zeros and the row above the matrix indicates the states as they are associated with the individual columns of the matrix.

The $M^{[X]}/M/1$ generator matrix can also be represented using submatrices for the elements within the various partitions. The advantage of doing so is that the generator $\boldsymbol{Q}$ can then be easily written for a general system with a maximum batch size of m customers. This gives

$$
\boldsymbol{Q} = \begin{bmatrix}
-\lambda & \overline{\boldsymbol{\lambda}} & & & & \\
\overline{\boldsymbol{\mu}} & \boldsymbol{A} & \boldsymbol{\Lambda}_0 & & & \\
 & \boldsymbol{M} & \boldsymbol{A} & \boldsymbol{\Lambda}_0 & & \\
 & & \boldsymbol{M} & \boldsymbol{A} & \boldsymbol{\Lambda}_0 & \\
 & & & \boldsymbol{M} & \boldsymbol{A} & \cdots \\
 & & & & \vdots & \ddots
\end{bmatrix}, \tag{10.1}
$$

where $\overline{\boldsymbol{\lambda}}$ is the row vector $(\lambda_1, \lambda_2, \cdots)$; $\overline{\boldsymbol{\mu}}$ is the column vector $(\mu, 0, 0, \cdots)$; $\boldsymbol{M}$ is a matrix of zeros, except for the top right element, which is μ; $\boldsymbol{A}$ is a matrix with $-(\lambda + \mu)$ on the diagonal, μ on the subdiagonal, and λ_k on the kth superdiagonal; and $\boldsymbol{\Lambda_0}$ is a matrix made up of zeros and the λ_k terms. (If the largest possible batch has m customers, then the kth column of $\boldsymbol{\Lambda}_0$ has $k - 1$ zeros and then $m - k + 1$ values of λ_i starting with λ_m and ending with λ_k.)

Our goal is to find the values for the steady-state probabilities denoted by $\boldsymbol{p} = (p_0 \mid p_1, p_2 \mid p_3, p_4 \mid \cdots) = (p_0, \boldsymbol{p}_1, \boldsymbol{p}_2, \cdots)$, where $\boldsymbol{p}_n = (p_{2n-1}, p_{2n})$ for $n = 1, 2, \cdots$. In other words, the vector $\boldsymbol{p}$ is partitioned in the same fashion as $\boldsymbol{Q}$. Combining Eq. (10.1) with the equation $\boldsymbol{pQ} = \boldsymbol{0}$ yields the following system of equations, written using the submatrices from our partitioning:

$$
\begin{aligned}
-p_0\lambda + \boldsymbol{p}_1\overline{\boldsymbol{\mu}} &= 0 \\
p_0\overline{\boldsymbol{\lambda}} + \boldsymbol{p}_1\boldsymbol{A} + \boldsymbol{p}_2\boldsymbol{M} &= 0 \\
\boldsymbol{p}_1\boldsymbol{\Lambda}_0 + \boldsymbol{p}_2\boldsymbol{A} + \boldsymbol{p}_3\boldsymbol{M} &= 0 \\
\boldsymbol{p}_2\boldsymbol{\Lambda}_0 + \boldsymbol{p}_3\boldsymbol{A} + \boldsymbol{p}_4\boldsymbol{M} &= 0 \\
&\vdots
\end{aligned} \tag{10.2}
$$

The above system of equations can be solved using the matrix geometric technique, which is simply the matrix analogue to the solution procedures of Section 10.1. The recursive section from the equations listed in Eq. (10.2) reduces to the following problem: Find the vectors $\boldsymbol{p}_1, \boldsymbol{p}_2, \cdots$ such that

$$\boldsymbol{p}_{n-1}\boldsymbol{\Lambda}_0 + \boldsymbol{p}_n\boldsymbol{A} + \boldsymbol{p}_{n+1}\boldsymbol{M} = 0$$

for $n = 2, 3, \cdots$. The first step in the solution procedure is to form the *matrix characteristic function* and find its root. That is, we must find the matrix $\boldsymbol{R}$ such that

$$\boldsymbol{\Lambda}_0 + \boldsymbol{RA} + \boldsymbol{R}^2\boldsymbol{M} = \boldsymbol{0}. \tag{10.3}$$

Once the characteristic root matrix $\boldsymbol{R}$ is found,[5] it follows that

$$\boldsymbol{p}_n = \boldsymbol{c}\boldsymbol{R}^n \tag{10.4}$$

for $n = 1, 2, \cdots$, where $\boldsymbol{c}$ is a vector of constants that must be determined from the boundary conditions—that is, the first two equations from Eq. (10.2).

10.2.1 Quasi–Birth-Death Processes

The characteristic function of Eq. (10.3) represents the main work of the matrix geometric method, so we pause momentarily from our discussion of the batch arrival queue and discuss the general solution method for this type of matrix geometric problem. Notice that the generator matrix in Eq. (10.1) is a tridiagonal block matrix. That is, the generator $\boldsymbol{Q}$ has submatrices along the diagonal, the subdiagonal, and the superdiagonal, and zeros everywhere else. Thus the general form for the matrix is

$$\boldsymbol{Q} = \begin{bmatrix} \boldsymbol{B}_{00} & \boldsymbol{B}_{01} & & & & \\ \boldsymbol{B}_{10} & \boldsymbol{B}_{11} & \boldsymbol{\Lambda}_0 & & & \\ & \boldsymbol{A}_2 & \boldsymbol{A}_1 & \boldsymbol{A}_0 & & \\ & & \boldsymbol{A}_2 & \boldsymbol{A}_1 & \boldsymbol{A}_0 & \\ & & & \boldsymbol{A}_2 & \boldsymbol{A}_1 & \cdots \\ & & & & \vdots & \ddots \end{bmatrix}. \tag{10.5}$$

Any queueing system that gives rise to a generator of this form is called a *quasi–birth-death process*. In finding the steady-state probabilities for a quasi–birth-death process, the characteristic function given by

$$\boldsymbol{A}_0 + \boldsymbol{R}\boldsymbol{A}_1 + \boldsymbol{R}^2\boldsymbol{A}_2 = \boldsymbol{0}. \tag{10.6}$$

will arise, where $\boldsymbol{R}$ is the unknown matrix and must be obtained. In some systems the matrices $\boldsymbol{A}_0$, $\boldsymbol{A}_1$, and $\boldsymbol{A}_2$ have simple structures, so that $\boldsymbol{R}$ can be obtained analytically; however, in many systems an expression for $\boldsymbol{R}$

[5]There are actually two matrices satisfying Eq. (10.3), but Neuts, *op. cit.*, showed that exactly one matrix exists that will satisfy the boundary (norming) conditions.

is impossible. What makes the matrix geometric approach attractive is the fact that numerical methods to compute $\boldsymbol{R}$ are readily available. To present one approach, let us rearrange the terms in Eq. (10.6) to obtain

$$\boldsymbol{R} = -(\boldsymbol{A}_0 + \boldsymbol{R}^2\boldsymbol{A}_2)\boldsymbol{A}_1^{-1}. \tag{10.7}$$

Equation (10.7) leads to a successive substitution scheme where a guess is made for $\boldsymbol{R}$; that guess is used in the right-hand side of the equation to produce a new estimate for $\boldsymbol{R}$; this new estimate is used again in the right-hand side to obtain another estimate of $\boldsymbol{R}$; and so on. Specifically, we have the following algorithm.

Algorithm 10.1 **Successive Substitution Algorithm.** Let $\boldsymbol{A}_0$, $\boldsymbol{A}_1$, and $\boldsymbol{A}_2$ be submatrices from the generator $\boldsymbol{Q}$ as defined by Eq. (10.5), and let $\boldsymbol{A}_s = \boldsymbol{A}_0 + \boldsymbol{A}_1 + \boldsymbol{A}_2$. Define the vector $\boldsymbol{\nu}$ such that $\boldsymbol{\nu}\boldsymbol{A}_s = \boldsymbol{0}$ and $\boldsymbol{\nu}\boldsymbol{1} = 1$, and define the traffic intensity as $\rho = \boldsymbol{\nu}\boldsymbol{A}_0\boldsymbol{1}/\boldsymbol{\nu}\boldsymbol{A}_2\boldsymbol{1}$. If $\rho < 1$, the following iteration procedure will yield an approximation to the matrix $\boldsymbol{R}$ that satisfies Eq. (10.6).

Step 1 Check to ensure that the steady-state conditions hold; that is, check that $\rho < 1$.

Step 2 Let $\boldsymbol{R}_0 = \boldsymbol{0}$ (i.e., a matrix of all zeros), let $n = 0$, and fix ϵ to be a small positive real value.

Step 3 Define the matrix $\boldsymbol{R}_{n+1}$ by

$$\boldsymbol{R}_{n+1} = -\left(\boldsymbol{A}_0 + \boldsymbol{R}_n^2\,\boldsymbol{A}_2\right)\boldsymbol{A}_1^{-1}.$$

Step 4 Define δ by

$$\delta = \max_{i,j}\{|R_{n+1}(i,j) - R_n(i,j)|\}.$$

Step 5 If $\delta < \epsilon$, let $\boldsymbol{R} = \boldsymbol{R}_{n+1}$ and stop; otherwise, increment n by one and return to step 3.

The successive substitution algorithm is often slow to converge, especially if the traffic intensity, ρ, is close to one; however, the algorithm is very easy to build into a computer code, so it becomes a convenient method for obtaining the matrix $\boldsymbol{R}$. Once the matrix $\boldsymbol{R}$ is determined, Eq. (10.4) gives an expression for the probability vectors $\boldsymbol{p}_n$ for $n \geq 1$. To obtain the values for the constant $\boldsymbol{c}$ and the probability vector $\boldsymbol{p}_0$, the norming equation and the boundary conditions must be used.

The boundary conditions are taken from the first columns of submatrices from the generator in Eq. (10.5). These equations are

$$\begin{aligned} \boldsymbol{p}_0\boldsymbol{B}_{00} + \boldsymbol{p}_1\boldsymbol{B}_{10} &= \boldsymbol{0} \\ \boldsymbol{p}_0\boldsymbol{B}_{01} + \boldsymbol{p}_1\boldsymbol{B}_{11} + \boldsymbol{p}_2\boldsymbol{A}_2 &= \boldsymbol{0}. \end{aligned} \tag{10.8}$$

The norming equation is

$$\sum_{n=0}^{\infty} \sum_{k} p_{nk} = \sum_{n=0}^{\infty} \boldsymbol{p}_n \mathbf{1} = 1.$$

A closed form expression for this equation is obtained by taking advantage of the geometric progression result for matrices[6] together with the matrix geometric form for the probabilities [Eq.(10.4)]. Thus, we have

$$\begin{aligned} 1 &= \boldsymbol{p}_0 \mathbf{1} + \sum_{n=1}^{\infty} \boldsymbol{p}_n \mathbf{1} \\ &= \boldsymbol{p}_0 \mathbf{1} + \sum_{n=1}^{\infty} \boldsymbol{c} \boldsymbol{R}^n \mathbf{1} \\ &= \boldsymbol{p}_0 \mathbf{1} + \boldsymbol{c} \boldsymbol{R} \left(\sum_{n=1}^{\infty} \boldsymbol{R}^{n-1} \right) \mathbf{1} \\ &= \boldsymbol{p}_0 \mathbf{1} + \boldsymbol{p}_1 (\boldsymbol{I} - \boldsymbol{R})^{-1} \mathbf{1}. \end{aligned} \qquad \textbf{(10.9)}$$

In conclusion, the long-run probabilities for quasi–birth-death processes are found by first finding the matrix $\boldsymbol{R}$, possibly using the successive substitution algorithm (Algorithm 10.1), and then finding the values for $\boldsymbol{p}_0$ and $\boldsymbol{p}_1$ by solving the system of equations

$$\begin{aligned} \boldsymbol{p}_0 \boldsymbol{B}_{00} + \boldsymbol{p}_1 \boldsymbol{B}_{10} &= \mathbf{0} \\ \boldsymbol{p}_0 \boldsymbol{B}_{01} + \boldsymbol{p}_1 \boldsymbol{B}_{11} + \boldsymbol{p}_1 \boldsymbol{R} \boldsymbol{A}_2 &= \mathbf{0} \\ \boldsymbol{p}_0 \mathbf{1} + \boldsymbol{p}_1 (\boldsymbol{I} - \boldsymbol{R})^{-1} \mathbf{1} &= 1. \end{aligned} \qquad \textbf{(10.10)}$$

Notice that the system of equations (10.10) is simply a combination of Eqs. (10.8) and (10.9) together with the substitution[7] $\boldsymbol{p}_2 = \boldsymbol{p}_1 \boldsymbol{R}$. As is common with all irreducible recurrent systems, one of the equations in the system of Eqs. (10.10) will be redundant.

10.2.2 Batch Arrivals (Continued)

Before proceeding with the batch arrival queueing system, we should check the steady-state conditions; namely, check that the traffic intensity is less than one (step 1 from Algorithm 10.1). For our example with a maximum batch size of two, the matrix $\boldsymbol{A}_s$ is

$$\boldsymbol{A}_s = \begin{bmatrix} -(\lambda_1 + \mu) & \lambda_1 + \mu \\ \lambda_1 + \mu & -(\lambda_1 + \mu) \end{bmatrix}$$

[6] If the absolute value of all eigenvalues for $\boldsymbol{R}$ is less than one, $\sum_{n=0}^{\infty} \boldsymbol{R}^n = (\boldsymbol{I} - \boldsymbol{R})^{-1}$.

[7] If the substitution is not obvious, observe that Eq. (10.4) implies that $\boldsymbol{p}_{n+1} = \boldsymbol{p}_n \boldsymbol{R}$.

and thus the distribution from $\boldsymbol{A}_s$ is $\nu_1 = 0.5$ and $\nu_2 = 0.5$. Therefore, the traffic intensity is given by

$$\begin{aligned}\rho &= \frac{\boldsymbol{\nu}\boldsymbol{\Lambda}_0\mathbf{1}}{\boldsymbol{\nu}\boldsymbol{M}\mathbf{1}}\\ &= \frac{0.5\lambda_2 + 0.5\lambda}{0.5\mu}\\ &= \frac{\lambda_1 + 2\lambda_2}{\mu}\\ &= \frac{\lambda E[X]}{\mu},\end{aligned}$$

where $E[X]$ is the mean batch size.

For purposes of a numerical example, let $\mu = 5$ per hour, $\lambda_1 = 2$ per hour, and $\lambda_2 = 1$ per hour. This yields $\rho = 0.8$, so that it makes sense to find the steady-state probabilities. The matrix $\boldsymbol{R}$ is the solution to the equation

$$\boldsymbol{R}^2\begin{bmatrix}0 & 5\\0 & 0\end{bmatrix} + \boldsymbol{R}\begin{bmatrix}-8 & 2\\5 & -8\end{bmatrix} + \begin{bmatrix}1 & 0\\2 & 1\end{bmatrix} = \mathbf{0},$$

and the successive substitution algorithm yields the following sequence:

$$\begin{aligned}\boldsymbol{R}_0 &= \begin{bmatrix}0 & 0\\0 & 0\end{bmatrix}\\ \boldsymbol{R}_1 &= \begin{bmatrix}0.148 & 0.037\\0.389 & 0.222\end{bmatrix}\\ \boldsymbol{R}_2 &= \begin{bmatrix}0.165 & 0.064\\0.456 & 0.329\end{bmatrix}\\ &\vdots\\ \boldsymbol{R} &= \begin{bmatrix}0.2 & 0.12\\0.6 & 0.56\end{bmatrix}.\end{aligned}$$

With the determination of $\boldsymbol{R}$, there are three unknowns that need to be obtained in order to fully define the steady-state probabilities—specifically, p_0 and the constant vector $\boldsymbol{c}$ from Eq. (10.4). Alternatively, we could say that there are three unknowns—p_0,p_1, and p_2—and these must be determined through the boundary conditions given by the system of equations (10.2) and the norming equation (10.9). (Remember that $\boldsymbol{p}_1 = (p_1, p_2)$ is a vector. Also, the vector $\boldsymbol{p}_0$ given in the general quasi–birth-death system equates to the scalar p_0 in the batch arrival system.)

The norming equation can be combined with the second equation from (10.2) to obtain a system defining the constant vector $\boldsymbol{c}$; thus,

$$1 - \boldsymbol{p}_1(\boldsymbol{I} - \boldsymbol{R})^{-1}\boldsymbol{1} = p_0$$
$$p_0\overline{\boldsymbol{\lambda}} + \boldsymbol{p}_1\boldsymbol{A} + \boldsymbol{p}_1\boldsymbol{R}\boldsymbol{M} = \boldsymbol{0}.$$

The above system is sufficient to determine the unknown probabilities; however, it can be simplified somewhat using a result based on Little's formula. All single-server systems that process customers one at a time have the same form for the probability that the system is empty; p_0 is always $1 - (\lambda_e/\mu)$ where λ_e is the mean arrival rate of customers *into* the system and μ is the mean service rate. Therefore,

$$p_0 = 1 - \frac{\lambda E[X]}{\mu}$$
$$\boldsymbol{p}_1 = -p_0\overline{\boldsymbol{\lambda}}(\boldsymbol{A} + \boldsymbol{R}\boldsymbol{M})^{-1},$$

where X is the random variable indicating the size of each batch.

Returning again to the numerical example for the batch arrival process, we have

$$\boldsymbol{A} + \boldsymbol{R}\boldsymbol{M} = \begin{bmatrix} -8.0 & 3.0 \\ 5.0 & -5.0 \end{bmatrix}$$

yielding $\boldsymbol{p}_1 = (0.12, 0.112)$.

Before going to another example of the use of the matrix geometric distribution, we present some background in phase-type distributions. These distributions will allow for a significant amount of generality over the exponential distribution, while taking advantage of the computational tractability afforded by the matrix geometric approach.

10.3 PHASE-TYPE DISTRIBUTIONS

The Erlang distribution defined in Chapter 1 is an example of a phase-type distribution because it is the sum of independent exponential distributions. That is, an Erlang random variable could be thought of as the length of time required to go through a sequence of phases or steps in which the time to go through each phase requires an exponentially distributed length of time. For example, suppose we wish to model a single-server queueing system with a Poisson arrival process, but all we know about its service distribution is that it has a mean service time of 10 minutes and a standard deviation of 5 minutes. The temptation might be to use an exponential distribution with parameter $\mu = 0.1$ for service times. However, such an assumption would clearly be wrong, because the standard deviation would be twice the correct size even though the mean would be correct. A much better modeling distribution would be the Erlang distribution [see Eq. (1.16)] with parameters $m = 4$ and $\lambda = 0.1$; such a distribution yields a mean of 10 and a standard deviation of 5. This Erlang distribution is the sum of four exponentials, each exponential having a mean of 2.5; thus, a schematic diagram for the Erlang is contained in Figure 10.1.

In other words, instead of modeling the Erlang server as a single system, we model it as a four-phase process (see Figure 10.1). To describe the dynamics of this phase-type model, assume that there is a queue outside the server and the server has just completed a service. Immediately, the customer at the head of the queue enters the phase 1 box. After an exponentially distributed length of time, the customer leaves phase 1 and enters phase 2 and remains in that box for another exponentially distributed length of time. This process is repeated until the customer passes through all four boxes, and as long as the customer is in any of the four boxes, no additional customer may enter the server. As soon as the customer leaves phase 4, the next customer in the queue enters phase 1. Thus, we are able to model the server using the exponential distribution, even though the server itself does not have an exponential but rather an Erlang distribution with a standard deviation half of its mean. In modeling such a system, we emphasize that the phases are *fictitious*. Whether or not customers actually enter individual physical phases is not relevant; the important question is whether or not the resultant distribution of the server adequately mimics the real process.

The exponential nature embedded within the Erlang distribution allows for a Markov process to be used for its distribution. To demonstrate, let $\{Y_t; t \geq 0\}$ be a Markov process with state space $E = \{1, 2, 3, 4, \Delta\}$, generator matrix

$$\boldsymbol{G} = \begin{bmatrix} -0.4 & 0.4 & 0.0 & 0.0 & 0.0 \\ 0.0 & -0.4 & 0.4 & 0.0 & 0.0 \\ 0.0 & 0.0 & -0.4 & 0.4 & 0.0 \\ 0.0 & 0.0 & 0.0 & -0.4 & 0.4 \\ 0.0 & 0.0 & 0.0 & 0.0 & 0.0 \end{bmatrix},$$

and initial probability vector $\boldsymbol{\alpha} = (1, 0, 0, 0, 0)$. In other words, Y always starts in state 1 and sequentially passes through all the states until being absorbed in the final state, Δ. Mathematically, this can be expressed by defining the random variable T as the first passage time to state Δ; that is,

$$T = \min\{t \geq 0 : Y_t = \Delta\}. \tag{10.11}$$

FIGURE 10.1 Schematic for the Erlang distribution.

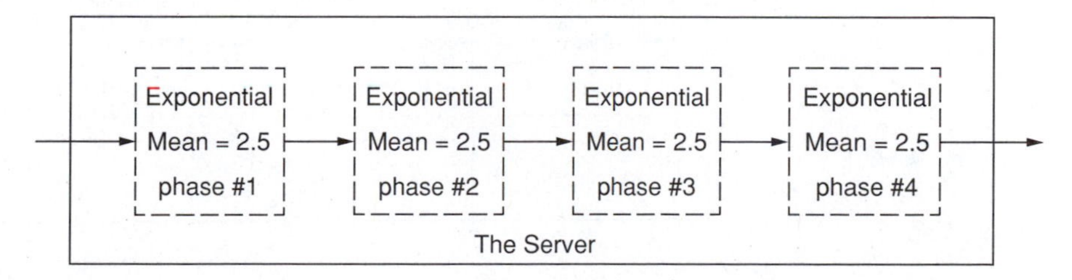

The distribution of T is the sum of four independent exponential distributions, each with mean 2.5; therefore, T has a four-phase Erlang distribution with mean 10. This example is an illustration of the property that any Erlang distribution can be represented as the first passage time for a Markov process.

An m-phase Erlang has the property that its mean divided by its standard deviation equals the square root of m. If we have a nonexponential server, one approach for finding an appropriate distribution for it is to take the "best" Erlang distribution based on the mean and standard deviation. In other words, if we square the ratio of the mean over the standard deviation and obtain something close to an integer, that gives the appropriate number of phases for the Erlang. For example, if the mean were 10 and standard deviation were 5.77, we would use a three-phase Erlang model [because $(10/5.77)^2 = 3.004$]. Or, as another example, consider a server that has a mean service time of 10 and standard deviation of 5.48 [observe that $(10/5.48)^2 = 3.33$]. We now have a dilemma, because the number of phases for the Erlang must be an integer. However, if you look again at Figure 10.1, you should realize that the assumptions for the Erlang are unnecessarily (from the perspective of modeling) restrictive. If the purpose of using the Erlang is to reduce everything to exponentials, we must ask the question, "Why do the means of each phase need to be the same?" The answer is, "They do not!" If in Figure 10.1 the first phase was exponential with mean 1, the second phase was exponential with mean 2, the third phase was exponential with mean 3, and the fourth phase was exponential with mean 4, then the overall mean would be 10 and the standard deviation would be 5.48 (review Property 1.23 for verifying the standard deviation).

Phase-type distributions require generalizing the Erlang concept to allow nonequal means for the phases and to allow more general paths than simply a "straight-through" path. For example, consider the phase-type distribution illustrated in Figure 10.2. The distribution of the service time illustrated in that figure can again be represented by a Markov process. Let $\{Y_t; t \geq 0\}$ be

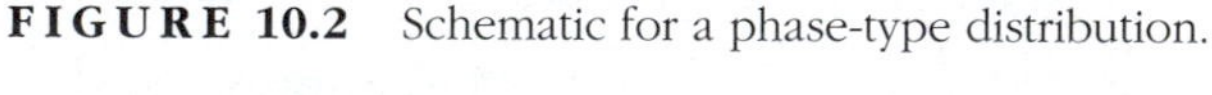
FIGURE 10.2 Schematic for a phase-type distribution.

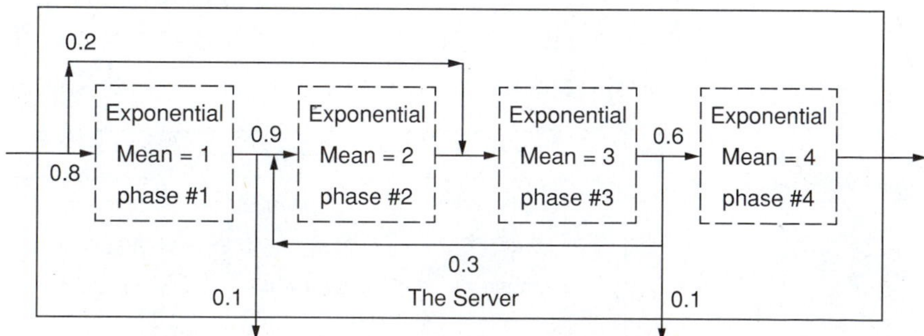

a Markov process with state space $E = \{1, 2, 3, 4, \Delta\}$, generator matrix

$$\boldsymbol{G} = \begin{bmatrix} -1.0 & 0.9 & 0.0 & 0.0 & 0.1 \\ 0.0 & -0.5 & 0.5 & 0.0 & 0.0 \\ 0.0 & 0.1 & -0.3333 & 0.2 & 0.0333 \\ 0.0 & 0.0 & 0.0 & -0.25 & 0.25 \\ 0.0 & 0.0 & 0.0 & 0.0 & 0.0 \end{bmatrix}, \tag{10.12}$$

and initial probability vector $\boldsymbol{\alpha} = (0.8, 0, 0.2, 0, 0)$. Consider a customer who, upon entering the server, begins in phase 1 with probability 0.8 and begins in phase 3 with probability 0.2. After entering a phase, the customer spends an exponential length of time in that phase and then moves according to a Markov chain as illustrated in Figure 10.2. In other words the length of time that the customer spends in the server, as illustrated in the figure, is equivalent to the first passage time to state Δ for the Markov process defined by the generator of Eq. (10.12). This furnishes the motivation for phase-type distributions; namely, any distribution that can be represented by a first passage time of a Markov process is called a *phase-type distribution*.

The generator matrix of Eq. (10.12) can be partitioned into four key submatrices as

$$\boldsymbol{G} = \left[\begin{array}{cccc|c} -1.0 & 0.9 & 0.0 & 0.0 & 0.1 \\ 0.0 & -0.5 & 0.5 & 0.0 & 0.0 \\ 0.0 & 0.1 & -0.3333 & 0.2 & 0.0333 \\ 0.0 & 0.0 & 0.0 & -0.25 & 0.25 \\ \hline 0.0 & 0.0 & 0.0 & 0.0 & 0.0 \end{array}\right]$$

$$= \left[\begin{array}{c|c} \boldsymbol{G}_* & \boldsymbol{G}_\Delta \\ \hline \boldsymbol{0} & 0 \end{array}\right]$$

and the initial probability vector can be partitioned as

$$\begin{aligned} \boldsymbol{\alpha} &= (0.8, 0.0, 0.2, 0.0 \mid 0.0) \\ &= (\boldsymbol{\alpha}_* \mid 0.0). \end{aligned}$$

Such partitioning is always possible for Markov processes that are to be used in defining phase-type distributions. The matrix $\boldsymbol{G}_*$ and the vector $\boldsymbol{\alpha}_*$ are sufficient to define a phase-type distribution, as is specified in the following definition.

Definition 10.2 *A random variable has a* Phase-type Distribution *with parameters m, $\boldsymbol{G}_*$, and $\boldsymbol{\alpha}_*$ if it can be expressed as a first-passage-time random variable [Eq. (10.11)] for a Markov process with state space $E = \{1, \cdots, m, \Delta\}$, generator*

$$\boldsymbol{G} = \left[\begin{array}{c|c} \boldsymbol{G}_* & \boldsymbol{G}_\Delta \\ \hline \boldsymbol{0} & 0 \end{array}\right],$$

and initial probability vector $\boldsymbol{\alpha} = (\boldsymbol{\alpha}_* \mid 0)$.

First-passage-time random variables have well-known properties that can be easily described mathematically, but not necessarily easy to compute. These properties are as follows.

Property 10.3 Let T be a random variable with a phase-type distribution from Definition 10.2. Its cumulative distribution function and first two moments are given by

$$F(t) = 1 - \boldsymbol{\alpha}_* \exp\{t\boldsymbol{G}_*\}\mathbf{1} \quad \text{for } t \geq 0,$$
$$E[T] = -\boldsymbol{\alpha}_*\boldsymbol{G}_*^{-1}\mathbf{1},$$
$$E[T^2] = 2\boldsymbol{\alpha}_*\boldsymbol{G}_*^{-2}\mathbf{1},$$

where $\boldsymbol{G}_*^{-2}$ indicates the square of the inverse and $\mathbf{1}$ is a vector of ones.

Property 10.3 gives an easy-to-calculate formula for determining the mean and variance; however, the distribution is somewhat more difficult. You should be familiar with the concept of scalar times a matrix; namely, the matrix $t\boldsymbol{G}_*$ is the matrix $\boldsymbol{G}_*$ with all components multiplied by t. However, a matrix in the exponent may be unfamiliar. Specifically, the matrix $e^{\boldsymbol{A}}$ is defined by the power series[8]

$$e^{\boldsymbol{A}} = \sum_{n=0}^{\infty} \frac{\boldsymbol{A}^n}{n!}.$$

Most linear algebra textbooks will discuss matrices of the form $e^{\boldsymbol{A}}$; therefore, we will not discuss its computational difficulties here. For the purposes of this text, we will only use the moment formulas from Property 10.3.

One caveat worth mentioning is that our example has focused only on the mean and standard deviation—as if those two properties adequately described the randomness inherent in the server. This is, of course, an oversimplification, although at times these may be the only data that the analyst has to work with. It is obviously best to try to fit the entire distribution to empirical data; however, goodness-of-fit tests require considerable data that are not always available. When data are not available to give a good estimate for the full distribution function, moments are often used. The first moment gives the central tendency, and the second moment is used to measure variability. The third moment is useful as a measure of skewness (see page 22). For phase-type distributions, the nth moment is given as

$$E[T^n] = (-1)^n n! \boldsymbol{\alpha}_*\boldsymbol{G}_*^{-n}\mathbf{1}.$$

As a final comment regarding the choice of distributions for modeling purposes, we note that it is often the tails of distributions that are the keys

[8]Remember, for scalars, the exponential power series is $e^a = \sum_{n=0}^{\infty} a^n/n!$.

to behavioral characteristics of queueing systems. Therefore, fitting the tail percentiles is an additional task that is worthwhile for the analyst in obtaining a proper distribution function for modeling purposes.

10.4 SYSTEMS WITH PHASE-TYPE SERVICE

The service distribution often describes a process designed and controlled by people who view variability as an undesirable feature; therefore, the exponential assumption for service times is too limiting because of its large variability. In this section we discuss generalizing the service time distribution. With single-server systems, using phase-type distributions presents little difficulty. With multiple-server systems, tractability becomes an important issue and places limitations on the usability of the matrix geometric methods.

10.4.1 The M/Ph/1 Queueing System

We begin our discussion by looking at a specific example of an M/Ph/1 system. Our example will use Poisson arrivals and a single server with a phase-type service distribution having two phases. The state space will have to be two-dimensional because we need to know not only the number of customers in the system, but also the phase number of the customer being served; therefore, let

$$E = \{0, (1,1), (1,2), (2,1), (2,2), (3,1), (3,2), (4,1), (4,2), \cdots\}.$$

If the queueing system is in state (n, k), there are n customers in the system with $n-1$ customers being in the queue, and the customer that is being served is in the kth phase. The mean arrival rate is denoted by λ, and the phase-type distribution has parameters $m = 2$, $\boldsymbol{\alpha}_*$, and $\boldsymbol{G}_*$. Thus, for $k, j = 1, 2$, we have the following interpretations: $\alpha_*(k)$ is the probability that a customer who enters the server will begin in phase k; $G_*(k, j)$ is the rate at which the customer in service moves from phase k to phase j; and $G_\Delta(k)$ is the rate at which a customer who is in phase k leaves the server. The state diagram for this system is given in Figure 10.3.

FIGURE 10.3 State diagram for an M/Ph/1 system. Rates associated with vertical arrows are $G_*(1,2)$ going down and $G_*(2,1)$ going up.

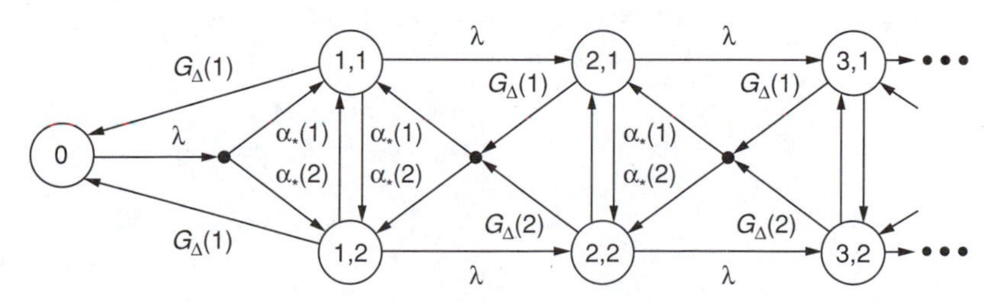

Thus, for an M/Ph/1 queueing system, an empty system will make a transition to a system containing one customer with a mean rate of λ; however, there are two states representing a system with one customer. When a customer enters the server, there is a probability of $\alpha_*(1)$ that the customer will begin in phase 1 and a probability of $\alpha_*(2)$ that the customer will begin in phase 2. Therefore, the rate of a transition from state 0 to state (1,1) is $\lambda\alpha_*(1)$, and the rate of a transition from state 0 to state (1,2) is $\lambda\alpha_*(2)$. If the process is in state (n, k), an arriving customer joins the queue so that the rate of transition from (n, k) to $(n+1, k)$ is λ. If there are two or more customers in the system, then a service completion occurs with rate $G_\Delta(k)$ if the customer who is in service is occupying phase k. Immediately after departure, the customer at the head of the queue enters service and joins phase k with probability $\alpha_*(k)$; therefore, the rate of transition from state (n, k) to $(n-1, j)$ is $G_\Delta(k)\alpha_*(j)$. With this description (and with the help of Figure 10.3), the generator matrix for the M/Ph/1 queueing model can be formed. We shall call it Q to keep it distinct from the generator G, which is associated with the description of the Markov process representing the server. Thus, Q is defined by

$$
\begin{array}{c|cc|cc|cc|c}
0 & (1,1) & (1,2) & (2,1) & (2,2) & (3,1) & (3,2) & \cdots \\
\hline
-\lambda & \lambda\alpha_1 & \lambda\alpha_2 & & & & & \\
\hline
G_\Delta(1) & g_{11}-\lambda & g_{12} & \lambda & 0 & & & \\
G_\Delta(2) & g_{21} & g_{22}-\lambda & 0 & \lambda & & & \\
\hline
 & G_\Delta(1)\alpha_1 & G_\Delta(1)\alpha_2 & g_{11}-\lambda & g_{12} & \lambda & 0 & \\
 & G_\Delta(2)\alpha_1 & G_\Delta(2)\alpha_2 & g_{21} & g_{22}-\lambda & 0 & \lambda & \\
\hline
 & & & G_\Delta(1)\alpha_1 & G_\Delta(1)\alpha_2 & g_{11}-\lambda & g_{12} & \cdots \\
 & & & G_\Delta(2)\alpha_1 & G_\Delta(2)\alpha_2 & g_{21} & g_{22}-\lambda & \cdots \\
\hline
 & & & & & \ddots & \ddots & \ddots
\end{array},
$$

where blanks within the matrix are interpreted as zeros and the row above the matrix indicates the states as they are associated with the individual columns of the matrix. (We have taken some notational liberties because of the large size of the matrix. The terms α_1 and α_2 are the components of $\boldsymbol{\alpha}_*$, and g_{ij} are components of $\boldsymbol{G}_*$.)

The M/Ph/1 generator can also be represented using submatrices for the elements within the various partitions. The advantage of doing so is that the matrix $\boldsymbol{Q}$ for the M/Ph/1 queue can be easily written for a general server with m phases. This gives

$$
\boldsymbol{Q} = \begin{bmatrix}
-\lambda & \lambda\boldsymbol{\alpha}_* & & & & \\
\boldsymbol{G}_\Delta & \boldsymbol{G}_* - \boldsymbol{\Lambda} & \boldsymbol{\Lambda} & & & \\
 & \boldsymbol{G}_\Delta\boldsymbol{\alpha}_* & \boldsymbol{G}_* - \boldsymbol{\Lambda} & \boldsymbol{\Lambda} & & \\
 & & \boldsymbol{G}_\Delta\boldsymbol{\alpha}_* & \boldsymbol{G}_* - \boldsymbol{\Lambda} & \boldsymbol{\Lambda} & \\
 & & & \boldsymbol{G}_\Delta\boldsymbol{\alpha}_* & \boldsymbol{G}_* - \boldsymbol{\Lambda} & \cdots \\
 & & & & \vdots & \ddots
\end{bmatrix}, \qquad \textbf{(10.13)}
$$

where $\boldsymbol{\Lambda} = \lambda\boldsymbol{I}$ (i.e., a matrix with λ's on the diagonal and zeros elsewhere). It should also be noted that $\boldsymbol{G}_\Delta$ is a *column* vector and $\boldsymbol{\alpha}_*$ is a *row* vector. Therefore, the matrix product of $\boldsymbol{G}_\Delta\boldsymbol{\alpha}_*$ is an m by m matrix, the i-j element of which is the product $G_\Delta(i)\alpha_*(j)$. Our goal is to find the values for the steady-state probabilities denoted by $\boldsymbol{p} = (p_0|p_{11}, p_{12}, \cdots|p_{21}, p_{22}, \cdots|\cdots) = (p_0, \boldsymbol{p}_1, \boldsymbol{p}_2, \cdots)$, where $\boldsymbol{p}_n = (p_{n1}, p_{n2}, \cdots)$ for $n = 1, 2, \cdots$. In other words the vector $\boldsymbol{p}$ is partitioned in the same fashion as the matrix $\boldsymbol{Q}$. Combining Eq. (10.13) with the equation $\boldsymbol{pQ} = \boldsymbol{0}$ yields the following system of equations, written using the submatrices from the partitioning of Eq. (10.13):

$$\begin{aligned} -\lambda p_0 + \boldsymbol{p}_1\boldsymbol{G}_\Delta &= 0 \\ \lambda\boldsymbol{\alpha}_* p_0 + \boldsymbol{p}_1(\boldsymbol{G}_* - \boldsymbol{\Lambda}) + \boldsymbol{p}_2\boldsymbol{G}_\Delta\boldsymbol{\alpha}_* &= \boldsymbol{0} \\ \boldsymbol{p}_1\boldsymbol{\Lambda} + \boldsymbol{p}_2(\boldsymbol{G}_* - \boldsymbol{\Lambda}) + \boldsymbol{p}_3\boldsymbol{G}_\Delta\boldsymbol{\alpha}_* &= \boldsymbol{0} \\ \boldsymbol{p}_2\boldsymbol{\Lambda} + \boldsymbol{p}_3(\boldsymbol{G}_* - \boldsymbol{\Lambda}) + \boldsymbol{p}_4\boldsymbol{G}_\Delta\boldsymbol{\alpha}_* &= \boldsymbol{0} \\ &\vdots \end{aligned} \tag{10.14}$$

The matrix in Eq. (10.13) is from a quasi–birth-death process, so we know the solution to the system given in Eq. (10.14) involves the characteristic matrix equation given by

$$\boldsymbol{\Lambda} + \boldsymbol{R}(\boldsymbol{G}_* - \boldsymbol{\Lambda}) + \boldsymbol{R}^2\boldsymbol{G}_\Delta\boldsymbol{\alpha}_* = \boldsymbol{0}. \tag{10.15}$$

Once the characteristic root matrix $\boldsymbol{R}$ is found, we can again take advantage of the matrix geometric form for the probabilities as we repeat Eq. (10.4); namely, we have

$$\boldsymbol{p}_n = \boldsymbol{c}\boldsymbol{R}^n \tag{10.16}$$

for $n = 1, 2, \cdots$, where $\boldsymbol{c}$ is a vector of constants.

Because this is a single-server system, traffic intensity is always given by $\rho = \lambda/\mu$, where $1/\mu$ is the mean service time. Thus, from Property 10.3, the traffic intensity is $\rho = -\lambda\boldsymbol{\alpha}_*\boldsymbol{G}_*^{-1}\boldsymbol{1}$. Assuming that ρ is less than one, it makes sense to find the matrix $\boldsymbol{R}$ that satisfies Eq. (10.15). It turns out that for this M/Ph/1 characteristic function, it is possible[9] to find a closed form expression to the equation and obtain

$$\boldsymbol{R} = \lambda(\boldsymbol{\Lambda} - \lambda\boldsymbol{1}\boldsymbol{\alpha}_* - \boldsymbol{G}_*)^{-1}, \tag{10.17}$$

where $\boldsymbol{1}$ is a column vector of ones so that the product $\boldsymbol{1}\boldsymbol{\alpha}_*$ is a matrix—each row of which is the vector $\boldsymbol{\alpha}_*$.

It can also be shown that the constant vector $\boldsymbol{c}$ from Eq. (10.16) is given by the vector $p_0\boldsymbol{\alpha}_*$ with $p_0 = 1 - \rho$, where $\rho = \lambda/\mu$. (To demonstrate this, we rewrite the second equation from the system of equations (10.14) using

[9] M. F. Neuts, p. 84.

Eq. (10.17) as a substitute for $\boldsymbol{p}_1$ and $\boldsymbol{p}_2$ and observe that $\lambda\boldsymbol{\alpha}_* = \boldsymbol{\alpha}_*\boldsymbol{\Lambda}$. This yields

$$\begin{aligned}\lambda\boldsymbol{\alpha}_* p_0 &+ \boldsymbol{p}_1(\boldsymbol{G}_* - \boldsymbol{\Lambda}) + \boldsymbol{p}_2\boldsymbol{G}_\Delta\boldsymbol{\alpha}_* \\ &= \boldsymbol{\alpha}_*\boldsymbol{\Lambda}p_0 + p_0\boldsymbol{\alpha}_*\boldsymbol{R}(\boldsymbol{G}_* - \boldsymbol{\Lambda}) + p_0\boldsymbol{\alpha}_*\boldsymbol{R}^2\boldsymbol{G}_\Delta\boldsymbol{\alpha}_* \\ &= \boldsymbol{\alpha}_* p_0\left[\boldsymbol{\Lambda} + \boldsymbol{R}(\boldsymbol{G}_* - \boldsymbol{\Lambda}) + \boldsymbol{R}^2\boldsymbol{G}_\Delta\boldsymbol{\alpha}_*\right] = 0,\end{aligned}$$

where the final equality is obtained by recognizing that the quantity in the square brackets is the same as Eq. (10.15). Thus, the M/Ph/1 queueing system with a mean arrival rate of λ and a service distribution defined by parameters m, $\boldsymbol{\alpha}_*$, and $\boldsymbol{G}_*$ has steady-state probabilities given, for $\rho < 1$, by

$$p_{n,k} = \begin{cases} 1-\rho & \text{for } n = 0, \\ (1-\rho)(\boldsymbol{\alpha}_*\boldsymbol{R}^n)(k) & \text{for } n = 1, 2, \cdots, \text{ and } k = 1, \cdots, m, \end{cases}$$

where $\rho = -\lambda\boldsymbol{\alpha}_*\boldsymbol{G}_*^{-1}\boldsymbol{1}$, and $\boldsymbol{R}$ is defined by Eq. (10.17). It also follows that the expected number of customers in the system and in the queue are given by

$$L = (1-\rho)\boldsymbol{\alpha}_*\boldsymbol{R}(\boldsymbol{I}-\boldsymbol{R})^{-2}\boldsymbol{1} \qquad \textbf{(10.18)}$$

$$L_q = (1-\rho)\boldsymbol{\alpha}_*\boldsymbol{R}^2(\boldsymbol{I}-\boldsymbol{R})^{-2}\boldsymbol{1} \qquad \textbf{(10.19)}$$

EXAMPLE 10.3 Let us consider an M/Ph/1 queueing system such that the arrival rate is $\lambda = 0.079$ and the distribution function governing the service time random variable is the one illustrated in Figure 10.2, with generator matrix given by Eq. (10.12). Thus the mean and the standard deviation of the service time are 10.126 and 7.365, respectively. Thus the queueing system has a traffic intensity of 0.8, which implies that a steady-state analysis is possible. The matrix that must be inverted [Eq. (10.17)] to obtain $\boldsymbol{R}$ is

$$\boldsymbol{\Lambda} - \lambda\boldsymbol{1}\boldsymbol{\alpha}_* - \boldsymbol{G}_* = \begin{bmatrix} 1.0158 & -0.9000 & -0.0158 & 0.0000 \\ -0.0632 & 0.5790 & -0.5158 & 0.0000 \\ -0.0632 & 0.1000 & 0.3965 & -0.2000 \\ -0.0632 & 0.0000 & -0.0158 & 0.3290 \end{bmatrix}.$$

Thus, the matrix $\boldsymbol{R}$ is

$$\boldsymbol{R} = \begin{bmatrix} 0.1298 & 0.2633 & 0.3563 & 0.2166 \\ 0.0579 & 0.2946 & 0.3951 & 0.2402 \\ 0.0490 & 0.1453 & 0.3999 & 0.2431 \\ 0.0273 & 0.0576 & 0.0876 & 0.2934 \end{bmatrix}.$$

The probability that the system is empty is 0.2; the probability that there is exactly one customer in the system is 0.188. The expected number in the system and the expected number in the queue are 3.24 and 2.44, respectively. As a practice in the use of Little's formula, we can also derive L_q knowing only that $L = 3.24$. With Little's formula we have that $W = 3.24/0.079 = 41.013$. Because the system has only one server, it follows that $W = W_q + 10.126$

or $W_q = 30.887$. Using Little's formula one more time yields $L_q = \lambda W_q = 0.079 \times 30.887 = 2.44$. ■

■ **EXAMPLE 10.4** **Simulation.** In order to provide as much help as possible in understanding the dynamics of a phase-type queueing system, we step through a simulation of Example 10.3. There will be three types of events, denoted as follows: "a" signifies an arrival, "p" signifies the completion of a sojourn time within a phase, and "q" signifies the movement of a queued entity into the server. (The example simulations of Chapter 6 did not have a separate event for the transfer from the queue to the server, but because of the extra complication of the movement within the server, it is easier to distinguish between the "p" event and the "q" event.) The random number sequence begins at row 4 of Table C.6. The transformation from the random number, R, in column 3 (Table 10.1) to the arrival time in column 4 is

$$\text{Clock time} - \ln(R)/0.079.$$

An arriving entity goes into phase 1 if the random number in column 5 is less than or equal to 0.8; otherwise, the entity goes into phase 3. An entity that is leaving phase 1 will enter phase 2 if the random number in column 5 is less than or equal to 0.9 (as at clock time 13.04 and 28.45 in Table 10.1); otherwise, the entity will leave the system (as at clock time 0.03, 6.52, etc.). An entity that is leaving phase 2 will always proceed into phase 3. An entity that is leaving phase 3 will enter phase 4 if the random number

TABLE 10.1 Simulation of an M/Ph/1 queueing system.

(1) Clock time	(2) Type	(3) RN	(4) Time next arrival	(5) RN	(6) Into which phase	(7) RN	(8) Within phase time	(9) Time leave phase	(10) No. in sys.
0	a	0.7156	4.24	0.5140	1	0.9734	0.03	0.03	1
0.03	p	—	—	0.9375	out	—	—	—	0
4.24	a	0.7269	8.28	0.2228	1	0.1020	2.28	6.52	1
6.52	p	—	—	0.9991	out	—	—	—	0
8.28	a	0.9537	8.88	0.6292	1	0.4340	0.83	9.11	1
8.88	a	0.2416	26.86	queued	—	—	—	—	2
9.11	p	—	—	0.9946	out	—	—	—	1
9.11	q	—	—	0.6799	1	0.0197	3.93	13.04	1
13.04	p	—	—	0.3988	2	0.4703	1.51	14.55	1
14.55	p	—	—	—	3	0.8872	0.36	14.91	1
14.91	p	—	—	0.5035	4	0.2335	5.82	20.73	1
20.73	p	—	—	—	out	—	—	—	0
26.86	a	0.6576	32.17	0.5989	1	0.2047	1.59	28.45	1
28.45	p	—	—	0.4328	2	0.2419	2.84	31.29	1
31.29	p	—	—	—	3	0.1215	6.32	37.61	1
32.17	a	0.6602	37.43	queued	—	—	—	—	2
37.43	a	0.7957	40.32	queued	—	—	—	—	3
37.61	p	—	—	0.9518	out	—	—	—	2
37.61	q	—	—	0.1589	1	0.5695	0.56	38.17	1

is less than or equal to 0.6; it will enter phase 2 if the random number is greater than 0.6 and less than or equal to 0.9; otherwise it will leave the system. An entity leaving phase 4 will always leave the system. The random number in column 7 is used to determine the random variate representing the sojourn time within a phase; therefore, its transformation depends on the phase in which the entity resides. The simulation results contained in Table 10.1 should contain enough of the dynamics of the system to illustrate these concepts. ■

10.4.2 The M/Ph/*c* Queueing System

There are two approaches for modeling the multiple-server phase-type system: (1) Let the state space contain information regarding the current phase number for each customer in each server, or (2) let the state space contain information regarding the number of customers in each phase. Consider an M/Ph/3 system with a phase-type distribution containing two phases. If we were to model this system using the first approach, the vector $\boldsymbol{p}_n$ would be defined by

$$\boldsymbol{p}_n = (p_{n111}, p_{n112}, p_{n121}, p_{n122}, p_{n211}, p_{n212}, p_{n221}, p_{n222}),$$

for $n \geq 3$. In this case the state $nijk$ refers to n customers in the system—the customer in the first server being in phase i, the customer in the second server being in phase j, and the customer in the third server being in phase k. If we were to model the three-server system with a two-phase service distribution using the second approach, the vector $\boldsymbol{p}_n$ would be defined by

$$\boldsymbol{p}_n = (p_{n30}, p_{n21}, p_{n12}, p_{n03}),$$

for $n \geq 3$. In this second case the state nij refers to there being n customers in the system—i customers being in their first phase, and j customers being in their second phase. For a system of c servers and a phase-type distribution of m phases, the first approach yields submatrices containing m^c rows and columns, and the second approach yields submatrices containing $(c+m-1)!/[c!(m-1)!]$ rows and columns. Thus the advantage of the second approach is that it can yield significantly smaller submatrices, especially when the number of phases is small. However, the first approach can be used to model a system that contains services with different service time distributions.

10.5 SYSTEMS WITH PHASE-TYPE ARRIVALS

Another example of a queueing system for which the matrix geometric procedure permits computational results for a problem otherwise difficult is a system with exponential service and phase-type arrivals.

For the phase-type arrival process, there is always a customer in an "arrival" phase. Consider first an Erlang type-2 system (see Figure 10.4). A customer (not yet in the queueing system) starts in phase 1 of the arrival

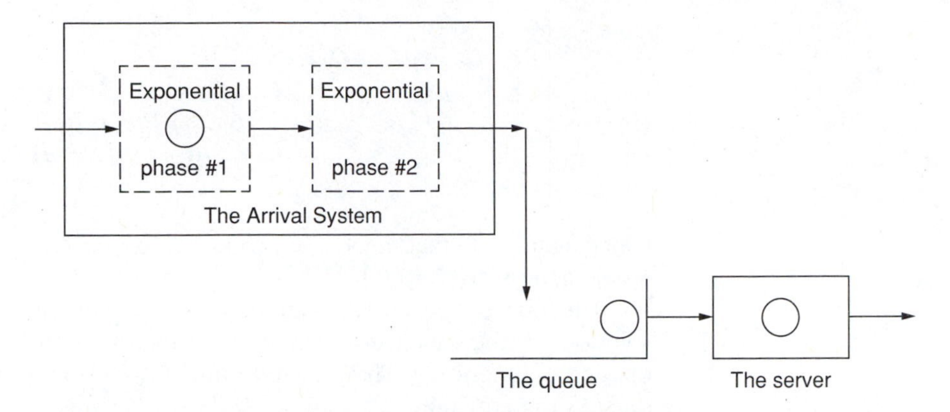

FIGURE 10.4 Schematic for the Erlang distribution showing two customers in the queueing system.

system. After an exponential length of time, the customer moves to phase 2. After an additional exponential length of time, the customer leaves phase 2 of the arrival system and enters the queueing system. As soon as the first customer enters the queueing system, the next customer instantaneously begins phase 1 in the arrival system.

We now generalize from Erlang distributions to phase-type distributions. Assume for illustrative purposes that the distribution governing interarrival times is a phase-type distribution with parameters $m = 2$, $\boldsymbol{\alpha}_*$, and $\boldsymbol{G}_*$. Also, let the mean service time be $1/\mu$. As soon as a customer leaves the arrival system and enters the queueing system, the next customer begins in phase k of the arrival system with probability $\alpha_*(k)$. The customer moves around the phases according to the rate matrix $\boldsymbol{G}_*$ and then enters the queueing system upon leaving the arrival system. Such a system gives rise to the following generator matrix:

$$
\begin{array}{c}
\begin{array}{cccccccc}
\quad\quad\quad & (1,0) & (2,0) & (1,1) & (2,1) & (1,2) & (2,2) & \cdots
\end{array}\\
Q = \left[\begin{array}{cc|cc|cc|c}
g_{11} & g_{12} & G_\Delta(1)\alpha_1 & G_\Delta(1)\alpha_2 & & & \\
g_{21} & g_{22} & G_\Delta(2)\alpha_1 & G_\Delta(2)\alpha_2 & & & \\
\hline
\mu & 0 & g_{11}-\mu & g_{12} & G_\Delta(1)\alpha_1 & G_\Delta(1)\alpha_2 & \\
0 & \mu & g_{21} & g_{22}-\mu & G_\Delta(2)\alpha_1 & G_\Delta(2)\alpha_2 & \\
\hline
 & & \mu & 0 & g_{11}-\mu & g_{12} & \cdots \\
 & & 0 & \mu & g_{21} & g_{22}-\mu & \cdots \\
\hline
 & & & & \ddots & \ddots & \ddots
\end{array}\right].
\end{array}
$$

The above representation of the Ph/M/1 system for a two-phase distribution provides the pattern to generalize the generator for the general phase-type distribution. In this case the state (k, n) indicates that there are n customers in the system, and the customer in the arrival system is in phase k. In matrix notation we have

$$Q = \begin{bmatrix} \boldsymbol{G}_* & \boldsymbol{G}_\Delta \boldsymbol{\alpha}_* & & & \\ \mu \boldsymbol{I} & \boldsymbol{G}_* - \mu \boldsymbol{I} & \boldsymbol{G}_\Delta \boldsymbol{\alpha}_* & & \\ & \mu \boldsymbol{I} & \boldsymbol{G}_* - \mu \boldsymbol{I} & \boldsymbol{G}_\Delta \boldsymbol{\alpha}_* & \\ & & \mu \boldsymbol{I} & \boldsymbol{G}_* - \mu \boldsymbol{I} & \boldsymbol{G}_\Delta \boldsymbol{\alpha}_* \\ & & & \mu \boldsymbol{I} & \boldsymbol{G}_* - \mu \boldsymbol{I} & \cdots \\ & & & & \vdots & \ddots \end{bmatrix}, \tag{10.20}$$

[Once again note that the vector product $\boldsymbol{G}_\Delta \boldsymbol{\alpha}_*$ yields a square matrix whose i-j element is $G_\Delta(i)\alpha_*(j)$.]

The phase-type arrival system does not yield itself to a closed form solution to the characteristic function; therefore, the matrix $\boldsymbol{R}$ must be obtained numerically. That is, Algorithm 10.1 can be used to find the $\boldsymbol{R}$ that satisfies the equation

$$\mu \boldsymbol{R}^2 + \boldsymbol{R}(\boldsymbol{G}_* - \mu \boldsymbol{I}) + \boldsymbol{G}_\Delta \boldsymbol{\alpha}_* = \boldsymbol{0}. \tag{10.21}$$

After $\boldsymbol{R}$ is found, the vector $\boldsymbol{p}_0$ can be found by solving the system of equations given by

$$\begin{aligned} \boldsymbol{p}_0(\boldsymbol{G}_* + \mu \boldsymbol{R}) &= \boldsymbol{0} \\ \boldsymbol{p}_0(\boldsymbol{I} - \boldsymbol{R})^{-1}\boldsymbol{1} &= 1. \end{aligned} \tag{10.22}$$

This chapter has touched only briefly on a rich topic for numerically analyzing queueing systems. There are many further examples of queueing systems that can be profitably analyzed through matrix geometric procedures. There are also many aspects of the matrix geometric method that we have not discussed; interested students should find this topic a rewarding pursuit. To help the student gain experience in modeling with these techniques, the exercises present several further queueing systems for which matrix geometric models must be given.

10.6 EXERCISES

To obtain the maximum benefit from these homework problems, it would be best to have a computer available. As a starting point, write a program implementing the successive substitution algorithm. The only difficult part is the matrix inversion procedure. If you do not have a subroutine available that inverts a matrix, consult the appendix for a description of an algorithm for matrix inversion.

10.1 Consider the following sequence of numbers: 0, 1, 2, 5, 12, 29, $\cdots$.

(a) What is the next number in the series?

(b) What is the characteristic function that defines the general expression for the numbers in the series?

(c) Without writing out the next several numbers, give the tenth number in the series.

10.2 Consider a batch arrival queueing system. The batches arrive according to a Poisson process with mean rate of one per hour, and every batch contains exactly two customers. Service is one at a time according to an exponential distribution, with a mean of 16 minutes. The purpose of this exercise is to practice the use of (scalar) difference equations, so you should analyze this queueing system without using the matrix geometric concepts.

(a) Form the generator matrix and write out the characteristic function that must be solved to obtain a general expression for p_n. (Note that the characteristic function is a cubic with one root equal to 1.)

(b) Show that

$$p_n = c_1 \left(\frac{2}{3}\right)^n + c_2 \left(-\frac{2}{5}\right)^n$$

for $n = 0, 1, \cdots$.

(c) Show that the probability that there are more than three in the system is 17.6%.

(d) What is the expected number in the system?

10.3 Consider a batch service system in which the arrival process is Poisson (i.e., one at a time), with a mean arrival rate of one per hour. The service mechanism is an oven that can hold two items at once, and service never starts unless two items are ready for service. In other words, if only one item is in the system, that item will wait in the queue until the second item arrives. The service time is exponentially distributed with a mean batch service rate of 0.8284 per hour. Again, this problem emphasizes the use of the difference equation methodology, so you should analyze this system without using the matrix geometric technique.

(a) Form the generator matrix and write out the characteristic function that must be solved to obtain a general expression for p_n. (Note that the characteristic function is a cubic with one root equal to 1.)

(b) Show that

$$p_n = p_1(0.7071)^{n-1}$$

for $n = 1, 2, \cdots$.

(c) Show that the probability that there are more than three in the system is approximately 30.2%.

(d) What is the expected number in the system?

10.4 Consider an $M^{[X]}/M/1$ queueing system with a mean arrival rate of 3 batches per hour and a mean service time of 7.5 minutes. The probability that the batch size is 1, 2, or 3 is 0.3, 0.5, or 0.2, respectively.

(a) What is the matrix characteristic equation for this system?

(b) Using Algorithm 10.1, determine the matrix $\boldsymbol{R}$.

(c) What is the probability that the queue is empty?

(d) What is the value of L?

10.5 Consider an $M^{[X]}/M/2$ queueing system with a mean arrival rate for batches of λ and a mean service rate of μ. Assume that the maximum batch size is 2 and show how this system can be analyzed using the matrix geometric methodology. In other words, give the relevant matrices for the matrix characteristic equation and the necessary boundary equations.

10.6 Consider the following batch service system. Arrivals occur one at a time, with mean rate $\lambda = 1$. The service mechanism is an oven, and, after service is started, it cannot be interrupted. If only one item is in the system, that item is placed in the oven, and service is immediately started. If two or more are in the system and the oven is empty, two items are placed in the oven and service is started. The length of time that the oven takes to complete its operation is an exponential random variable with mean rate $\mu = 0.8284$—whether one or two items are in the oven.

(a) Write the generator matrix using a two-dimensional state space, where the state (n, k) denotes n items in the queue and k items in the oven. Also, give the (cubic) matrix characteristic equation.

(b) Solve for $\boldsymbol{R}$ analytically—that is, without using the successive substitution algorithm. (*Hint:* Observe that, because all the matrices in the characteristic equation are upper-triangular, the matrix $\boldsymbol{R}$ will be upper-triangular; thus,

$$\boldsymbol{R}^3 = \begin{bmatrix} r_{11}^3 & r_{12}^{(3)} \\ 0 & r_{22}^3 \end{bmatrix},$$

where $r_{12}^{(3)}$ can be written in terms of the elements of $\boldsymbol{R}$, and r_{ii}^3 is the cube of the r_{ii} term.)

(c) What is the probability that the system is empty?

10.7 A two-channel communication node can take two types of calls—data calls and voice calls. Data calls arrive according to a Poisson process with mean rate λ_d, and the length of time a data call uses a channel is exponentially distributed with a mean time of $1/\mu_d$. If both channels are busy, there is a buffer that can hold the data calls until a channel is free. The buffer has an unlimited capacity. Voice calls arrive according to a Poisson process with mean rate λ_v, and the length of time a voice call uses a channel is exponentially distributed with a mean time of $1/\mu_v$. However, if both channels are busy when a voice call arrives, the call is lost.

(a) This system can be modeled as a quasi–birth-death process. Give the matrix characteristic equation that must be solved to find $\boldsymbol{R}$, and write out the boundary equations.

(b) Data calls come in at a rate of 1 per minute and the average service time is 1 minute. Voice calls come in at a rate of 4 per hour and last an average of 12 minutes. Assume that voice calls are *nonpreemptive*. Find the numerical values for the matrix $\boldsymbol{R}$ analytically instead of using the successive substitution algorithm. (*Hint:* Because the matrices $\boldsymbol{A}_0$, $\boldsymbol{A}_1$, and $\boldsymbol{A}_2$ are triangular, the matrix $\boldsymbol{R}$ will also be triangular; thus the diagonal elements of $\boldsymbol{R}^2$ are the square of the diagonal elements of $\boldsymbol{R}$.)

(c) What is the expected number of data calls in the system?

(d) What is the probability that both channels are occupied by voice calls?

(e) Given that both channels are occupied by voice calls, what is the expected number of data calls in the system?

10.8 Answer the questions of Exercise 10.7, assuming now that the voice calls have *preemptive priority*. In other words, if a voice call arrives and both channels are busy with at least one channel being occupied by a data call, then that data call is placed back in the buffer and the voice call takes over the channel. If both channels are occupied by voice calls, the newly arrived voice call is lost. [For part (b) use Algorithm 10.1 to find the matrix $\boldsymbol{R}$.]

10.9 Consider an M/M/1 system with a mean arrival rate of 10 per hour and a mean service time of 5 minutes. A cost is incurred for each item that is in the system at a rate of $50 per hour, and an operating cost of $100 per hour is incurred whenever the server is busy. Through partial automation, the variability of the service time can be reduced, although the mean will be unaffected by the automation. The engineering department claims that they will be able to cut the standard deviation of the service time in half. What are the potential savings due to this reduction?

10.10 Develop a model for an M/E_2/2 queueing system.

10.11 Consider the E_2/M/1 model with a mean arrival rate of 10 per hour and a mean service time of 5 minutes.

(a) Find the matrix $\boldsymbol{R}$ using the successive substitution algorithm.

(b) What is the probability that a queue is present?

(c) What is the expected number in the system?

10.12 Consider a Ph/M/2 queueing system with a mean arrival rate of λ and a mean service rate of μ. Show how this system can be analyzed using the matrix geometric methodology. In other words, give the relevant matrices for the matrix characteristic equation and the necessary boundary equations.

10.13 Let $X = \{X_t; t \geq 0\}$ denote a Markov process with state space $E = \{\ell, m, h\}$ and generator matrix

$$G = \begin{bmatrix} -0.5 & 0.5 & 0 \\ 1 & -2 & 1 \\ 0 & 1 & -1 \end{bmatrix}.$$

The process X is called the *environmental* process, and the three states refer to low-traffic conditions, medium-traffic conditions, and high-traffic conditions.

Let $N = \{N_t; t \geq 0\}$ denote a single-server queueing system. The arrivals to the system form a Poisson process whose mean arrival rate is dependent on the state of the environmental process. The mean arrival rate at time t is 2 per hour during periods of low traffic, 5 per hour during periods of medium traffic, and 10 per hour during heavy traffic conditions. (In other words, at time t the mean arrival rate is 2, 5, or 10 per hour, depending on whether X_t is ℓ, m, or h, respectively.) The mean service time is 10 minutes, independent of the state of the environment.

(a) Give the generator matrix for this system, where a state (n, k) denotes that there are n customers in the system and that the environmental process is in state k.

(b) Give the matrix characteristic equation and the boundary equations for this system.

(c) Show that this system satisfies the steady-state conditions.

(d) Derive an expression for L, the mean number in the system.

(e) Derive an expression for L_k, the *conditional* mean number in the system, given that the environment is in state k.

(f) Use Algorithm 10.1 to obtain numerical values for L and L_k.

APPENDIX A

Matrices

Within the body of this textbook, we have assumed that students are familiar with matrices and their basic operations. This appendix has been written for those students who might benefit from a review.

Definition A.1 *An $m \times n$ Matrix is an array of numbers in m rows and n columns as follows:*

$$A = \begin{bmatrix} a_{11} & a_{12} & \cdots & a_{1n} \\ a_{21} & a_{22} & \cdots & a_{2n} \\ \vdots & \vdots & & \vdots \\ a_{m1} & a_{m2} & \cdots & a_{mn} \end{bmatrix}.$$

Notice that the row index is always mentioned first, so the a_{ij} element refers to the element in row i and column j. A matrix that consists of only one row is called a *row vector*. Similarly, a matrix that consists of only one column is called a *column vector*.

Definition A.2 *If $m = n$, the matrix is a Square Matrix of order n.*

Definition A.3 *The Identity Matrix is a square matrix with ones on the diagonal and zeros elsewhere. That is,*

$$I = \begin{bmatrix} 1 & 0 & \cdots & 0 \\ 0 & 1 & \cdots & 0 \\ \vdots & \vdots & & \vdots \\ 0 & 0 & \cdots & 1 \end{bmatrix}.$$

A.1 MATRIX ADDITION AND SUBTRACTION

Two matrices can be added or subtracted if the number of rows and columns is the same for both matrices. The resulting matrix has the same number of rows and columns as the matrices being added or subtracted.

DEFINITION A.4 *The* SUM (DIFFERENCE) *of two $m \times n$ matrices* **A** *and* **B** *is an $m \times n$ matrix* **C** *with elements $c_{ij} = a_{ij} + b_{ij}$ $(c_{ij} = a_{ij} - b_{ij})$. Thus, to obtain* **A** + **B** (**A** − **B**) *we add (subtract) corresponding elements. Thus,*

$$\boldsymbol{A} + \boldsymbol{B} = \boldsymbol{C} = \begin{bmatrix} a_{11} + b_{11} & a_{12} + b_{12} & \cdots & a_{1n} + b_{1n} \\ a_{21} + b_{21} & a_{22} + b_{22} & \cdots & a_{2n} + b_{2n} \\ \vdots & \vdots & & \vdots \\ a_{m1} + b_{m1} & a_{m2} + b_{m2} & \cdots & a_{mn} + b_{mn} \end{bmatrix};$$

$$\boldsymbol{A} - \boldsymbol{B} = \boldsymbol{C} = \begin{bmatrix} a_{11} - b_{11} & a_{12} - b_{12} & \cdots & a_{1n} - b_{1n} \\ a_{21} - b_{21} & a_{22} - b_{22} & \cdots & a_{2n} - b_{2n} \\ \vdots & \vdots & & \vdots \\ a_{m1} - b_{m1} & a_{m2} - b_{m2} & \cdots & a_{mn} - b_{mn} \end{bmatrix}.$$

A.2 MATRIX MULTIPLICATION

Two matrices can be multiplied if the number of columns of the first matrix is equal to the number of rows of the second matrix. The resulting matrix then has the same number of rows as the first matrix and the same number of columns as the second matrix.

DEFINITION A.5 *The* PRODUCT ***AB*** *of an $m \times n$ matrix* **A** *and an $n \times l$ matrix* **B** *results in an $m \times l$ matrix* **C** *with elements defined by*

$$c_{ij} = \sum_{k=1}^{n} a_{ik} b_{kj} = a_{i1} b_{1j} + a_{i2} b_{2j} + \cdots + a_{in} b_{nj}$$

for $i = 1, 2, \cdots, m$ and $j = 1, 2, \cdots, l$.

Note that if $m \neq l$, then, even though ***AB*** is possible, the product ***BA*** makes no sense and is undefined. Even for square matrices of the same order, the product ***AB*** is *not* necessarily equal to ***BA***.

To illustrate, consider a row vector ***v*** with three components (i.e., a 1×3 matrix) multiplied with a 3×2 matrix ***P***; namely,

$$\boldsymbol{v} = \boldsymbol{vP} = [5 \quad 10 \quad 20] \begin{bmatrix} 1 & 12 \\ 3 & 13 \\ 7 & 17 \end{bmatrix}.$$

The resulting matrix is a row vector with two elements given by

$$\begin{aligned} \boldsymbol{v} &= [5 \times 1 + 10 \times 3 + 20 \times 7, 5 \times 12 + 10 \times 13 + 20 \times 17] \\ &= [175, 530]. \end{aligned}$$

However, the product of $\boldsymbol{Pv}$ is *not* possible, because the number of columns of $\boldsymbol{P}$ is different from the number of rows of $\boldsymbol{v}$. For a further example, let the column vector $\boldsymbol{f}$ denote the first column of $\boldsymbol{P}$. The product $\boldsymbol{vf}$ is a 1×1 matrix, usually written as a scalar as

$$\boldsymbol{vf} = 175,$$

and $\boldsymbol{fv}$ is a 3×3 matrix given by

$$\boldsymbol{fv} = \begin{bmatrix} 5 & 10 & 20 \\ 15 & 30 & 60 \\ 35 & 70 & 140 \end{bmatrix}.$$

A.3 DETERMINANTS

The determinant is a scalar associated with every square matrix. Determinants play a role in obtaining the inverse of a matrix and in solving systems of linear equations. We give a theoretical definition of the determinant of a matrix in this section and a more computationally oriented definition in the next section.

Before presenting the definition of determinants, it is important to understand permutations. Take, for example, the integers 1 through 6; their natural order is (1,2,3,4,5,6). Any rearrangement of this natural order is called a *permutation;* thus, (1,5,2,3,6,4) is a permutation. For the natural order, each number is smaller than any number to its right; however, in permutations, some numbers are larger than other numbers to their right. Whenever a number is larger than another one to its right, it is called an *inversion*. For example, in the permutation (1,5,2,3,6,4) there are four inversions—namely, (5,3), (5,4), (5,2), and (6,4). A permutation is called *even* if it has an even number of inversions; it is called *odd* if it has an odd number of inversions.

DEFINITION A.6 *The* DETERMINANT *of a square matrix* $\boldsymbol{A}$ *of order* n *is given by*

$$|\boldsymbol{A}| = \sum (\pm) a_{1j_1} a_{2j_2} \cdots a_{nj_n},$$

where the summation is over all possible permutations $j_1 j_2 \cdots j_n$ *of the set* $\{1, 2, \cdots, n\}$. *The sign is positive* $(+)$ *if the permutation* $j_1 j_2 \cdots j_n$ *is even and negative* $(-)$ *if the permutation is odd.*

A permutation $j_1 j_2 \cdots j_n$ contains exactly one element from each row and each column. Thus, a square matrix of order n has $n!$ permutations. For example, the matrix

$$A = \begin{bmatrix} a_{11} & a_{12} & a_{13} \\ a_{21} & a_{22} & a_{23} \\ a_{31} & a_{32} & a_{33} \end{bmatrix}$$

has 3!, or 6, permutations, and the determinant is given by

$$|A| = a_{11}a_{22}a_{33} - a_{11}a_{23}a_{32} + a_{12}a_{23}a_{31} - a_{12}a_{21}a_{33} + a_{13}a_{21}a_{32} - a_{13}a_{22}a_{31}$$

where, for example, $a_{13}a_{22}a_{31}$ has a negative sign because the sequence of the second subscripts $(3, 2, 1)$ is a permutation with three inversions—namely, (3,2), (3,1), and (2,1). Similarly, the determinant of a square matrix A of order 2 is

$$|A| = \begin{vmatrix} a_{11} & a_{12} \\ a_{21} & a_{22} \end{vmatrix} = a_{11}a_{22} - a_{12}a_{21}.$$

(In the next section this 2×2 determinant will form the basic building block on which the determinant for a general $n \times n$ matrix is obtained.)

The main properties of determinants can be summarized as follows:

1. If A is a square matrix, then $|A| = |A^T|$, where A^T is the transpose matrix of A. The transpose matrix A^T has elements $a_{ij}^T = a_{ji}$.
2. If any two rows (or columns) of a square matrix A are interchanged, resulting in a matrix B, then $|B| = -|A|$.
3. If any two rows or columns of a square matrix A are identical, then $|A| = 0$.
4. If any row (column) of a square matrix A can be written as a linear combination of the other rows (columns), then $|A| = 0$.
5. If any row or column of a square matrix A has all zeros, then $|A| = 0$.
6. If any row or column of a square matrix A is multiplied by a constant k, resulting in a matrix B, then $|B| = k|A|$.
7. If matrix B results from replacing the uth row (column) of a square matrix A by that uth row (column) plus a constant times the vth $(v \neq u)$ row (column), then $|B| = |A|$.
8. If A and B are two $n \times n$ matrices, then $|AB| = |A||B|$.
9. If a square matrix A of order n is upper- or lower-triangular, then its determinant is the product of its diagonal elements; in other words, $|A| = a_{11}a_{22}, \cdots, a_{nn}$.

A.4 DETERMINANTS BY COFACTOR EXPANSION

Evaluating the determinant using Definition A.6 is difficult for medium- or large-sized matrices. A systematic way of finding the determinant of a matrix is given by the cofactor expansion method. This method evaluates the determinant of a square matrix of order n by reducing the problem to the evaluation of determinants of square matrices of order $(n - 1)$. The procedure is then repeated by reducing the square matrices of order $(n-1)$ to finding determinants of square matrices of order $(n-2)$. This continues until determinants for square matrices of order 2 are evaluated.

DEFINITION A.7 *The* MINOR *of a_{ij} (element of the square matrix $\boldsymbol{A}$ of order n) is the square matrix $\boldsymbol{M}_{ij}$ of order $n-1$ that results from deleting the ith row and the jth column from matrix $\boldsymbol{A}$.*

DEFINITION A.8 *The* COFACTOR *c_{ij} of a_{ij} is defined as $c_{ij} = (-1)^{i+j}|\boldsymbol{M}_{ij}|$.*

To illustrate, consider the following 3×3 matrix $\boldsymbol{A}$:

$$\boldsymbol{A} = \begin{bmatrix} 1 & 2 & 3 \\ 4 & 5 & 6 \\ 7 & 8 & 9 \end{bmatrix}.$$

The minor of a_{21} is given by

$$\boldsymbol{M}_{21} = \begin{bmatrix} 2 & 3 \\ 8 & 9 \end{bmatrix},$$

and the cofactor of a_{21} is given by

$$c_{21} = (-1)^{2+1}|\boldsymbol{M}_{21}| = -(2 \times 9 - 3 \times 8) = 6.$$

Property A.9 Let $\boldsymbol{A}$ be a square matrix of order n with elements a_{ij}. Then the determinant of matrix $\boldsymbol{A}$ can be obtained by expanding the determinant with cofactors c_{ij} about the ith row,

$$|\boldsymbol{A}| = \sum_{j=1}^{n} a_{ij} c_{ij},$$

or by expanding the determinant about the jth column,

$$|\boldsymbol{A}| = \sum_{i=1}^{n} a_{ij} c_{ij}.$$

For example, to evaluate the determinant of a square matrix of order 3, we could consider the following expansion of the first row:

$$\begin{aligned} |\boldsymbol{A}| &= (-1)^{1+1} a_{11} \begin{vmatrix} a_{22} & a_{23} \\ a_{32} & a_{33} \end{vmatrix} + (-1)^{1+2} a_{12} \begin{vmatrix} a_{21} & a_{23} \\ a_{31} & a_{33} \end{vmatrix} \\ &\quad + (-1)^{1+3} a_{13} \begin{vmatrix} a_{21} & a_{22} \\ a_{31} & a_{32} \end{vmatrix} \\ &= a_{11}(a_{22}a_{33} - a_{23}a_{32}) - a_{12}(a_{21}a_{33} - a_{23}a_{31}) \\ &\quad + a_{13}(a_{21}a_{32} - a_{22}a_{31}). \end{aligned} \qquad \textbf{(A.1)}$$

Notice that the determinant is the same if an expansion of any other row or column is used.

The cofactor expansion method can be used recursively, so that the original $n \times n$ determinant problem is reduced to computing determinants of 2×2 matrices. For example, a square matrix of order 4 could be calculated by

$$|\boldsymbol{A}| = a_{11}|\boldsymbol{M}_{11}| - a_{12}|\boldsymbol{M}_{12}| + a_{13}|\boldsymbol{M}_{13}| - a_{14}|\boldsymbol{M}_{14}|.$$

Because each of the determinants above involves a 3×3 matrix, Eq. (A.1) can then be used to evaluate each $|\boldsymbol{M}_{ij}|$ using determinants of 2×2 matrices.

A.5 NONSINGULAR MATRICES

Another important concept in the algebra of matrices is that of singularity. Square matrices can be classified as singular or nonsingular. Nonsingular matrices represent a set of linearly independent equations, whereas singular matrices represent a set in which at least one row (column) can be expressed as a linear combination of other rows (columns).

DEFINITION A.10 *A square matrix $\boldsymbol{A}$ of order n that has a multiplicative inverse is called a* NONSINGULAR MATRIX. *That is, there exists a square matrix, denoted by $\boldsymbol{A}^{-1}$, of order n such that $\boldsymbol{A}\boldsymbol{A}^{-1} = \boldsymbol{A}^{-1}\boldsymbol{A} = \boldsymbol{I}$. If no inverse exists, the matrix is called* SINGULAR. *For a nonsingular matrix, the matrix $\boldsymbol{A}^{-1}$ is called the* INVERSE *of $\boldsymbol{A}$.*

Property A.11 If $\boldsymbol{A}$ is nonsingular, the matrix $\boldsymbol{A}^{-1}$ is unique and the determinants of $\boldsymbol{A}$ and $\boldsymbol{A}^{-1}$ are nonzero. Furthermore, $|\boldsymbol{A}^{-1}| = 1/|\boldsymbol{A}|$.

From the above property it is easy to see that a matrix is nonsingular if and only if the determinant of the matrix is nonzero. For example, the following Markov matrix,

$$\boldsymbol{P} = \begin{bmatrix} 0.1 & 0.3 & 0.6 & 0.0 \\ 0.0 & 0.2 & 0.5 & 0.3 \\ 0.0 & 0.1 & 0.2 & 0.7 \\ 0.1 & 0.3 & 0.6 & 0.0 \end{bmatrix},$$

is singular because the determinant of $\boldsymbol{P}$ is zero. (Use the third property given on page 286.)

A.6 INVERSION IN PLACE

When a square matrix is nonsingular and the elements of the matrix are all numeric, several algorithms that have been developed can be used to find its inverse. The best known method for numerical inversion is the Gauss-Jordan elimination procedure. Conceptually, the Gauss-Jordan algorithm appends an identity matrix to the original matrix. Then, through a series of elementary row operations, the original matrix is transformed to an identity matrix. When the original matrix is transformed to an identity, the appended identity matrix is transformed to the inverse. This method has the advantage that it is intuitive but has the disadvantage that it is wasteful of computer space when implemented.

Methods that use memory more efficiently are those that use a procedure called *inversion in place*. The operations to perform inversion in place are identical to Gauss-Jordan elimination, but savings in storage requirements are accomplished by not keeping an identity matrix in memory. The inversion in place requires a vector with the same dimension as the matrix to keep track of the row changes in the operations. This is necessary because the identity matrix is not saved in memory, and row interchanges for the inversion procedure need to be reflected as column changes in the implicitly appended identity matrix.

We shall first give an overview of the mathematical transformation required for the inversion-in-place procedure and then give the specific steps of the algorithm. The algorithm uses a pivot element on the diagonal, denoted as a_{kk}, and transforms the original matrix using the elements on the pivot row and pivot column. Let a_{ij} denote an element in the original matrix, and let b_{ij} denote an element in the transformed matrix. The equations describing the transformation are:

Pivot element

$$b_{kk} = 1/a_{kk}$$

Pivot row

$$b_{kj} = a_{kj}/a_{kk} \qquad \text{for } j = 1, 2, \cdots, k-1, k+1, \cdots, n$$

Pivot column

$$b_{ik} = -a_{ik}/a_{kk} \qquad \text{for } i = 1, 2, \cdots, k-1, k+1, \cdots, n$$

General transformation

$$b_{ij} = a_{ij} - a_{kj}a_{ik}/a_{kk} \quad \text{for} \begin{cases} i = 1, 2, \cdots, k-1, k+1, \cdots, n \\ j = 1, 2, \cdots, k-1, k+1, \cdots, n \end{cases}$$

The above equations are applied iteratively to the matrix for each $k = 1, \cdots, n$. After finishing the transformation using pivot element a_{kk}, the matrix (a_{ij}) is replaced by the matrix (b_{ij}) and the value of k is incremented by one. Notice, however, that the iteration could be done in place. That is, the equations presented above could be written with a_{ij} values instead of b_{ij} values, as long as the operations are performed in the correct sequence.

The only additional problem that may arise occurs when the pivot element, a_{kk}, is zero. In such a case the equations cannot be applied, because the pivot element is in the denominator. To solve this problem, the pivot row is interchanged with a row that has not yet been pivoted. (If a suitable row for the pivot cannot be found for an interchange, the matrix is singular.) The interchange must be recorded so that, when the iterations are finished, the inverted matrix can be reordered.

To diminish round-off error, pivots are selected as large as possible. Thus the inversion-in-place algorithm interchanges rows whenever the current candidate pivot is smaller in absolute value than other candidate pivots. In the description of the inversion-in-place algorithm, the following conventions will be followed: The matrix $\boldsymbol{A}$ is a square matrix of order n; the index k indicates the current row; the index *irow* indicates the row containing the maximum pivot element; the vector *newpos* is of dimension n, where *newpos*(k) is the row used for the kth pivot; *piv* is the reciprocal of the pivot element; and *temp* is a temporarily held value. It should also be emphasized that, because this is an inversion-in-place algorithm, as soon as a value of the matrix a_{ij} is changed, the next step in the algorithm will use the changed value.

Step 1 For each $k = 1, \cdots, n$ do **Steps 2** through **6**.

Step 2 Let *irow* be the index in the set $\{k, k+1, \cdots, n\}$ that gives the row that has the largest (in absolute value) element of column k; that is,

$$irow = \text{argmax}\,\{|a_{ik}| : i = k, \cdots, n\}.$$

(Remember, k is fixed.)

Step 3 If the largest (in absolute value) element on the kth column of rows $k, k+1, \cdots, n$ is zero, the matrix is singular and the program should stop. In other words, if $|a_{irow,k}| = 0$, stop; no inverse exists.

Step 4 Update the vector that keeps track of row interchanges by setting $newpos(k) = irow$.

Step 5 If the pivot row, *irow*, is not equal to the current row, k, interchange rows k and *irow* of the matrix.

Step 6 Apply the operations for inversion in place using the following sequence:

Operation A Set *piv* equal to $1/a_{kk}$, and then set a_{kk} equal to 1.

Operation B For each $j = 1, \cdots, n$ do

$$a_{kj} = a_{kj} \times piv$$

Operation C For each $i = 1, 2, \cdots, k-1, k+1, \cdots, n$ do

Operation C.i Set *temp* equal to a_{ik}, and then set a_{ik} equal to 0.

Operation C.ii For $j = 1, \cdots, n$ do

$$a_{ij} = a_{ij} - a_{kj} \times temp.$$

Step 7 For each $j = n, n-1, \cdots, 1$, check to determine if $j \neq newpos(j)$. If $j \neq newpos(j)$, then interchange column j with column $newpos(j)$. ■

Step 3 of the algorithm is to determine if the original matrix is nonsingular or singular. When implementing the algorithm on a computer, you must specify some error tolerance. For example, set ε equal to 0.0001 and stop the algorithm if the absolute value of the pivot element is less than ε. (It is much better to check for a value that is "too" small than to check for an exact equality to zero.) The inverse of the matrix is obtained after steps 2 through 6 of the algorithm are completed for each k, but the columns may not be in the correct order. Step 7 is a reordering of the matrix using the information in the vector *newpos*. Notice that the reordering must be done in reverse order to "undo" the row changes made in step 5.

If a matrix is upper- or lower-triangular, the inversion-in-place algorithm can be made much more efficient. (An upper-triangular matrix is one in which all elements below the diagonal are zero; that is, $a_{ij} = 0$ if $i > j$. A lower-triangular matrix has zeros above the diagonal.) *For a triangular matrix,* make the following modifications to the inversion-in-place algorithm.

1. Check for any zeros on the diagonal. Stop if there are any zeros on the diagonal, because the matrix is singular. If there are no zeros on the diagonal, the matrix is nonsingular. (See the ninth property given on page 286.)
2. The vector *newpos* is not needed and steps 2 through 5 can be skipped.
3. For an upper-triangular matrix, the instruction for operation B should be, "For each $j = k, k + 1, \cdots, n$ do". For a lower-triangular matrix, the instruction for operation B should be, "For each $j = 1, \cdots, k$ do".
4. For an upper-triangular matrix, the instruction for operation C should be, "For each $i = 1, \cdots, k - 1$ do"; for $k = 1$, operation C is skipped. For a lower-triangular matrix, the instruction for operation C should be, "For each $i = k + 1, \cdots, n$ do"; for $k = n$, operation C is skipped.

Consider the following example, in which the inverse of a square matrix of order 3 is to be obtained using the inversion-in-place algorithm.

$$\begin{bmatrix} 2.0 & 4.0 & 4.0 \\ 1.0 & 2.0 & 3.0 \\ 2.0 & 3.0 & 2.0 \end{bmatrix}$$

We begin by looking at row 1. Because the first element of the first row is the largest element in the first column, there is no need to do row interchanges. [In other words, $newpos(1) = 1$.] The new matrix after performing operations A through C is:

$$\begin{bmatrix} 0.5 & 2.0 & 2.0 \\ -0.5 & 0.0 & 1.0 \\ -1.0 & -1.0 & -2.0 \end{bmatrix}$$

Next we look at the elements in column 2, starting at row 2, and observe that the largest (in absolute value) element among the candidate values is the a_{32} element. (Note that $|a_{12}| > |a_{32}|$, but because a pivot operation has already been performed on row 1, the element in row 1 is not a candidate value.) Thus, *irow* = 3 and *newpos*(2) = 3, which indicates that row 2 and row 3 must be interchanged before any transformation can be performed. This interchange results in:

$$\begin{bmatrix} 0.5 & 2.0 & 2.0 \\ -1.0 & -1.0 & -2.0 \\ -0.5 & 0.0 & 1.0 \end{bmatrix}.$$

After performing operations A through C in the matrix above we have:

$$\begin{bmatrix} -1.5 & 2.0 & -2.0 \\ 1.0 & -1.0 & 2.0 \\ -0.5 & 0.0 & 1.0 \end{bmatrix}.$$

The final pivot element is a_{33}. The resulting matrix, after performing operations A through C, is

$$\begin{bmatrix} -2.5 & 2.0 & 2.0 \\ 2.0 & -1.0 & -2.0 \\ -0.5 & 0.0 & 1.0 \end{bmatrix}.$$

The matrix has been inverted. However, the second and third rows were interchanged during the inversion process; thus the columns in the above matrix cannot be in the correct order. The inverse of the original matrix is thus obtained by "undoing" the row interchanges by changing the corresponding columns on the last matrix. [Or, more pedantically, we first observe that *newpos*(3) = 3, so no change is necessary to start with; *newpos*(2) = 3, so columns 2 and 3 must be interchanged; finally, *newpos*(1) = 1, so no change is necessary for the last step.) The inverse to the original matrix is thus given by:

$$\begin{bmatrix} -2.5 & 2.0 & 2.0 \\ 2.0 & -2.0 & -1.0 \\ -0.5 & 1.0 & 0.0 \end{bmatrix}.$$

A.7 DERIVATIVES

Matrix equations sometimes occur in optimization problems, so it may be necessary to take derivatives of expressions involving matrices. For example, suppose that the elements of a matrix $\boldsymbol{Q}$ have been written as a function of some parameter, for example,

$$\boldsymbol{Q} = \begin{bmatrix} 1-\theta & \theta \\ 0 & 1-\theta^2 \end{bmatrix}. \quad \textbf{(A.2)}$$

The *derivative* of a matrix entails the term-by-term evaluation of the matrix:

$$\frac{d\boldsymbol{Q}}{d\theta} = \begin{bmatrix} -1 & 1 \\ 0 & -2\theta \end{bmatrix}.$$

The difficulty comes when the inverse of a matrix is part of an objective function. For example, a common matrix expression when dealing with Markov chains is the matrix $\boldsymbol{R}$, whose (i, j) term gives the expected number of visits to state j starting from state i. Assume that, for the matrix of Eq. (A.2), $\theta < 1$ and the matrix represents the transition probabilities associated with the transient states of a Markov chain. Then the matrix giving the expected number of visits to the two states is

$$\boldsymbol{R} = \left(\begin{bmatrix} 1 & 0 \\ 0 & 1 \end{bmatrix} - \begin{bmatrix} 1-\theta & \theta \\ 0 & 1-\theta^2 \end{bmatrix} \right)^{-1}$$
$$= \frac{1}{\theta^2} \begin{bmatrix} \theta & 1 \\ 0 & 1 \end{bmatrix}.$$

As long as $\boldsymbol{R}$ can be written analytically, it is usually not difficult to give its derivative. In this case it would be

$$\frac{d\boldsymbol{R}}{d\theta} = -\frac{1}{\theta^3} \begin{bmatrix} \theta & 2 \\ 0 & 2 \end{bmatrix}.$$

However, when the matrix is larger than a 2×2, it may be impossible to express $\boldsymbol{R}$ analytically; therefore, we present the following property,[1] by which the derivative can be calculated for the more general case.

Property A.12 Let $\boldsymbol{Q}$ be a matrix that is a function of the parameter θ and assume that the potential of $\boldsymbol{Q}$, defined by

$$\boldsymbol{R} = (\boldsymbol{I} - \boldsymbol{Q})^{-1},$$

exists. Then the derivative of $\boldsymbol{R}$ with respect to the parameter is given by

$$\frac{d\boldsymbol{R}}{d\theta} = \boldsymbol{R}\frac{d\boldsymbol{Q}}{d\theta}\boldsymbol{R}.$$

It is interesting to observe that the matrix result is similar to the scalar result: Let $f(t) = [1 - x(t)]^{-1}$, then $df/dt = f(dx/dt)f$.

[1]See R. R. Hocking, *The Analysis of Linear Models* (Monterey, Calif.: Brooks/Cole Publishing Co., 1985).

APPENDIX B

Numerical Analysis

After models have been developed, it becomes necessary to perform the calculations required by the models. Sometimes this process is straightforward, and sometimes it is not. The intent in this appendix is not to present a chapter on classical numerical analysis or any of its theory, but to present some numerical techniques that may be useful when writing computer code to evaluate your models. All of this material can be found in the *Handbook of Mathematical Functions.*[1] Therefore, if you need more accuracy than is possible by the limited number of parameters given in this appendix, or if you need to evaluate functions not listed in the appendix, use the *Handbook of Mathematical Functions*. Another valuable resource is the book *Numerical Recipes,*[2] which is a book (and accompanying software) containing description and computer code for a wide variety of numerical analysis techniques. There are three separate books—one for FORTRAN, one for Pascal, and one for C.

B.1 NUMERICAL INTEGRATION

For a reasonably accurate and easy integration scheme, we suggest Gaussian integration for definite integrals. Gaussian integration uses x values and their associated w values. The basic equation used for evaluation is

$$\int_{-1}^{1} f(x)\,dx \approx \sum_{i=1}^{n} w_i f(x_i),$$

[1] *Handbook of Mathematical Functions,* M. Abramowitz and I. A. Stegun, editors, (New York, N.Y.: Dover Publications, Inc., 1970).

[2] W. H. Press, B. P. Flannery, S. A. Teukolsky, and W. T. Vetterling, *Numerical Recipes: The Art of Scientific Computing,* (New York, N.Y.: Cambridge University Press, 1986).

where the value of n determines the number of x values used in the summation and the accuracy of the integral. In order to apply the above to an arbitrary definite integral, a change of variable is needed to expand the range of the integral. Because the x values are symmetric around zero, the following can be used if n is an even number and an integral from a to b needs to be evaluated:

$$x_m = \frac{b+a}{2}$$

$$\Delta = \frac{b-a}{2}$$

$$\int_a^b f(x)dx \approx \Delta \sum_{i=1}^{n/2} w_i \left[f(x_m - x_i\Delta) + f(x_m + x_i\Delta) \right].$$

The x values and w values for $n/2 = 4$ are as follows:[3]

x values	w values
0.18343 46424	0.36268 37833
0.52553 24099	0.31370 66458
0.79666 64774	0.22238 10344
0.96028 98564	0.10122 85362

B.2 FUNCTIONAL EVALUATION

Two common functions often need to be evaluated—the gamma function and the cumulative standard normal distribution.

B.2.1 The Gamma Function

The gamma function is denoted as $\Gamma(\cdot)$ and is defined, for $x > 0$, by

$$\Gamma(x) = \int_0^\infty e^{-s} s^{x-1}\, ds.$$

For integer values the value of the gamma function is a simple factorial; that is, for n a positive integer, $\Gamma(n) = (n-1)!$. However, for noninteger values the gamma function is difficult to evaluate, so approximations are commonly used.

A polynomial approximation for the gamma distribution is

$$\Gamma(1+x) \approx 1 + a_1 x + a_2 x^2 + \cdots + a_5 x^5 \quad \text{for } 0 \le x \le 1,$$

where the constants[4] are defined by $a_1 = -0.5748646$, $a_2 = 0.9512363$, $a_3 = -0.6998588$, $a_4 = 0.4245549$, and $a_5 = -0.1010678$. If it is necessary

[3]Taken from P. Davis and P. Rabinowitz, "Abscissas and Weights for Gaussian Quadratures of High Order," *Journal of Research of the National Bureau of Standards,* **56** (1956), pp. 35–37.

[4]C. Hastings, Jr., *Approximations for Digital Computers* (Princeton, N.J.: Princeton University Press, 1955), p. 155.

to evaluate $\Gamma(x)$ for $x < 1$, then use the relationship

$$\Gamma(x) = \frac{1}{x}\Gamma(1 + x).$$

If it is necessary to evaluate $\Gamma(n + x)$ for $n > 1$ and $0 \le x \le 1$, then use the relationship

$$\Gamma(n + x) = (n - 1 + x)(n - 2 + x) \cdots (1 + x)\Gamma(1 + x).$$

(See Table C.1 for an example.)

B.2.2 The Normal Distribution

If Z is a normally distributed random variable with mean of zero and standard deviation of one, its distribution function is defined, for all x, by

$$\Pr\{Z \le x\} = \int_{-\infty}^{x} \frac{1}{\sqrt{2\pi}} e^{-(1/2)s^2}\, ds.$$

As with the gamma function, the above integral is difficult to evaluate, so approximations are often used. Because the distribution is symmetric around zero, it is sufficient to give an approximation for the positive portion of the domain. In particular, for $x > 0$ we have

$$\Pr\{Z \le x\} \approx 1.0 - \frac{1}{\sqrt{\pi}} e^{-(1/2)x^2}(a_1\eta + a_2\eta^2 + a_3\eta^3 + a_4\eta^4),$$

where

$$\eta = \frac{1}{1 + 0.27009x}$$

and the constants[5] are defined by $a_1 = 0.12771538$, $a_2 = 0.54107939$, $a_3 = -0.53859539$, and $a_4 = 0.75602755$. To evaluate the normal distribution for negative values, we take advantage of its symmetric nature and use the relationship $\Pr\{Z \le -x\} = 1 - \Pr\{Z > x\}$.

B.3 RANDOM NUMBER GENERATION

The implementation of the random number generator discussed in Chapter 3 is more difficult than simply writing a computer code to perform the operations listed. The reason for the difficulty is an "overflow" problem. Consider the example on page 80, in which the largest integer was assumed

[5]Based on C. Hastings, Jr., *Approximations for Digital Computers* (Princeton, N.J.: Princeton University Press, 1955), p. 168; and *Handbook of Mathematical Functions*, M. Abramowitz and I. A. Stegun, editors, (New York, N.Y.: Dover Publications, Inc., 1970), p. 298, Eq. 7.1.22.

to be 32,767. In the calculation of S_2, a value of 34,065,047 was obtained, which would produce a fatal overflow error by many compilers. Therefore, if a random number generator is needed for your programs, we suggest a procedure given by Park and Miller,[6] as listed below. It is designed to avoid overflows completely (assuming 32-bit integers) and should therefore be acceptable to most compilers.

```
main code
          seed   =   some_value
                 ⋮
            rn   =   random()
subprogram random()
        factor   =   1.0/2147483647.0
             i   =   seed/127773
             j   =   seed MOD 127773
             k   =   j * 16807 - i * 2836
         if  k > 0   then seed = k
         if  k ≤ 0   then seed = k + 2147483647
  return value   =   seed * factor
```

The variables *seed*, i, j, and k must be 32-bit integers. [Double-precision integers (or long integers) on micros are often 32-bits. The largest possible 32-bit integer is $2^{31} - 1 = 2{,}147{,}483{,}647$.] The division in the second step of the subprogram assumes truncation. The variables *rn* and *value* are floating-point variables, and the variable *factor* is a double-precision floating point variable. The variable *seed* is a global variable; all other variables are local to the subprogram. The seed can be initialized to any value between 1 and 2,147,483,646.

B.4 MISCELLANEOUS

If λ is small and t is large, the two expressions $\lambda^\alpha t^\alpha$ and $(\lambda t)^\alpha$ may not be equal, because of round-off error. To avoid unnecessary accuracy problems when evaluating expressions like $(\lambda t)^\alpha$, the product λt should be evaluated first before being raised to a power.

Another expression that occurs is of the form $r^n/n!$. It is much better to evaluate that expression iteratively instead of evaluating the numerator separate from the denominator. For example, do not use the equality $r^4/4! = (r*r*r*r)/(4*3*2*1)$. Use instead $r^4/4! = r*(r/2)*(r/3)*(r/4)$. In

[6] S. K. Park and K. W. Miller, "Random Number Generators: Good Ones are Hard to Find," *Communications of the ACM* **31** (1988), pp. 1192–1201.

other words, to code the evaluation of $w = r^n/n!$, the following iterative scheme could be used:

$$\begin{aligned} w_0 &= 1 \\ w_i &= w_{i-1} * r/i \quad \text{for } i = 1, \cdots, n. \end{aligned}$$

Polynomials are often evaluated iteratively, instead of as they are written. For example, consider the polynomial given as

$$f(x) = a_0 + a_1 x^1 + a_2 x^2 + a_3 x^3 .$$

It is more efficient to evaluate this polynomial written as

$$f(x) = a_0 + x[a_1 + x(a_2 + a_3 x)].$$

In other words, to code the evaluation of an nth degree polynomial, the following iterative scheme could be used:

$$\begin{aligned} \hat{f}_0 &= 0 \\ \hat{f}_i &= (\hat{f}_{i-1} + a_{n-i+1}) * x \quad \text{for } i = 1, \cdots, n \\ f_n &= \hat{f}_n + a_0 . \end{aligned}$$

APPENDIX C

Statistical Tables

C.1 GAMMA FUNCTION

$$\Gamma(x) = \int_0^\infty s^{x-1} e^{-s}\, ds.$$

If $0 < x < 1$, use the fact that $\Gamma(x) = \Gamma(1 + x)/x$. For example, $\Gamma(0.7) = 0.90864/0.7 = 1.298$. If $x > 2$, use the fact that $\Gamma(x) = (x - 1)\Gamma(x - 1)$. For example, $\Gamma(5.7) = 4.7 \times 3.7 \times 2.7 \times 1.7 \times 0.90864 = 72.5277$.

x	$\Gamma(x)$	x	$\Gamma(x)$
1.00	1.00000	1.05	0.97350
1.10	0.95135	1.15	0.93304
1.20	0.91817	1.25	0.90640
1.30	0.89747	1.35	0.89115
1.40	0.88726	1.45	0.88565
1.50	0.88623	1.55	0.88887
1.60	0.89352	1.65	0.90012
1.70	0.90864	1.75	0.91906
1.80	0.93138	1.85	0.94561
1.90	0.96177	1.95	0.97988
2.00	1.00000		

Numbers derived from the approximation in *Handbook of Mathematical Functions,* M. Abramowitz and I. A. Stegun, editors (New York, N.Y.: Dover Publications, Inc., 1970), p. 257, Eq. 6.1.36.

C.2 NORMAL DISTRIBUTION

$P\{Z \leq z\}$ where Z is normal with mean 0 and variance 1.

The tabular values give the "left-hand" probabilities. The values denoted by z_α refer to "right-hand" probabilities. The value of z_α is the z such that $P\{Z \leq z\} = 1 - \alpha$. For example, to find $z_{0.025}$, the tabular value of 0.975 must be found; thus, $z_{0.025} = 1.96$.

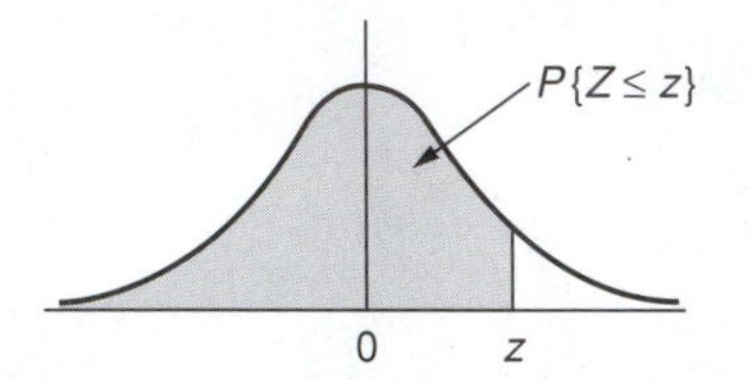

z	0.00	0.01	0.02	0.03	0.04	0.05	0.06	0.07	0.08	0.09
0.0	.5000	.5040	.5080	.5120	.5160	.5199	.5239	.5279	.5319	.5359
0.1	.5398	.5438	.5478	.5517	.5557	.5596	.5636	.5675	.5714	.5753
0.2	.5793	.5832	.5871	.5910	.5948	.5987	.6026	.6064	.6103	.6141
0.3	.6179	.6217	.6255	.6293	.6331	.6368	.6406	.6443	.6480	.6517
0.4	.6554	.6591	.6628	.6664	.6700	.6736	.6772	.6808	.6844	.6879
0.5	.6915	.6950	.6985	.7019	.7054	.7088	.7123	.7157	.7190	.7224
0.6	.7257	.7291	.7324	.7357	.7389	.7422	.7454	.7486	.7517	.7549
0.7	.7580	.7611	.7642	.7673	.7704	.7734	.7764	.7794	.7823	.7852
0.8	.7881	.7910	.7939	.7967	.7995	.8023	.8051	.8078	.8106	.8133
0.9	.8159	.8186	.8212	.8238	.8264	.8289	.8315	.8340	.8365	.8389
1.0	.8413	.8438	.8461	.8485	.8508	.8531	.8554	.8577	.8599	.8621
1.1	.8643	.8665	.8686	.8708	.8729	.8749	.8770	.8790	.8810	.8830
1.2	.8849	.8869	.8888	.8907	.8925	.8944	.8962	.8980	.8997	.9015
1.3	.9032	.9049	.9066	.9082	.9099	.9115	.9131	.9147	.9162	.9177
1.4	.9192	.9207	.9222	.9236	.9251	.9265	.9279	.9292	.9306	.9319
1.5	.9332	.9345	.9357	.9370	.9382	.9394	.9406	.9418	.9429	.9441
1.6	.9452	.9463	.9474	.9484	.9495	.9505	.9515	.9525	.9535	.9545
1.7	.9554	.9564	.9573	.9582	.9591	.9599	.9608	.9616	.9625	.9633
1.8	.9641	.9649	.9656	.9664	.9671	.9678	.9686	.9693	.9699	.9706
1.9	.9713	.9719	.9726	.9732	.9738	.9744	.9750	.9756	.9761	.9767
2.0	.9772	.9778	.9783	.9788	.9793	.9798	.9803	.9808	.9812	.9817
2.1	.9821	.9826	.9830	.9834	.9838	.9842	.9846	.9850	.9854	.9857
2.2	.9861	.9864	.9868	.9871	.9875	.9878	.9881	.9884	.9887	.9890
2.3	.9893	.9896	.9898	.9901	.9904	.9906	.9909	.9911	.9913	.9916
2.4	.9918	.9920	.9922	.9925	.9927	.9929	.9931	.9932	.9934	.9936
2.5	.9938	.9940	.9941	.9943	.9945	.9946	.9948	.9949	.9951	.9952
2.6	.9953	.9955	.9956	.9957	.9959	.9960	.9961	.9962	.9963	.9964
2.7	.9965	.9966	.9967	.9968	.9969	.9970	.9971	.9972	.9973	.9974
2.8	.9974	.9975	.9976	.9977	.9977	.9978	.9979	.9979	.9980	.9981
2.9	.9981	.9982	.9982	.9983	.9984	.9984	.9985	.9985	.9986	.9986
3.0	.9987									
3.5	.9998									
4.0	1.000									

Numbers derived from the approximation in *Handbook of Mathematical Functions*, M. Abramowitz and I. A. Stegun, editors (New York, N.Y.: Dover Publications, Inc., 1970), p. 932, Eq. 26.2.17.

C.3 STUDENT'S t DISTRIBUTION

$P\{T_n > t\} = \alpha$ where T_n has n degrees of freedom.

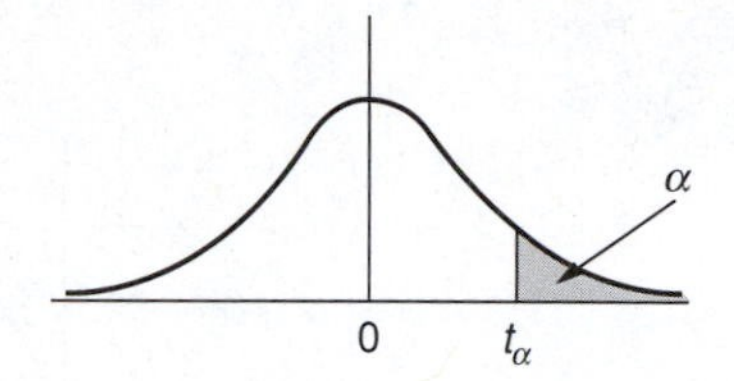

n	$\alpha = .10$	$\alpha = .05$	$\alpha = .025$	$\alpha = .01$	$\alpha = .005$
1	3.078	6.314	12.706	31.820	63.656
2	1.886	2.920	4.303	6.964	9.925
3	1.638	2.353	3.182	4.541	5.841
4	1.533	2.132	2.776	3.747	4.604
5	1.476	2.015	2.571	3.365	4.032
6	1.440	1.943	2.447	3.143	3.707
7	1.415	1.895	2.365	2.998	3.499
8	1.397	1.860	2.306	2.896	3.355
9	1.383	1.833	2.262	2.821	3.250
10	1.372	1.813	2.228	2.764	3.169
11	1.363	1.796	2.201	2.718	3.106
12	1.356	1.782	2.179	2.681	3.055
13	1.350	1.771	2.160	2.650	3.012
14	1.345	1.761	2.145	2.624	2.977
15	1.341	1.753	2.131	2.602	2.947
16	1.337	1.746	2.120	2.583	2.921
17	1.333	1.740	2.110	2.567	2.898
18	1.330	1.734	2.101	2.552	2.878
19	1.328	1.729	2.093	2.539	2.861
20	1.325	1.725	2.086	2.528	2.845
21	1.323	1.721	2.080	2.518	2.831
22	1.321	1.717	2.074	2.508	2.819
23	1.320	1.714	2.069	2.500	2.807
24	1.318	1.711	2.064	2.492	2.797
25	1.316	1.708	2.060	2.485	2.787
26	1.315	1.706	2.056	2.479	2.779
27	1.314	1.703	2.052	2.473	2.771
28	1.313	1.701	2.048	2.467	2.763
29	1.312	1.699	2.045	2.462	2.756
30	1.310	1.697	2.042	2.457	2.750
40	1.303	1.684	2.021	2.423	2.704
60	1.296	1.671	2.000	2.390	2.660
120	1.289	1.658	1.980	2.358	2.617
∞	1.282	1.645	1.960	2.327	2.576

Numbers derived from the distributions given in *Handbook of Mathematical Functions,* M. Abramowitz and I. A. Stegun, editors, (New York, N.Y.: Dover Publications, Inc., 1970), p. 948, Eqs. 26.7.3 and 26.7.4.

C.4 CHI-SQUARE DISTRIBUTION

$$P\{\chi_n^2 > t\} = \alpha \qquad \text{where } \chi_n^2 \text{ has } n \text{ degrees of freedom.}$$

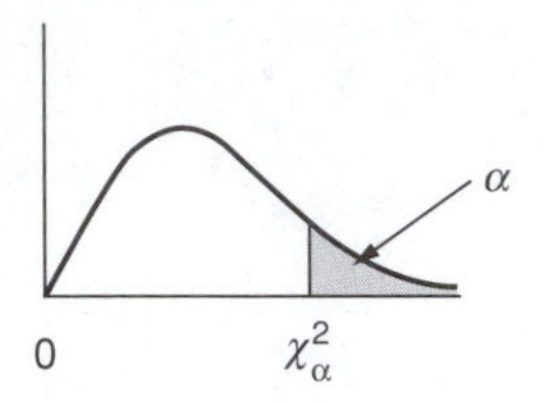

n	$\alpha = .99$	$\alpha = .975$	$\alpha = .95$	$\alpha = .90$	$\alpha = .10$	$\alpha = .05$	$\alpha = .025$	$\alpha = .01$
1	.000157	.000982	.00393	.0158	2.71	3.84	5.02	6.63
2	.0201	.0506	.103	.211	4.61	5.99	7.38	9.21
3	.115	.216	.352	.584	6.25	7.81	9.35	11.35
4	.297	.484	.711	1.06	7.78	9.49	11.14	13.28
5	.554	.831	1.15	1.61	9.24	11.07	12.83	15.09
6	.872	1.24	1.64	2.20	10.65	12.59	14.44	16.81
7	1.24	1.69	2.17	2.83	12.02	14.07	16.01	18.48
8	1.65	2.18	2.73	3.49	13.36	15.51	17.53	20.09
9	2.09	2.70	3.33	4.17	14.68	16.92	19.02	21.67
10	2.56	3.25	3.94	4.87	15.99	18.31	20.48	23.21
11	3.05	3.82	4.57	5.58	17.27	19.68	21.92	24.73
12	3.57	4.40	5.23	6.30	18.55	21.03	23.34	26.22
13	4.11	5.01	5.89	7.04	19.81	22.36	24.74	27.69
14	4.66	5.63	6.57	7.79	21.06	23.69	26.12	29.14
15	5.23	6.26	7.26	8.55	22.31	25.00	27.49	30.58
16	5.81	6.91	7.96	9.31	23.54	26.30	28.85	32.00
17	6.41	7.56	8.67	10.09	24.77	27.59	30.19	33.41
18	7.01	8.23	9.39	10.87	25.99	28.87	31.53	34.81
19	7.63	8.91	10.12	11.65	27.20	30.14	32.85	36.19
20	8.26	9.59	10.85	12.44	28.41	31.41	34.17	37.57
21	8.90	10.28	11.59	13.24	29.62	32.67	35.48	38.93
22	9.54	10.98	12.34	14.04	30.81	33.92	36.78	40.29
23	10.20	11.69	13.09	14.85	32.01	35.17	38.08	41.64
24	10.86	12.40	13.85	15.66	33.20	36.41	39.36	42.98
25	11.52	13.12	14.61	16.47	34.38	37.65	40.65	44.31
26	12.20	13.84	15.38	17.29	35.56	38.89	41.92	45.64
27	12.88	14.57	16.15	18.11	36.74	40.11	43.19	46.96
28	13.56	15.31	16.93	18.94	37.92	41.34	44.46	48.28
29	14.26	16.05	17.71	19.77	39.09	42.56	45.72	49.59
30	14.95	16.79	18.49	20.60	40.26	43.77	46.98	50.89

Numbers derived from the distributions given in *Handbook of Mathematical Functions,* M. Abramowitz and I. A. Stegun, editors (New York, N.Y.: Dover Publications, Inc., 1970), p. 941, Eqs. 26.6.4 and 26.6.5.

C.5 CRITICAL VALUES FOR THE KOLMOGOROV-SMIRNOV TEST

$$P\{D_n > t\} = \alpha$$

where D_n is the largest difference (in absolute value) between theoretical and empirical cumulative distribution functions.

For $\alpha = .01$ and .05, asymptotic formulas give values that are slightly high.

n = Sample size	$\alpha = .20$	$\alpha = .15$	$\alpha = .10$	$\alpha = .05$	$\alpha = .01$
1	.900	.925	.950	.975	.995
2	.684	.726	.776	.842	.929
3	.565	.597	.642	.708	.829
4	.494	.525	.564	.624	.734
5	.446	.474	.510	.563	.669
6	.410	.436	.470	.521	.618
7	.381	.405	.438	.486	.577
8	.358	.381	.411	.457	.543
9	.339	.360	.388	.432	.514
10	.322	.342	.368	.409	.486
11	.307	.326	.352	.391	.468
12	.295	.313	.338	.375	.450
13	.284	.302	.325	.361	.433
14	.274	.292	.314	.349	.418
15	.266	.283	.304	.338	.404
16	.258	.274	.295	.328	.391
17	.250	.266	.286	.318	.380
18	.244	.259	.278	.309	.370
19	.237	.252	.272	.301	.361
20	.231	.246	.264	.294	.352
25	.21	.22	.24	.264	.32
30	.19	.20	.22	.242	.29
35	.18	.19	.21	.23	.27
40				.21	.25
50				.19	.23
60				.17	.21
70				.16	.19
80				.15	.18
90				.14	
100				.14	
Asymptotic formula:	$1.07/\sqrt{n}$	$1.14/\sqrt{n}$	$1.22/\sqrt{n}$	$1.36/\sqrt{n}$	$1.63/\sqrt{n}$

This table is adapted from Z. W. Birnbaum, "Numerical Tabulation of the Distribution of Kolmogorov's Statistic for Finite Sample Size," *Journal of the American Statistical Association* **47** (1952), pp. 425–441. Reprinted with permission from the

C.6 RANDOM NUMBERS

1	0.0140	0.5319	0.2443	0.7732	0.9240	0.4042	0.8806	0.4158
2	0.1855	0.5574	0.7842	0.1194	0.6112	0.6421	0.3221	0.4554
3	0.6594	0.7259	0.6301	0.1797	0.3775	0.5157	0.1454	0.9683
4	0.7156	0.5140	0.9734	0.9375	0.7269	0.2228	0.1020	0.9991
5	0.9537	0.6292	0.4340	0.2416	0.9946	0.6799	0.0197	0.3988
6	0.4703	0.8872	0.5035	0.2335	0.6576	0.5989	0.2047	0.4328
7	0.2419	0.1215	0.6602	0.7957	0.9518	0.1589	0.5695	0.8823
8	0.4605	0.5212	0.1180	0.9862	0.9965	0.5643	0.5486	0.8357
9	0.1829	0.7604	0.5357	0.7768	0.8637	0.4946	0.7607	0.9254
10	0.0368	0.3365	0.5396	0.1069	0.7673	0.2997	0.7799	0.2747
11	0.9982	0.5265	0.2921	0.5158	0.0846	0.4759	0.6883	0.4096
12	0.0334	0.3243	0.7865	0.1718	0.8602	0.1670	0.9555	0.3851
13	0.7400	0.4135	0.1927	0.6486	0.0130	0.2967	0.0997	0.9583
14	0.0827	0.2605	0.6894	0.9348	0.9241	0.7901	0.4082	0.2058
15	0.8230	0.1849	0.1063	0.3566	0.8655	0.3360	0.3279	0.5518
16	0.2358	0.5907	0.3483	0.6849	0.9204	0.6083	0.2709	0.7610
17	0.5374	0.1730	0.5044	0.6206	0.0474	0.8403	0.9076	0.3143
18	0.0383	0.0513	0.9537	0.6614	0.1637	0.4939	0.6086	0.1542
19	0.2330	0.8310	0.8647	0.2143	0.5414	0.0214	0.3664	0.6326
20	0.7129	0.1671	0.2222	0.9869	0.9278	0.1854	0.3735	0.8792
21	0.9019	0.4779	0.1472	0.3953	0.5929	0.6838	0.4449	0.2235
22	0.3689	0.5881	0.9092	0.1110	0.3456	0.2000	0.7063	0.3995
23	0.2013	0.9000	0.3621	0.8129	0.2016	0.4512	0.3482	0.1755
24	0.8849	0.7879	0.7132	0.6077	0.9739	0.9131	0.2607	0.1193
25	0.5020	0.7121	0.6690	0.1828	0.2882	0.1731	0.3260	0.4354
26	0.4530	0.0345	0.0804	0.7218	0.8138	0.8508	0.2301	0.2370
27	0.4227	0.2128	0.2858	0.9167	0.5889	0.1401	0.8972	0.9903
28	0.6843	0.7528	0.4794	0.0418	0.7935	0.0525	0.1044	0.2407
29	0.0320	0.7857	0.4718	0.0936	0.8061	0.3582	0.8688	0.9033
30	0.8598	0.1847	0.9695	0.0333	0.2935	0.2598	0.9509	0.6378

Numbers derived using the method described at the end of Appendix B, with an initial seed of 1783.

Index